International Series in Operations Research & Management Science

Volume 301

More information about this series at http://www.springer.com/series/6161

Chialin Chen • Yihsu Chen
Vaidyanathan Jayaraman
Editors

Pursuing Sustainability

OR/MS Applications in Sustainable Design, Manufacturing, Logistics, and Resource Management

Springer

Editors
Chialin Chen
Department of Business Administration
College of Management
National Taiwan University
Taipei, Taiwan

Yihsu Chen
Jack Baskin School of Engineering
University of California, Santa Cruz
Santa Cruz, CA, USA

Vaidyanathan Jayaraman
Great Lakes Institute of Management
Chennai, India

ISSN 0884-8289 ISSN 2214-7934 (electronic)
International Series in Operations Research & Management Science
ISBN 978-3-030-58022-3 ISBN 978-3-030-58023-0 (eBook)
https://doi.org/10.1007/978-3-030-58023-0

This Springer imprint is published by the registered company Springer Nature Switzerland AG
The registered company address is: Gewerbestrasse 11, 6330 Cham, Switzerland

Contents

1 **Pursuing Sustainability: An Interdisciplinary Perspective** 1
Chialin Chen, Vaidy Jayaraman, and Yihsu Chen

Part I Design, Innovation, and Technology

2 **Green Product Development: Price Competition, Quality Choice, and First-Mover Advantage** 15
Chialin Chen and Kilsun Kim

3 **Decision Support for Sustainable Supply Chain Management** 43
Hendrik Reefke and David Sundaram

4 **DEA Application in Sustainability 1996–2019: The Origins, Development, and Future Directions** 71
Haibo Zhou, Yi Yang, Yao Chen, Joe Zhu, and Yu Shi

5 **Sustainable Manufacturing and Technology: The Development and Evaluation** 111
Panitas Sureeyatanapas and Jian-Bo Yang

6 **Circular Economy and New Research Directions in Sustainability** 141
Hwong-Wen Ma, Hsiu-Ching Shih, and Meng-I Liao

Part II Manufacturing, Logistics, and Value Creation

7 **Sustainable Supply Chain Management: Research Pathways Based on Empirical Evidence from Chinese Automotive Companies** 171
Lin Wu and Nachiappan Subramanian

8 **Inventory Management in Multi-echelon After-Sales Service Networks** 199
Sandeep Srivathsan

9 Towards a Unified Understanding and Management of Closed Loop Operations 219
Santosh Mahapatra, Dwayne Cole, Raktim Pal, and Scott Webster

10 Orchestrating Sustainable Stakeholder Value Creation: A Product Life Cycle Extension Perspective 239
Suresh Srinivasan and Vaidyanathan Jayaraman

11 Exercise Your Rs! You Never Know When You May Need Them: Revisiting and Extending Modes of Product Life for the Future 255
Jonathan D. Linton and Vaidyanathan Jayaraman

Part III Sustainable Natural Resource Management

12 Economic Management of Electric Power Systems 279
Alberto J. Lamadrid L.

13 Reservoir Capacity Planning Using Stochastic Multiobjective Programming Integrated with MCMC Technique 315
Ho Wen Chen and Kieu Lan Phuong Nguyen

14 High-Resolution Multiobjective Optimization of Sustainable Supply Chains for a Large-Scale Lignocellulosic Biofuel Industry . . 341
T. Edward Yu, Burton C. English, Jia Zhong, James A. Larson, Joshua S. Fu, Lixia He-Lambert, and Brad Wilson

15 Mathematical Models for Evolving Natural Gas Markets 359
Felipe Feijoo, Sriram Sankaranarayanan, Charalampos Avraam, and Sauleh A. Siddiqui

16 Future Research Directions for Sustainable Natural Resource Management 387
Yihsu Chen, Antonio J. Conejo, and Andrew L. Liu

Index 403

Chapter 1
Pursuing Sustainability: An Interdisciplinary Perspective

Chialin Chen, Vaidy Jayaraman, and Yihsu Chen

1.1 Introduction

Sustainability is generally defined as managing the triple bottom line—a process by which companies and governments manage their economic, social, and environmental risks, obligations and opportunities. These three impacts are sometimes referred to as profits, people, and planet. Pursuing sustainability requires these entities to manage products, processes, value chains, and resources in a way that meets the needs of the present without compromising the ability of future generations to meet their own needs (Elkington 1997). On the same basis, the 2030 Agenda for Sustainable Development, adopted by all United Nations (2015), call for a shared set of goals into the future. At its heart are the 17 sustainable development goals (SDGs), an urgent call for action by all countries around the world from an interdisciplinary perspective. Consistent with the goals for pursuing sustainability, the purpose of the book is to provide a holistic review of the state-of-the-art OR/MS research confronting and enhancing sustainability as well as to identify future research directions through bridging differences and assimilating similarities among

C. Chen (✉)
Department of Business Administration, College of Management, National Taiwan University, Taipei, Taiwan, ROC
e-mail: cchen026@ntu.edu.tw

V. Jayaraman
Supply Chain Operations, Analytics, Great Lakes Institute of Management, Chennai, Tamil Nadu, India

Y. Chen
Jack Baskin School of Engineering, University of California, Santa Cruz, Santa Cruz, CA, USA
e-mail: yihsuchen@ucsc.edu

C. Chen et al. (eds.), *Pursuing Sustainability*, International Series in Operations Research & Management Science 301,
https://doi.org/10.1007/978-3-030-58023-0_1

three domains: (1) Design, Innovation, and Technology, (2) Manufacturing, Logistics, and Value Creation, and (3) Sustainable Natural Resource Management.

The first part of the book (Chaps. 2–6) focuses on the creation and development of sustainable products and services from a systems design perspective. Key areas covered include Green Design and Development, Technology and Sustainable Manufacturing, Decision Support for Sustainable Supply Chain Management, Data Envelopment Analysis and Performance Evaluation, and Circular Economy. The second part of the book (Chaps. 7–11) concentrates on the major supply-chain and logistic issues faced by today's industries in delivering sustainable products and services. Major areas covered include Remanufacturing, Reverse Logistics, Closed-Loop Management, and Sustainable Value Creation. The third part of the proposed book (Chaps. 12–16) centers on the applications of analytical tools, such as dynamic optimization, risk analysis, and geographic information system analysis, to sustainable natural resource management. Key areas covered include Electric Power Systems, Water Resource, Biofuel Infrastructure, Natural Gas, and Global Climate Models.

1.2 Part 1: Design, Innovation, and Technology

The focus of the first part of the book is on the creation and development of new products, services, and systems which allows firms to address sustainability issues and/or to capture new opportunities in emerging markets. This part of the book starts with a general competitive model and its extensions for designing green (environmentally friendly) products and services under different market, technological, and regulatory settings. In the next three chapters, guidelines and decision-supporting tools are proposed for a number of key issues, such as sustainable manufacturing and technology management, green supply chain management, and sustainable performance evaluation with data analytics. It concludes with a chapter that integrates key concepts and practices of circular economy based on which future research directions are identified and discussed.

Based on the pioneering work in green product design by Chen (2001), in Chap. 2, the authors (Chen and Kim) present a general game-theoretic model to analyze the entry, quality, and pricing decisions for green product development in competitive environments, and investigate the resulting economic and environmental consequences. In the simultaneous game, the conditions for a subgame perfect equilibrium is first identified in which a subset of the competitive firms in the market enter with distinct products for multiple segments to reach a price equilibrium. It is shown that, at equilibrium, the overall environmental quality in products sold to all the customers in the competitive market is higher than that under monopoly, a result that differs from the popular view that "the monopolist is the conservationist's friend." In sequential games, firm's pricing decisions under exogenous arrangements of price leadership are first analyzed, and then a role-choosing model is used to identify the endogenous arrangements of price leadership. Based on the analytical results, the chapter further explores the first-mover advantage with sustainable product development in a competitive market. It is shown that the first mover in quality choice will introduce a green product to a sufficiently large green segment even with less than 50% of green consumers in a market in order to obtain a desirable arrangement of price leadership. A number of managerial and policy implications concerning how to manage and regulate green product development in competitive environments are also discussed in this chapter.

With a focus on supply chains as a crucial component for sustainable development, in Chap. 3, the authors (Reefke and Sundaram) start with an overview of the definitions of sustainable supply chain management (SSCM) and illustrate key characteristics of SSCM. The chapter then synthesizes research recommendations from a selection of seminal articles in the field with regard to decision support in SSCM. Based on seminal management approaches, the authors then present two models aimed at transforming SC towards SSCM and developing SSCM towards higher levels of maturity. These models are further elaborated within a case study of Mattel, an American multinational company that designs, manufactures, and markets toys worldwide, to illustrate their applications in SSCM. The chapter is then concluded with discussions reflecting on how the findings contribute to the understanding of SSCM and how they can be leveraged by scholars in sustainable supply chain management.

As data envelopment analysis (DEA) has become one of the popular tools for sustainability performance evaluation, in Chap. 4, the authors (Zhou, Yang, Chen, and Zhu) conduct an extensive review of DEA applications in sustainability using citation-based approaches. A directional network is constructed based on citation relationships among DEA papers published in journals indexed by the Web of Science database from 1996 to 2019. They first draw the citation chronological graph to present a complete picture of literature development trajectory since 1996 to identify the local main DEA development paths in sustainability research. Based on the Kamada-Kawai layout algorithm, four research clusters in the literature are identified, which include corporate sustainability assessment, regional sustainability assessment, sustainability composite indicator construction, and sustainability performance analysis. For each of the clusters, the authors further identify the key articles based on citation network and local citation scores, demonstrate the developmental trajectory of the literature, and explore future research directions.

From the perspective of guideline development and evaluation, in Chap. 5, the authors (Sureeyatanapas and Yang) propose guidelines for how a manufacturing company can develop and evaluate its sustainability in a practical manner. They first explore the definitions and the components of sustainable manufacturing to develop a general framework for sustainable manufacturing. Examples of operational strategies with corporate social responsibility implications are also provided to identify relevant aspects based on which key steps in performance evaluation are identified. The authors then review a number of performance analysis techniques, and a hierarchical framework of corporate sustainability assessment with a list of key performance indicators is proposed. A case study of sugar manufacturing is presented to show the applications of the proposed framework and evaluation techniques.

Circular economy has recently received considerable attention from both researchers and practitioners. In Chap. 6, the authors (Ma, Shih, and Liao) present frameworks and analytical tools for developing circular economy. The chapter starts with a review of key concepts and principles of circular economy and their association with Sustainable Development Goals (SDGs) set by the United Nations. The authors then review literature related to the goals and strategies for developing circular economy and business models for transition toward circular economy. Several assessment tools, including material flow analysis and life cycle assessment, are presented with processes and detailed steps of transition toward circular economy. A case study of the transition paths of polypropylene in the automotive industry toward circular economy is presented to demonstrate the applications of the proposed framework. The chapter is concluded with a discussion on implementation issues and future research directions.

1.3 Part 2: Manufacturing, Logistics, and Value Creation

The second part of the book focuses on sustainable supply chain management practices that include several product life extension strategies including reverse logistics, remanufacturing and closed-loop management. This section of the book describes four chapters that cover some of the breadth and depth of sustainable supply chain strategies. It concludes with a fifth chapter where the authors encourage multiple stakeholders, including suppliers, producers, distributors, retailers and customers, to be mindful of the 12 Rs related to product life extension strategies and how this could be a profitable venture for one and all.

From a sustainable manufacturing perspective, in Chap. 7, the authors (Wu and Subramanian), through a set of eight case studies provide insights on the status quo of sustainable development in the context of manufacturing firms located in China. The eight companies were drawn from a population that consists of SMEs as well as large companies. Through a systematic review of theories and empirical studies from the literature, they clarified the research pathways on sustainability. For the sample of eight companies, both primary data (collected from semi-structured interviews with managers and senior staff members) and secondary data (company annual reports, CSR reports, and other internal company documentation that are pertinent to the study) were obtained and triangulated to enhance the reliability of their research findings. Under the umbrella of sustainable manufacturing practices, the authors found sufficient empirical evidence that practices, such as lean, green and social management, were implemented by the eight automotive manufacturing firms. The authors, based on their case study findings, propose a causal relationship between the implementation of sustainable practices and the development of operational capabilities through a series of propositions that include factors such as the size and age of the firm and their commitment to sustainable management implementation.

From a reverse logistics perspective, in Chap. 8, the author (Srivathsan) investigates a scenario in which firms offer service contracts that guarantee a quick turnaround time for repairing faulty components while ensuring full functionality of their returned products. These contracts are viewed from an inventory management scenario in a multi-echelon service supply chain. The inventory management problem that is covered in this chapter is a subset of the broad area of reverse logistics and viewed from an after-sales service system that is closed related to the repairable item inventory situation with product returns and remanufacturing. The author models the problem using an M/M/1/K queuing system and derive properties for the total cost function.

From a closed-loop management perspective, in Chap. 9, the authors (Mahapatra, Cole, Paul, and Webster) present a unified understanding and management of closed loop supply chain operations. They remark that given the position of closed loop supply chains to reach the goal of "cradle-to-cradle" operations, it would be fruitful to investigate its significance in light of sustainable supply chains. The authors emphasize the importance of closed-loop supply chain operations, analyze its significance to both business and environmental performance and suggest consideration for managing the operations in a cost effective fashion. The term closed-loop supply chain involves a supply chain where there it includes a traditional "forward" supply chain (typically from the Supplier all the way to the end customer) and "reverse" supply chain that handles post-consumer products that travels backward from the customer all the way back to the manufacturer or their associated third party provider. In a sustainable environment where there is a focus on reduction in pollution, waste, energy, and materials in process; in a closed-loop environment, there is an attempt to minimize environmental impact of products through its life cycle design, manufacturing, distribution, use, and end-of-life disposal. The authors recognized that the key challenge in analyzing the effectiveness of various environmental practices is how to quantify the cost incurred and the associated benefits that are accrued. The book chapter identifies three principles that companies need to pay attention: (a) the selection and use of environmentally benign materials that reduce waste generation and promote upcyclability has to be carefully investigated; (b) the promotion of cost effective remanufacturing strategy wherein companies can make the case that it is a profitable venture; and (c) the need for collaborative and mutually rewarding engagement among both internal and external partners along a supply chain to promote and embrace sustainability practices.

From a sustainable value creation perspective, in Chap. 10, the authors (Srinivasan and Jayaraman) discuss the need to orchestrate a business model for organizations that identifies "sweet spots" of value creation for all stakeholders with an objective of creating a sustainable competitive advantage. By bringing together a value creation framework that is grounded in the resource-based theory, the triple bottom line and CSR framework, the authors argue that product life cycle extensions are an important paradigm to achieve shared value in terms of sustainable competitive advantages for firms. The genesis of this chapter is to postulate that produce life cycle extensions are a critical paradigm to achieve shared and sustainable competitive advantage for organizations. The chapter explores the challenges for companies to implement such product life extensions to achieve competitive advantages. The authors remark that by avoiding or ignoring the efficient return, refurbishment, or disposal of products, companies miss out on significant return of investment. It is indeed possible that applying product life extension strategies, companies can create a competitive advantage and convert it to a profitable venture. From a customer standpoint, more research needs to be done to explore the customer's behavior and their willingness to pay for remanufactured products vis-à-vis new products. From a company perspective, the risk of cannibalization is also low. The authors conclude with a series of challenges in implementing a circular economy and raise valid questions that perhaps research scholars could investigate in their quest toward sustainable stakeholder value creation.

In the last chapter of Part 2, new research directions are provided for researchers, practitioners and organizations to pursue as a way forward. In Chap. 11, the authors (Linton and Jayaraman) make a passionate plea to have organizations focus on the 12 Rs to do with product life extension modes under the umbrella of sustainable supply chain management. The idea is to move beyond recycling, reuse or reduction and focus on other product life extension models so as to extract value from the products and by-products that result from post-consumer activities. Building upon the principles of industrial ecology that suggests the idea of a food chain in which companies can exploit unutilized resources of by-products and thereby increase resource utilization, the authors address post-consumer waste, where the source of waste is the consumer. Issues such as closed-loop supply chains, remanufacturing, reverse logistics, and recycling are considered in this book chapter. The authors indicate that the current pandemic of 2020 shows us that even with tremendous care taken by organizations to look at greenhouse gas emissions and other measures of pollution, it's important for organizations to look at their supply chain in relation to the 12 Rs. They conclude by proposing that companies that do not engage in these product life cycle extension modes will face challenges in the near future while companies that incorporate these strategies will not only avoid future difficulties but also reduce their cost and increase their profitability.

1.4 Part 3: Sustainable Natural Resource Management

The third part of the book focuses on illustrating studies that apply operations research (OR) tools to sustainable natural resource management. Compared to the first two parts, OR applications to sustainably managing natural resource is difficult and challenging to study for at least three reasons. First, studying natural resource management generally requires an explicit incorporation of engineering systems so as to account for the processes by which the natural resources are extracted, processed, and delivered to end-users or consumers. Second, management of natural resources also entails understanding the impacts by physical environment, for example, climate change, or its resulting impacts on the natural environment, for example, pollution. Third, it also needs to develop a representation to consider supply-and-demand relationship of natural resources through market transactions or induced by climate or energy policies. Toward this end, this part of the book collects four chapters, illustrating OR applications in managing renewable energy or energy in general, water resource, biofuel infrastructure, and natural gas. A conclusion chapter discusses emerging challenges and lays out future research directions for interested readers.

Chapter 12 introduces tools related to dynamic optimization and optimal control, which are commonly used in engineering fields. However, the focus is on the energy sector. The chapter starts with a review of the Lagrangian Method, the intuition behind the formulation, for example, shadow prices. Then, the chapter reviews the basic dynamic optimization problem, starting with deterministic formulations of the discrete and the continuous problem and present an example applied to the replacement of an energy storage system (ESS). Next, the chapter presents the stochastic formulation based on geometric Brownian motion (GBM) with an example of an agent optimizing the use of an ESS. Finally, the chapter discusses the formulation of stochastic programs and decision-making under uncertainty. An example of a unit commitment problem is presented wherein potential ways to deal with the curse of dimensionality are discussed. These concepts are for early graduate students, as a toolset in their graduate studies and beyond.

Chapter 13 presents an optimization-based risk analysis framework formulated by a chance-constrained problem to size a new reservoir in a river basin with a focus on sustainable development. Sizing a new reservoir is a long-run and irreversible investment. An undersized reservoir is unlikely to provide needed functions, while an oversized reservoir is doubtfully to be economically desirable. Moreover, uneven streamflow during dry and wet seasons, coupled with urban development in adjacent areas, increasing economic activities, and impacts induced by climate change, make the problem hardly deterministic. Thus, siting new reservoir requires a stochastic-based and risk-informed decision approach. The scope of risk management, which is increasingly being recognized today, should consider the resulting economic benefits and the local unique hydrological conditions while maintaining necessary water quality in the river. An incorporation of environmental and economic factors into the reservoir capacity planning process is essential.

The framework developed in the chapter begins by decomposing a water basin into a number of river reaches based on their hydraulic characteristics. Within each river reach, several segments with a finer length are defined for modeling purpose. Between these segments, flow- and mass-balancing relationship is established based on finite element approach. The Streeter–Phelps model is then applied to model changes in dissolved oxygen (DO) and biochemical oxygen demand (BOD) of dry and wet seasons by calibrating collected data with governing equations (Harrison 1980). The total cost includes capital/fixed cost of initial project construction, and operation/variable costs. The chapter combines cost information from the existing reservoirs to derive fixed capacity construction cost. The variable cost of water supply is estimated based on contract cost and on cost of sea water desalinization when the demand is more than contracted quantity. Two objectives are considered, including maximization of the monthly water supply from reservoir and minimization of the total cost incurred from constructing and operating facilities with various constraints that are imposed to model mass balances of the flow as well as maintenance of water quality. A Markov Chain Monte Carlo (MCMC) approach is used to modeling the transition probability of streamflow between months. A case study of Hou-Lung River Basin in Taiwan is used to illustrate the capability of the framework. The framework presented in the chapter is designed for a multidisciplinary assessment of reservoir sizing, simultaneously considering natural patterns of stream flows, adequate water supply, and pristine water quality in the river watershed.

Chapter 14 focuses on study the development of biofuel infrastructure to meet policy-induced biofuel demand. The determination for reducing reliance on fossil fuels and greenhouse gas (GHG) emissions has driven the development of a biofuel sector in the USA. The Renewable Fuel Standard 2 (RFS 2) in the Energy Independence and Security Act of 2007 mandated 136 billion liters of biofuels to be produced annually in the USA by 2022, with at least 79 billion liters of advanced biofuels (The U.S. Congress 2007). Advanced biofuels generated from the lignocellulosic biomass (LCB), such as agricultural and forest residues, short-rotation woody crops, and herbaceous grasses, has been advocated by the US Environmental Protection Agency (EPA) as a strategy to reduce GHG emissions (The U.S. Environmental Protection Agency 2014).

The chapter introduces an integrative framework that couples GIS (Geographic Information System) tools, a biophysical model, enterprise budgeting tools, GHG emission models, and optimization mechanism to determine a sustainable switchgrass biofuel supply chain network in a multiple-objective (cost and GHG emissions) framework formulated as a mixed integer problem. The supply chain consists of framers, biorefineries, blending facilities. The location and production area of feedstock, also the location and capacity of biorefineries, is simultaneously determined endogenously by solving the optimization problems. The high-resolution spatial GIS data are used as inputs in a biophysical model to simulate the effects of cropping system change that covert from original crops to switchgrass and related land management practices on soil carbon cycling and stock. The simulated ecosystem outputs are then employed to estimate GHG emissions associated with land use. The spatial explicit data of resource availability and physical characteristics are also used to constraint resources in the optimization model to determine facility locations. Cost information is obtained from an enterprise budget model (Larson et al. 2010) that accounts for capital investment and annual operational cost of farms and of biorefineries.

The framework is applied to a case study of replacing 30% of gasoline used in transportation in Tennessee. The results indicate that land use choice makes substantial impacts on the deployment of the supply chains under different objectives. On the one hand, when considering only the private cost, hay and pasture land concentrated in the east and central Tennessee will be the major source for switchgrass production. On the other hand, if minimizing GHG emissions, more than 500 thousand hectares of the state's cropland is converted to switchgrass for biofuel production. This highlights the tradeoff between cost and GHG emissions in the supply chains that entails marginal rate of substitution between total cost and GHG emissions on the frontier curve increases at an accelerating rate.

Chapter 15 addresses modeling of natural gas market in north American, a regional-scale model. Owing to the concerns about climate change, natural gas has expanded its significance as an energy source mainly due to its low carbon emissions and low competitive prices as result of new technologies. In the meantime, the regional and global natural gas markets have also been significantly influenced by domestic and international socioeconomic conditions and politics. This chapter discusses (1) the main drivers behind the evolution of natural gas markets and (2) the modeling approaches that have been taken to understand such evolutions at regional and global levels.

A gas market model typically considers three levels of the supply chain, including the upstream sector where natural gas is produced, the midstream sector where natural gas is either transformed into secondary energy forms (e.g., electricity), stored or delivered, and finally the downstream sector where natural gas is sold to the different demand sectors (e.g., residential, commercial, and industrial). The chapter presents the optimization problem faced by the entities in each layer of the natural gas supply chain and solve them jointly by the collection of their first-order conditions. The resulting problem is an example of Mixed Complementarity Problem (MCP), where can be solving commercial solver, for example, PATH (Munson and Ferris 2000). A case study using North American Natural Gas Model (NANGAM) is presented to illustrate the impacts of increase in demand and of discovery of cheap shale gas at Marcellus region. The chapter concludes with an outlook of future research and examples of historical energy system transformations due to the appearance of natural gas as a competitive energy source.

Finally, a conclusion chapter is presented in Chap. 16 that discusses three directions for future research in sustainable natural resource management. First, we believe that future sustainability and resilience in natural resource management can be enhanced through better utilization of outcomes from process- or physical-based climate change models. One example is Global Climate Models or CGE that simulates global climate with standard emission scenarios, for example, RCP (Representative Concentration Pathway) (van Vuuren et al. 2011). We illustrate here use of downscaled hourly sea level rise data at the level of weather stations to identify and to locate the natural gas facilities that are prone to the risk of sea level rise. The stochasticity of water depth at which each station submerge under the water allows evaluating timing of implementing resilience options while considering their costs and benefits of avoided damage. Second, directly related to the electric power sectors is to better harnessing variability and unpredictability via flexibility, transmission, and storage. This is especially relevant with increasing penetration of both utility-scale and residential renewable energy, distributed energy resource, such as electric cars, that could serve as energy storage, provide demand response to offset the energy requirement from the bulk energy market, or other much needed ancillary services that help maintain the reliability of the grid. Third, a better natural resource management should consider interdependence of multiple systems, such as power, natural gas and water systems, through co-optimization of the systems. The

U.S. Department of Homeland Security (2019) identifies that there are at least 16 such sectors that touch upon every aspect of our daily lives, and such sectors are all intrinsically linked. For example, meeting growing electricity demand could require more fresh water for power plant cooling, causing tension on water supply. On the hand, energy is essential for abstraction, conveyance, distribution, and treatment of freshwater or wastewater treatment. For instance, in California, the State Water Project (SWP), one of the major water projects, delivers water originated from the Northern California region over more than 400 miles to the demand location in the Southern California via San Joaquin Valley through a combination of natural rivers and man-made aqueducts (California Department of Water Resources 2019). The delivery of water is not entirely by gravity, and a series of pumping stations were built to lift water for more than 3500 ft beyond which the gravity flow takes place. Therefore, properly managing these sectors need a holistic approach that accounts for the interdependence among them so that scarcity of natural resources can be accurately priced in different system. Eventually, this is expected to lead to an efficient allocation of resources across these interdependent systems.

1.5 Summary

Pursuing sustainability requires collective and interdisciplinary efforts from research communities and industries around the world. We envision that the book will bridge the three main OR/MS research domains in terms of the creation and development of sustainable products and services through design, innovation, and system integration, the delivery of sustainable products and services through manufacturing, logistics, and value chain management, and the management of natural resources such as energy, water, biofuel, and climate, for sustainable development. We hope that this book can serve as a reference and guide for researchers and practitioners to fulfill their needs in understanding the state-of-the-art OR/MS research in sustainability and potential future research directions as well as in how to incorporate sustainability in OR/MS research in the three main domains and their interconnections.

References

California Department of Water Resources. (2019). *State Water Project*. Retrieved from https://water.ca.gov/Programs/State-Water-Project.

Chen, C. (2001). Design for the environment: A quality-based model for green product development. *Management Science, 47*(2), 250–263.

Elkington, J. (1997). *Cannibals with forks. The triple bottom line of 21st century business*. Oxford: Capstone.

Harrison, G. W. (1980). A stream pollution model with intervals for the rate coefficients. *Mathematical Biosciences, 49*, 111–120.

Larson, J. A., Yu, T. E., English, B. C., Mooney, D. F., & Wang, C. (2010). Cost evaluation of alternative switchgrass producing, harvesting, storing, and transporting systems and their logistics in the Southeastern USA. *Agricultural Finance Review, 70*, 184–200.

Munson, T. S., & Ferris, M. C. (2000). Complementarity problems in GAMS and the PATH solver. *Journal of Economics Dynamic Control, 24*, 165–188.

The U.S. Congress. (2007). *Energy Independence and Security Act of 2007*. Retrieved from http://frwebgate.access.gpo.gov/cgi-bin/getdoc.cgidbname=110_cong_bills&docid=f:h6enr.txt.pdf.

The U.S. Department of Homeland Security. (2019). *Critical infrastructure design and adaptive resilient systems*. Retrieved from https://www.dhs.gov/science-and-technology/cidars.

The U.S. Environmental Protection Agency. (2014). *Carbon pollution emission guidelines for existing stationary sources: Electric utility generating units*. Retrieved from http://yosemite.epa.gov/opei/rulegate.nsf/byRIN/2060-AR33.

United Nations. (2015). *Transforming our world: The 2030 agenda for sustainable development*. Retrieved from https://sustainabledevelopment.un.org/content/documents/21252030%20Agenda%20for%20Sustainable%20Development%20web.pdf.

van Vuuren, D. P., Edmonds, J., Kainuma, M., Riahi, K., Thomson, A., Hibbard, K., et al. (2011). The representative concentration pathways: An overview. *Climatic Change, 109*(1), 5–31.

Part I
Design, Innovation, and Technology

Chapter 2
Green Product Development: Price Competition, Quality Choice, and First-Mover Advantage

Chialin Chen and Kilsun Kim

2.1 Introduction

Green product design, which aims to improve a product's environmental benefits and/or to reduce its environmental impacts through product design and development, was originally introduced in the early 1990s. One notable approach for green product design was the Design for the Environment (DfE) program initiated by the United States Environmental Protection Agency (EPA) in 1992. The United Nations Environmental Programme, Division of Technology Industry and Economics (UNEP) was also a key player in this field in producing publications and collaborate on product sustainability projects. Over the years, green product design has evolved from a cleaner production method to a comprehensive design approach for improving the sustainability performance of products including social, economic, and environmental elements (Clark et al. 2009).

Developing environmentally friendly (green) products to create competitive advantages has become one of the biggest challenges and opportunities for industry. Many design decisions for environmentally friendly products, however, involve inevitable trade-offs between traditional and environmental attributes, such as finding the best combination of recyclable and non-recyclable materials and dealing with the trade-off between a vehicle's size and fuel efficiency (see, e.g., Field and Clark 1997; Calcott and Walls 2000; Chen and Zhang 2009; Chen and Ren 2010; Chen et al. 2012). While some companies have actively introduced environmentally

C. Chen (✉)
Department of Business Administration, College of Management,
National Taiwan University, Taipei, Taiwan
e-mail: cchen026@ntu.edu.tw

K. Kim
College of Business Administration, Sogang University, Seoul, South Korea
e-mail: kilsunkim@sogang.ac.kr

C. Chen et al. (eds.), *Pursuing Sustainability*, International Series in Operations Research & Management Science 301,
https://doi.org/10.1007/978-3-030-58023-0_2

friendly products, such as 100% recycled paper and alternative fuel vehicles, to the so-called green consumers who are willing to pay a premium for these products, many other companies have decided to adopt a wait-and-see policy because they still consider the green segment a niche market that is not sufficiently large. For example, it is well documented that American automakers' ignorance of the market segment for small, highly fuel-efficient vehicles allowed for Japanese companies, such as Toyota and Honda, to establish a strong and growing presence in North America in the 1970s. Even in the 1990s, many top executives of the Big Three dismissed the idea of introducing alternative-fuel vehicles with better environmental performance but poor traditional performance attributes due to market considerations (Christensen 1997; Christensen and Overdorf 2000). Not until the "surprising" growth of the market segments for small and alternative fuel vehicles due to the sky-high gasoline price and global environmental protection drives had most automakers realized that they should not have ignored the "small" number of green customers in the first place (*Wall Street Journal* 2006). According to White et al. (2019), some of the recently observed trends, such as millennials' strong interest in embracing sustainability, have provided great opportunities for today's firms to create competitive advantages through green product development.

With the presence of green consumers in a competitive market, the different strategies used by companies for new product development, as exemplified by the above anecdotes, raise a number of questions. How do companies design and price green and non-green products under different competitive settings? What are the economic and environmental consequences of green product development with competition? Can competition benefit the environment? What is the condition that favors the introduction of a green product? Is there any first mover advantage in green product development? To answer these questions, we will develop a game theoretical model to analyze the entry, quality, and pricing decisions for green product development in competitive environments, and investigate the resulting economic and environmental consequences. Specifically, we will first analyze the simultaneous games to establish some basic properties of the model, and then analyze the sequential games to derive important insights to the first mover advantage in green product development. The remainder of the paper is organized as follows. In Sect. 2.2, we review relevant literature. In Sect. 2.3, we present the basic model. In Sects. 2.4 and 2.5, we analyze the entry, quality, and pricing decision for green product development in simultaneous and sequential games. Managerial and policy implications and concluding remarks are given in Sect. 2.6.

2.2 Literature Review

Reinhardt (1998) proposes one of the early definitions of green product design which states that "environmental product differentiation takes place when a business creates products that provide greater environmental benefits, or that impose smaller environmental costs, than similar products." Commission of the European

Communities (2001) defines green products as products that "use less resource, have lower impacts and risks to the environment and prevent waste generation already at the conception stage." Chen (2001) treats green product design as a product development process, driven by consumer demand and governmental regulations, which addresses environmental issues through product design and innovation as opposed to the traditional end-of-pipe-control approach. Through a comprehensive literature review, Sdrolia and Zarotiadis (2019) recently define green product as "a product (tangible or intangible) that minimizes its environmental impact (direct and indirect) during its whole life-cycle, subject to the present technological and scientific status." Given the above definitions, we envision green product design as a product development approach for improving a product's environmental performances with greater environmental benefits and/or lower environmental damages, taking into considerations of consumer demand, technology capability, and governmental regulations.

There exists a rich literature on quantitative models for sustainable operations (see, e.g., Savaskan et al. 2004; Heese et al. 2005; Jayaraman et al. 2003; Ferrer 2003; Su et al. 2017; Zhu and He 2017; Bernard 2018; Xu et al. 2018). Good reviews are given in Brandenburg et al. (2014) and Bouchery et al. (2017). Extensive surveys of quantitative models in these areas are also given in Bloemhof-Ruwaard et al. (1995), Fleischmann et al. (1997), Dekker et al. (2004), Guide and van Wassenhove (2006), and Corbett and Klassen (2006). Majumder and Groenevelt (2001) and Corbett and DeCroix (2001) use game theory to analyze some interesting problems in reverse logistics and supply chain management. Debo et al. (2005) analyze the pricing and technology decisions for remanufacturing in an infinite-horizon setting, and Ferrer and Swaminathan (2006) consider a problem of offering a remanufactured and an all-new product in a duopoly and multi-period setting. Debo et al. (2006) consider joint life-cycle dynamics of new and remanufactured products through diffusion and substitution under different competitive settings, and Ferguson and Toktay (2006) address a question of how to position a remanufactured product in the presence of competition and analyze two entry-deterrent strategies: entering remanufacturing market and preemptive collection. Geyer et al. (2007) study the economics of remanufacturing considering limited component durability and finite product life cycles. Studies of reverse logistics network design are also active. Fleischmann et al. (2001) consider logistics network design in a reverse logistics context; Guide et al. (2006) incorporate a notion of marginal value of time in supply chain design choice; and Salema et al. (2007) study the design of a capacitated multi-product reverse logistics network with uncertainty. Most of the above quantitative models, however, are the so-called operations models which only treat environmental elements as an added-on in the operations and/or supply-chain management decision-making processes as opposed to focusing on the product creation and development processes with environmental performance evaluation in addition to the traditional economic performance metrics.

A considerable literature exists on economic models that analyze the interactions between market structures and the state of the natural environment. Hotelling (1931) first discusses the importance of particular market structures linked to the conservation of natural resources. His studies lead to the conclusion that monopoly markets exploit resources on a smaller scale than their competitive counterparts, a situation which Solow (1974) captures by stating that "the monopolist is the conservationist's friend." Most of the economic models have shown fairly consistent results that competition will generally lead to overuse of natural resources and thus is harmful to the environment (Tietenberg 2004). Fullerton and Wu (1998), Calcott and Walls (2000), Eichner and Pethig (2001) and Eichner and Pethig (2003) model different design and production decisions for consumer products with recycled/recyclable contents under different regulatory and policy settings. Studies by Smith et al. (2002), Bovenberg et al. (2005), and Subramanian et al. (2007) show that firms' profits can increase with stricter environmental standards.

Excellent reviews of quantitative models for product development decisions and managing product variety are given in Krishnan and Ulrich (2001) and Ramdas (2003). Moorthy and Png (1992), Desai et al. (2001), and Kim and Chhajed (2002) develop quality-based models for new product development with multiple market segments under different marketing and manufacturing considerations. Chen (2001) develops a quality-based model that analyzes the design decisions of new products for ordinary and green market segments with technological and regulatory constraints. Chen and Zhang (2013) present a general framework for optimal design decisions with different functional forms of technology efficient frontiers and a novel solution method based on trigonometric functions. These quality-based models, however, are under the monopoly assumption without the consideration of the impact of competition on new product development. Moorthy (1985, 1988), Tyagi (1999, 2000), Eriksson (2004), and Bernard (2018) develop quality-based models for new product development with multiple market segments under single and multiple quality dimensions. In terms of competitive green design models with evaluation of environmental performance, Chen and Liu (2014) develop a game-theoretic model for green product designed under price leadership in competitive environments. Murali et al. (2019) present a framework for studying the triple bottom line impact of voluntary ecolabels and mandatory environmental regulation on green product development among competing firms. In the section that follows, we will present the model structure for analyzing price competition, quality choice, and first mover advantage in green product development.

2.3 Model Structure

Our model framework is based on the N-firm, three-stage, non-cooperative game proposed by Shaked and Sutton (1982). In the first stage, firms choose whether to enter the market. If a firm decides to enter the market, it will incur a fixed cost δ for introducing any product type. In the second stage, each firm that enters the market in the first stage chooses (designs) the quality of its product by identifying the best location (position) in the quality dimension. In the last stage, each firm chooses its price. The profit of a firm is equal to the revenue from selling its product minus the production and fixed costs. If a firm decides not to enter the market in the first stage, its profit is equal to zero. We also adopt the framework for green product design by Chen (2001). Specifically, we consider a specific class of durable products with two competing qualities, the traditional and environmental qualities, q_t and q_e, where $q_t + q_e = 1$. For example, q_t and q_e can represent the non-recyclable and recyclable material contents of a product (whose sum is equal to 1 or 100%), or any other traditional and environmental attributes with competing relationships, such as vehicle safety (or size) and fuel efficiency as well as material consistency and recycled material content (see, e.g., Crandall and Graham 1989; Malloy 1996; Field and Clark 1997; Eichner and Pethig 2001; Eichner and Pethig 2003; Chen and Liu 2014).

On the demand side, there are M market segments, labeled $m = 1, 2, \ldots, M$, where $M \geq 2$. While customers in all the segments agree on the valuation for traditional attribute of a product, they differ in their valuations for the environmental attribute. Specifically, given a set of quality levels (q_t, q_e), the valuation of the traditional quality is equal to $v_t q_t$ for all segments, where v_t is the consumers' marginal valuation of the traditional quality. On the other hand, the valuation of the environmental quality by the customers in segments 1, 2, …, M are equal to $v_1 q_e$, $v_2 q_e$, …, $v_M q_e$, respectively, where $0 = v_1 < v_2 < \ldots < v_M$. While customers in segment 1 can be viewed as the ordinary customers who do not value the environmental quality at all (i.e., $v_1 = 0$), segments 2, …, M can be viewed as the sub-segments of the so-called green segment where customers receive positive values from the environmental quality of a product. This classification scheme of market segments resembles that used by RoperASW in its annual surveys which classifies all the consumers according to their different willingness-to-pay for environmentally friendly products (Ginsberg and Bloom 2004). As in many choice models and game-theoretical models based on Hotelling's (1929) framework, we assume that each customer will buy exactly one unit of the product offered by a firm. Also let n_1, n_2, …, n_M denote the sizes of the M segments. It should be noted that, compared to the competitive models of product design with continuous distributions of customer tastes (i.e., $M \to \infty$) whose tractability depends heavily on the assumption of uniformly distributed customer tastes, the above valuation structure allows us to analyze the effects of different preferential strengths and sizes of multiple segments on firms' design decisions. In practice, the discrete definition of market segments is also consistent with standard approaches in marketing research for product design and market segmentation, such as cluster analysis and discriminant analysis (see, e.g., Urban and Hauser 1993; Lilien and Rangaswamy 2003).

On the supply side, we consider N firms in a market where $N > M$. Note that the assumption that the number of firms is greater than that of market segments is needed in order to analyze the first mover advantage with the existence of potential entrants to the market. Following the Hotelling (1929) tradition, we assume that each firm introduces a single product under the identical cost structure (see, e.g., Shaked and Sutton 1982; Moorthy 1988). The unit cost of a product with qualities q_t and q_e is $c_t q_t^2 + c_e q_e^2$, where c_t and c_e are positive cost coefficients for the traditional and environmental qualities, respectively (Moorthy and Png 1992; Chen 2001). Note that the assumption of single product offering by each firm allows us to focus our analysis on the effects of competition (among firms) without the interfering factors of cannibalization (competition among a firm's own products). Also note that the assumption of the identical cost structure can rule out some trivial explanations of product differentiation, namely, technological differences (Moorthy 1988). We label all the products offered in the market by an index $k = 1, \ldots, N$. Firm k offers product k with quality set $s_k = \left(q_t^k, q_e^k\right)$ and price p_k, where q_t^k and q_e^k are the traditional and environmental qualities in product k. To simplify our exposition, we introduce the following functions for any product k with quality set s_k:

$C(s_k)$ = variable cost to produce one unit of product k.

$V_m(s_k)$ = the valuation of a customer in segment m for product k.

$U_m(s_k) = V_m(s_k) - C(s_k)$ = utility received by a customer in segment m from product k priced at the variable cost.

It should be noted that our assumption of competing attributes ($q_t + q_e = 1$) is due to the fact that one major challenge concerning the development of environmentally friendly products is how to deal with the trade-off between the traditional and environmental attributes (see, e.g., Porter and van der Linde 1995). Because of the multi-objective nature of green product development, very often the improvement in one attribute can only be accomplished at the expense of another (Keeney and Raiffa 1993). Theoretically, our modeling of the trade-off relationship is based on the theory of multiobjective decision making with the Pareto efficient frontier between competing product performance attributes, which stems from the neoclassical conception of production function of technology (Sahal 1981; de Neufville 1990). Also notice that the posited linear form of trade-off between the two attributes is widely adopted in engineering design and decision analysis with multiple attributes, and has been shown to be parsimonious, tractable, and can be expected to provide a reasonable approximation over a moderate range (see, e.g., Juran 1988; Keeney and Raiffa 1993).

The model we study in this paper does not have a pure-strategy Nash–Bertrand price equilibrium for the simultaneous game under certain situations. Therefore, we assume that each firm will try to achieve the maximin value for the simultaneous game under the novel price equilibrium concept proposed by Davis et al. (2004), which is based on the assumption that sellers announce prices from which they cannot renege by raising price whereas buyers always have the right to purchase at a previously announced price. This assumption leads to the equilibrium condition, referred to as "price cut immune," in which neither seller has an incentive to cut the price in order to steal its rivals' customers whereas consumers are content with their purchase decisions. Note that the maximin value is defined as the best payoff against the worst collective strategy (price cutting) that other firms could use against the firm. In addition, to simplify our analyses, we also assume that, when it makes no difference in the profit, a firm will always price or position its product at the most competitive level. We also assume that, although corner solutions of quality choice may exist, we will focus our attention on the interior solutions in all the analyses throughout the paper. Note that for all the competitive quality levels to be analyzed in the paper, the condition $-2c_t < v_m - v_t < 2c_e$ for all m is required to ensure the existence of interior quality solutions.

Given the model framework, it can be shown that the efficient quality sets, denoted by s_m^*, which maximize the differences between customer valuations and production costs for segments 1, …, M, are

$$s_m^* = \left(\frac{2c_e + v_t - v_m}{2(c_t + c_e)}, \frac{2c_t - v_t + v_m}{2(c_t + c_e)} \right), \quad \text{where} \quad m = 1, \ldots, M. \tag{2.1}$$

Figure 2.1 shows the cost and valuation structures of the model and the locations of the efficient quality sets for an example with three market segments (M = 3). According to the figure, while customers in segment 1 do not value the environmental quality of a product at all, customers in segments 2 receive positive but lower value from the environmental quality than customers in segment 3. Note that the figure is based on the assumptions $v_t > c_t$, $c_e > v_M$, and $v_3 > v_t > v_2 > v_1 = 0$. Given different sets of model parameters, the cost and valuation functions may take different forms.

Before analyzing the competition among multiple firms, we will identify the optimal quality sets under monopoly as the benchmarks. Specifically, if a monopolist intends to use the market-segmentation strategy, as defined in Moorthy (1984), it will introduce M products to the market, one for each segment. Let s_m^0 denote the

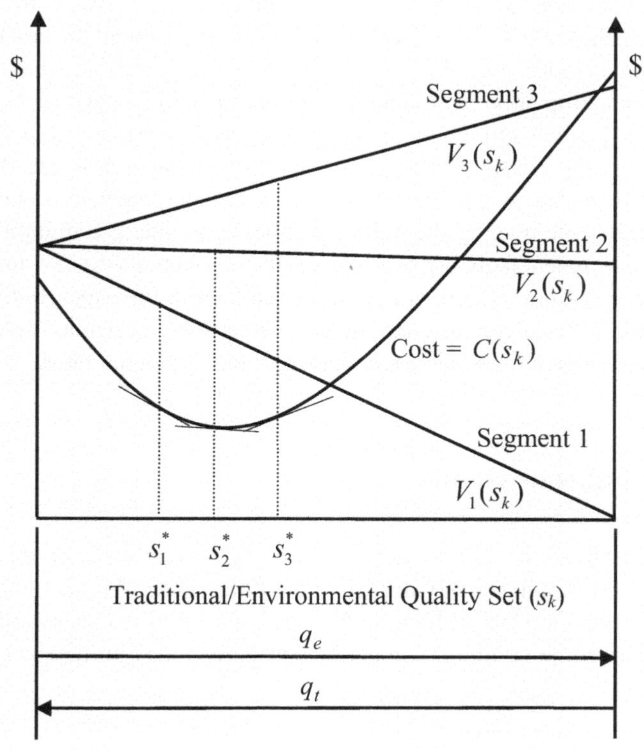

Fig. 2.1 Cost and valuation structures

quality set introduced by the monopolist for segment m. Also let $A_m = n_m + \cdots + n_M$. The monopolist's design decisions can then be characterized by the following proposition.

Proposition 1 *The quality levels of the products introduced by a monopolist for the M segments are*

$$s_m^0 = \left(\frac{n_m\left(2c_e + v_t - v_m\right) + A_{m+1}\left(v_{m+1} - v_m\right)}{2n_m\left(c_t + c_e\right)}, \frac{n_m\left(2c_t - v_t + v_m\right) - A_{m+1}\left(v_{m+1} - v_m\right)}{2n_m\left(c_t + c_e\right)} \right), \quad (2.2)$$

for $m = 1, \ldots, M-1$,

and

$$s_m^0 = \left(\frac{2c_e + v_t - v_m}{2\left(c_t + c_e\right)}, \frac{2c_t - v_t + v_m}{2\left(c_t + c_e\right)} \right), \quad \text{for } m = M. \quad (2.3)$$

The proof of the proposition is in Appendix 1. The above result is due to the fact that only the downward self-selection constrains are binding as discussed in Moorthy (1984). The reasoning is that the monopolist will make sure that the participation constraint for segment M (with the highest willingness-to-pay) will be binding while reducing the environmental qualities of the products for all the other segments to avoid cannibalization. As a result, only customers in segment M were offered the segment's efficient quality set, as can be observed from (2.1) and (2.3), while the environmental qualities offered to all the other segments are reduced by the amount $A_{m+1}(v_{m+1} - v_m)$ to ensure that customers in each segment will buy the product designed for them but not the one designed for another segment. Such a market segmentation scheme used by the monopolist has some interesting implications in the overall environmental quality of all the products sold in the market under different competitive settings, which will be discussed later in the paper. In the sections that follow, we will analyze the entry, quality, and pricing decisions under two different situations where firms make simultaneous and sequential moves.

2.4 Simultaneous Games

We will first analyze the simultaneous games where firms make entry, quality, or pricing decisions simultaneously in each of the three stages. As a common practice in game-theoretical models with multiple stages, we will use a backward induction process to derive the conditions under which perfect equilibrium exists for the three stages.

2.4.1 *Equilibria with Simultaneous Pricing, Quality, and Entry Decisions*

Starting from the last stage, we first analyze the price competition among products with quality sets s_k, $k = 1, \ldots, N$.. Let $-i$ denote the entire firm set that excludes only firm i. (The subscript $-i$ stands for "except i" or $\forall k \neq i$). As mentioned previously, to derive price equilibrium for the simultaneous game (in which the traditional Nash price equilibrium does not exist under certain situations), we will use the maximin rule under the price cut immune (PCI) concept proposed by Davis et al. (2004) with the following two definitions. (See Appendix 2 for the technical definitions of PCI.)

Definition 1 *A pricing outcome is PCI if no firm in a market can increase its profit by cutting price.*

Definition 2 *At equilibrium, each firm chooses its price from the set of PCI to maximize its profit.*

While Definition 1 ensures that price-cutting will not lead to higher profit for any firm in the market, Definition 2 ensures that each firm restricts its announcement to the most profitable price in the set of PCI. The price equilibrium defined by the above two conditions is thus sustainable since none of the firms finds it profitable to revise its offer (cut the price) given the price announced by its rivals, which is also a characteristic of Bertrand outcomes in standard models according to Davis et al. (2004).

Let $i(m)$ be the index of the product that offers the highest utility for segment m among the N products, where $m = 1, \ldots, M$. Here we assume that a single distinct product offers the highest utility for each segment. (Note that the major analytical results do not depend on this assumption which is only used to simplify our exposition.) At equilibrium, the prices of all the products in the market can be characterized in the next proposition.

Lemma 1 *At equilibrium, all the products will be priced at their marginal costs except for products $i(m)$, whose prices are equal to*

$$p_{i(m)} = V_m\left(s_{i(m)}\right) - \max\left\{U_m\left(s_{-i(m)}\right)\right\}, \quad \text{where} \quad m = 1, \ldots, M. \tag{2.4}$$

First note that, given the above price, customers in segment m will buy product $i(m)$ because firm $i(m)$ can undercut the prices of all its rivals by setting the price at $p_{i(m)}$ based on the Bertrand argument and the maximin rule. All the firms, including firm $i(m)$, cannot further cut their prices to increase their profits from segment m. This also holds true for all the other segments. Definition 1 above is thus verified. Since each firm $i(m)$ maximizes its profit by setting the price at $p_{i(m)}$ to undercut the price of the rival that offers customers in segment m with the second highest utility, $\max\{U_m(s_{-i(m)})\}$, and all the other firms set their prices at the most competitive levels (i.e., the marginal costs), as assumed previously, Definition 2 above is also verified. Because customers are allowed to purchase at a previously announced price, raising prices is not an option of any firm either even in the absence of pure-strategy Nash price equilibrium (Davis et al. 2004).

Based on the above result of price equilibrium, the next proposition establishes the equilibrium condition for the competition of quality choice in the second stage of the game.

Lemma 2 *At equilibrium of the game of quality choice,* (1) *each firm offers one of the M efficient quality sets; and* (2) *each of the M efficient qualities is taken by at least one firm.*

To show the above result, first note that, given $p_{i(m)}$ in (2.3), the profit of firm $i(m)$ is equal to

$$\pi_{i(m)} = n_m\left[p_{i(m)} - C\left(s_{i(m)}\right)\right] - \delta = n_m\begin{bmatrix} V_m\left(s_{i(m)}\right) \\ -\max\left\{U_m\left(s_{-i(m)}\right)\right\} - C\left(s_{i(m)}\right)\end{bmatrix} - \delta. \qquad (2.5)$$

Taking the first-order conditions of $\pi_{i(m)}$ with respective to $s_{i(m)}$, it can be easily verified that the efficient quality sets s_m^* in (2.1) are the maximizers of $\pi_{i(m)}$. It follows that all the firms in the market will try to take one of the M efficient quality sets for profit maximization. When each of the M efficient qualities is taken by at least one firm, none of the N firms in the market can further increase its profit by choosing another quality set since any firm offering identical products will receive zero profit from sales based on the Bertrand argument.

We now consider the analysis of the entry decisions in the first stage of the game. Recall that any firm that decides to enter the market will incur δ as the fixed cost. The following proposition concerns the entry decisions by the N potential entrants in the market.

Proposition 2 *There exists an equilibrium in which M firms enter; and in which they offer distinct products with the efficient quality sets* s_m^* *for segment m, where* $m = 1, \ldots, M$.

To show the above result, first note that, according to Lemmas 1 and 2, at most M firms can survive with positive profits and market shares at equilibrium with the efficient quality sets for the M segments. Given any group of M firms drawn from N potential entrants, if these M firms decide to enter the market, the payoff of each of the other firms from not entering is zero, while the payoff from entering is $-\delta$. As a result, only M firms enter. The M firms that enter will each earns a positive profit if δ is sufficiently small. Combining the above results, we have a subgame perfect equilibrium for the three-stage game in which M firms enter with distinct efficient quality sets given in (2.1) and the prices given in (2.4). The decision-making process based on the subgame perfect equilibrium is described as follows. To reach a sustainable set of price equilibrium under the concept of price cut immune in the last stage, each firm will offer the price that is just low enough to undercut the price of another firm that offers the second highest utility to a targeted segment. To maximize the equilibrium prices, firms will try to occupy one of the M efficient quality sets to better position themselves in the second stage. Knowing that all the firms will try to take the efficient quality sets and that at most M of them will make a positive profit, only M firms will enter the market in the first stage.

2.4.2 Can Competition Benefit the Environment?

To investigate the environmental consequences, let's assume that the total environmental quality is linear additive across all the products sold in the market. The index of the total environmental quality represents the improvement in environmental quality or reduction in environmental damage as a result of the use of all the products in the market. For example, if the environmental quality of interest is vehicle fuel efficiency, the overall environmental quality can be translated into the total amount of gasoline consumption with a particular product design. Given the optimal quality sets s_m^0 for the M segments under monopoly in (2.2) and (2.3), the overall environmental quality offered by a monopolist, denoted by TG_0, is equal to

$$TG_0 = \sum_{m=1}^{M-1} \frac{n_m\left(2c_\mathrm{t} - v_\mathrm{t} + v_m\right) - A_{m+1}\left(v_{m+1} - v_m\right)}{2\left(c_\mathrm{t} + c_\mathrm{e}\right)} + \frac{n_M\left(2c_\mathrm{t} - v_\mathrm{t} + v_M\right)}{2\left(c_\mathrm{t} + c_\mathrm{e}\right)} \tag{2.6}$$

where $A_m = n_m + \cdots + n_M$. With the presence of N potential entrants in the market, however, the equilibrium qualities are the efficient quality sets s_m^* in (2.1) for the M segments, according to Proposition 2. The overall environmental quality of all the products sold in the market with N firms, denoted by TG_N, is then equal to

$$TG_N = \sum_{m=1}^{M} \frac{n_m\left(2c_\mathrm{t} - v_\mathrm{t} + v_m\right)}{2\left(c_\mathrm{t} + c_\mathrm{e}\right)} \tag{2.7}$$

Since $v_m \le v_{m+1}$, comparing TG_0 and TG_N leads to the following proposition.

Proposition 3 *The overall environmental quality under competition is higher than that under monopoly. Specifically, the improvement in overall environmental quality due to competition is equal to*

$$TG_N - TG_0 = \sum_{m=1}^{M-1} \frac{A_{m+1}\left(v_{m+1} - v_m\right)}{2\left(c_\mathrm{t} + c_\mathrm{e}\right)}, \quad \text{where} \quad A_m = n_m + \cdots + n_M \tag{2.8}$$

The major reason behind this proposition is that competition can help correct the distortion of product qualities under monopoly. Recall from our previous discussion that the environmental qualities of some products offered by a monopolist are reduced from the efficient levels in order to prevent cannibalization. With the presence of new and potential entrants in the market, however, such a market segmentation scheme no longer works since firms now need to take the efficient qualities to better position themselves in the competitive market in order to maximize their profits during the price competition in the third stage of the game. The result that competition can benefit the environment by inducing competitive firms to take more environmentally friendly product positions differs from the popular view that competition is harmful to the environment since it will usually lead to overuse of natural resources. It should be noted that our analytical results are based on the assumption that the total demand quantity in a market is fixed. If the total quantity sold in the market could also depend on the price and quality levels, the resulting improvement in environmental quality could be different. One interesting implication of the above proposition is that, if the monopolist is allowed to adjust the quality levels of its product offerings to prevent potential entry, it will "voluntarily" improve the environmental qualities of all the products by moving from the monopoly levels (s_m^0) to the efficient levels (s_m^*). That is, the pressure of competition would be enough to induce a firm to become greener even without the regulatory pressure. Also note that the above result about the overall environmental quality differs from those derived from traditional duopoly models where firms enjoy more freedom for product differentiation in the absence of potential entrants (Chen and Liu 2014).

2.5 The First-Mover Advantage in Sequential Games

We will now analyze the sequential games where firms take turn to make their decisions in each stage, which resemble the "games with perfect information" discussed in Tirole (1988). The main objective of our analysis is to depict the way competition evolves in an emerging market through analyzing the impacts of the first mover advantage and price leadership on green product development. In particular, we are interested in the condition(s) under which the first mover that enters a market will choose to become a "green" firm through introducing a single green product to the market due to competitive considerations. Such an analysis is of importance since, as mentioned previously, many of today's firms are facing the dilemma of whether to stay with the ordinary segment or to pursue the new opportunity in the green segment.

To allow us to use the framework presented previously to analyze the issues concerning the first mover advantage and price leadership in green product development, we will consider a simplified version of the problem with two market segments: the ordinary and green segments, denoted by segment o and segment g (i.e., $M = 2$), with sizes n_o and n_g. Let n_o^g denote the ratio of market sizes. Note that it is a common practice in standard models for analyzing the first mover advantage to adopt the two-segment assumption due to the consideration of tractability (see, e.g., Johnson and Myatt 2003; Davis et al. 2004; Adner and Zemsky 2005). In practice, detailed information about customer valuation in an emerging market is usually difficult to obtain (Christensen 1997; Christensen and Overdorf 2000). Therefore, it is reasonable to assume that firms will adopt a more general definition of market segments (i.e., green and non-green) when making their entry/design decisions. We thus assume that all the customers with positive willingness-to-pay for the environmental quality can be combined into a green (environmentally friendly) segment with an overall marginal valuation, denoted by $v_e > 0$, for the environmental quality, while the marginal valuation for the environmental quality for the ordinary segment is equal to zero. We will also combine stages 1 and 2 so that when a firm decides to enter the market, it has to make the quality decision in the same stage. This assumption allows us to analyze the first mover advantage since the first mover is also given the "right" to choose the quality first. In this modified two-stage game, firms take turn to enter the market with specific quality sets in the first stage, and then take turns to set their prices (under a price leadership arrangement) in the second stage.

Based on the previous analytical results and the additional assumptions above, it can be easily verified that, among the N potential entrants ($N > M = 2$), the first two firms that enter the market will take turn to occupy one of the two distinct efficient quality levels for the ordinary and green segments since any firm that deviates from the efficient quality sets will be priced out of the market by the third entrant that will take the untaken efficient quality set. With this unwanted consequence in mind, the first two entrants will take the two efficient quality positions, and all the other firms will then choose not to enter the market given the fixed entry cost $\delta > 0$. There are still two major issues, however, that need to be resolved. First, which efficient quality set, ordinary or green, should the first mover offer to preempt the desirable market segment? Also, what kind of arrangement of price leadership can be reached between the two firms with distinct quality positions? These questions will be answered with our analyses that follow.

2.5.1 Price Competition and Role-Choosing Games

We will now analyze the sequential game of price competition where the two firms try to reach the Stackelberg equilibrium with one firm assuming the role of price leader. As discussed in Tyagi (1999), arrangements of price leadership can be either exogenously or endogenously determined. Therefore, we will first analyze the pricing decisions of the two firms under prespecified (exogenous) leadership arrangements, and then try to identify the endogenously determined outcomes.

To distinguish the two firms which take turn to offer the efficient quality sets for the two segments in the sequential game, we will call the firms which offer the efficient ordinary quality set (s_o^*) and the efficient green quality set (s_g^*) the ordinary and green firms, respectively, and call their products ordinary and green products. Also let p_o and p_g denote the prices of the ordinary and green products, respectively. Based on the efficient quality levels given in (2.1), we can first derive the following values of customer valuations and production costs to be used later in our analysis.

$$V_o\left(s_o^*\right)=\frac{v_t^2+2c_e v_t}{2\left(c_t+c_e\right)},\ V_o\left(s_g^*\right)=\frac{v_t^2+2c_e v_t-v_t v_e}{2\left(c_t+c_e\right)},\ V_g\left(s_o^*\right)=\frac{v_t^2+2c_e v_t+2c_t v_e-v_t v_e}{2\left(c_t+c_e\right)},$$
$$V_g\left(s_g^*\right)=\frac{v_t^2+2c_e v_t+2c_t v_e-2v_t v_e+v_e^2}{2\left(c_t+c_e\right)},\ C\left(s_o^*\right)=\frac{4c_t c_e+v_t^2}{4\left(c_t+c_e\right)},\ \text{and} \quad (2.9)$$
$$C\left(s_g^*\right)=\frac{4c_t c_e-2v_t v_e+v_t^2+v_e^2}{4\left(c_t+c_e\right)}.$$

Let's start with the situation where the ordinary firm sets its price first as the price leader and the green firm sets its price next as the follower. Note that the analysis will be similar if the two firms reverse their roles. For any price set by the ordinary firm as the price leader, the green firm (the follower) must decide to protect (the green segment) or to attack (the ordinary segments). To protect, it will set the price of the green product just a little lower than the level where customers in the green segment are indifferent between the green and ordinary products. To attack, the green firm will set the price of the green product just a little lower than the level where customers in the ordinary segment are indifferent between the green and ordinary products. Note that the price to attack is lower than that to protect. Therefore, when the green firm attacks, the green segment will be protected as well. Let us use Fig. 2.2 as an example. Suppose that the ordinary product is priced at p_1. The green firm will have to decide whether to attack (with p_a) or to protect (with p_b). It can be easily shown that, given any feasible set of p_a and p_b, the difference between the prices to protect and to attack, denoted by $K(=p_b - p_a)$, is a constant. As shown in Fig. 2.2, the value of K is equal to

$$K=V_g\left(s_g^*\right)-V_o\left(s_g^*\right)-V_g\left(s_o^*\right)+V_o\left(s_o^*\right)=\frac{v_e^2}{2\left(c_t+c_e\right)}. \quad (2.10)$$

To make the decision between to attack or to protect, the green firm will compare the profits resulted from the two options. Let π_a and π_b be the profits of the attacking and protecting options, respectively. Given any leader's price p_1, the follower's profits from the attacking and protecting options are $\pi_a=\left(n_o+n_g\right)\left[p_a-C\left(s_g^*\right)\right]-\delta$ and $\pi_b=n_g\left[p_b-C\left(s_g^*\right)\right]-\delta=n_g\left[p_a+K-C\left(s_g^*\right)\right]-\delta$. It can then be shown that there exists a unique critical price p_o^* set by the ordinary firm (the leader) where the green firm (the follower) is indifferenwt between attacking and protecting, that is, $\pi_a = \pi_b$. Notice that there also exists a similar critical price set by the green firm where the ordinary firm is indifferent between attacking and protecting. It can also be shown that, as the leader's price reduces from a higher level, the follower's profit from attacking reduces faster than that from protecting (i.e., $\partial\pi_a/\partial p_o \geq \partial\pi_b/\partial p_o > 0$). It follows that, at equilibrium, the leader will set its critical price at (or, precisely, a little lower than) the level where the follower is indifferent between attaching and protecting, and the follower will then set the price to protect accordingly. The results under the two different price leadership arrangements are summarized in the following proposition. (See Appendix 3 for proof.)

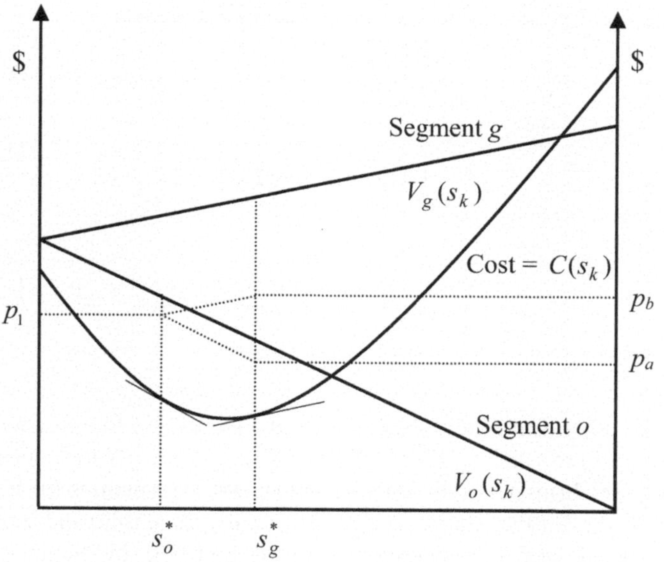

Fig. 2.2 Critical price levels

Proposition 4 *If the ordinary and green firms act as the price leader and follower, respectively, the prices of the ordinary and green products, denoted by* p_o^l *and* p_g^f*, are*

$$p_o^l = \frac{v_t^2 + 2n_o^g v_e^2 + 4c_t c_e + v_e^2}{4(c_t + c_e)} \quad \text{and} \quad p_g^f = \frac{v_t^2 - 2v_t v_e + 4c_t c_e + 2n_o^g v_e^2 + 3v_e^2}{4(c_t + c_e)}. \tag{2.11}$$

If the ordinary and green firms act as the price follower and leader, respectively, the prices of the ordinary and green products, denoted by p_o^f and p_g^l*, are*

$$p_o^f = \frac{n_o^g v_t^2 + 4n_o^g c_t c_e + 2n_o^g v_e^2 + 2v_e^2}{4n_o^g (c_t + c_e)} \quad \text{and}$$
$$p_g^l = \frac{n_o^g v_t^2 - 2n_o^g v_t v_e + 4n_o^g c_t c_e + 2n_o^g v_e^2 + 2v_e^2}{4n_o^g (c_t + c_e)}. \tag{2.12}$$

To identify the endogenous (mutually acceptable) arrangements of price leadership, we will use the role-choosing game proposed by Dowrick (1986). Table 2.1 describes the profits of the two firms under the four possible arrangements of price leadership in the role-choosing game where π_o and π_g denote the profits of the ordinary and green firms, respectively. In Outcomes I and II where one firm acts as the

Table 2.1 The Role-Choosing Game between the Ordinary and Green Firms

	Green firm	
Ordinary firm	Leader	Follower
Leader	$\pi_o = n_o\left[p_o^l - C\left(s_o^*\right)\right] - \delta$ $\pi_g = n_g\left[p_g^l - C\left(s_g^*\right)\right] - \delta$ III	$\pi_o = n_o\left[p_o^l - C\left(s_o^*\right)\right] - \delta$ $\pi_g = n_g\left[p_g^f - C\left(s_g^*\right)\right] - \delta$ II
Follower	$\pi_o = n_o\left[p_o^f - C\left(s_o^*\right)\right] - \delta$ $\pi_g = n_g\left[p_g^l - C\left(s_g^*\right)\right] - \delta$ I	$\pi_o = n_o\left[p_o - C\left(s_o^*\right)\right] - \delta$ $\pi_g = n_g\left[p_g - C\left(s_g^*\right)\right] - \delta$ IV

leader and the other firm acts as the follower, the leader and follower will set their prices at the p_o^l or p_g^l and p_g^f or p_o^f, respectively, accordingly to Proposition 4. In Outcome III, both firms act as the leader, and they will both assume the leader's role (without knowing that the other firm also chooses the same role) and set their leader's prices (i.e., p_o^l and p_g^l) according to Dowrick (1986). In Outcome IV where both firms act as the followers, the equilibrium result of the simultaneous game (in which there is no apparent price leader) can be used to identify the price levels (Dowrick 1986). Let p_o and p_g denote the equilibrium prices of the ordinary and green firms, respectively, for the simultaneous game. Based on (2.4), we have $p_o = V_o\left(s_o^*\right) - V_o\left(s_g^*\right) + C\left(s_g^*\right)$ and $p_g = V_g\left(s_g^*\right) - V_g\left(s_o^*\right) + C\left(s_o^*\right)$. Given the customer valuations and production costs in (2.9), the values of p_o and p_g are equal to

$$p_o = \frac{v_t^2 + 4c_t c_e + v_e^2}{4\left(c_t + c_e\right)} \quad \text{and} \quad p_g = \frac{v_t^2 - 2v_t v_e + 4c_t c_e + 2v_e^2}{4\left(c_t + c_e\right)} \tag{2.13}$$

To identify the acceptable outcomes for both firms, we will look for the Nash equilibrium solutions for the role-choosing game described in Table 2.1, which leads to the following proposition.

Proposition 5 (1) *If* $n_o^g \le 0.78$, *the green and ordinary firm will act as the price leader and follower, respectively, with prices given in* (2.12). (2) *If* $n_o^g > 1.28$, *the ordinary and green firms will act as the price leader and follower, respectively, with prices given in* (2.11). (3) *If* $0.78 < n_o^g \le 1.28$, *one firm will act as the price leader and the other will act as the follower with prices given in* (2.11) *or* (2.12).

The proof of the proposition is given in Appendix 4. According to the proposition, if the ratio of the numbers of green to ordinary customers is relatively small (i.e., $n_o^g \le 0.78$), the green firm will act as the price leader and the ordinary firm will act as the follower. When the ratio is relatively large (i.e., $n_o^g \ge 1.28$), the green firm will act as the price follower, and the ordinary firm will act as the leader. This result is quite surprising because, as the green firm "owns" more customers in the green segment, one would expect that the green firm should assume the role of price leader. The reasoning behind this result is as follows. The green firm will choose to lead when the ordinary segment is so large that, if the ordinary firm leads, it will set the price of the ordinary product at a very low level (to protect the large ordinary segment), which leads to low profits for both firms. Since the green segment is relatively small, the green firm's critical price (p_g^l) is relatively high. The green firm's being the price leader will lead to higher profits for both firms. On the other hand, as the number of green customers becomes large, it is the green firm that has too much to lose. Therefore, letting the ordinary firm be the price leader will lead to higher profits for both firms.

The above proposition also indicates that, when the numbers of green and ordinary customers are relatively even ($0.78 < n_o^g < 1.28$), both firms will be better off if they choose different roles in price leadership. Which firm will eventually become the price leader, however, is yet to be determined because there is a conflict between the two firms' best interests (of being a follower). As discussed in Dowrick (1986), some sort of coordination mechanisms or implicit/explicit collusion for reaching a mutually acceptable arrangement between the two firms can be expected under such a situation since, compared to the other two outcomes (where both firms are leaders or followers), it is still more profitable for one of the firms to play the "second best" role as the price leader.

2.5.2 *The First Mover Advantage in Green Product Development*

Let's now analyze the situation where the two firms make sequential entry/quality decisions. For ease of discussion, we will call the firm that makes the entry/quality decisions first the first mover and the firm that makes the decisions next the second mover. Since the second mover has the opportunity to observe the first mover's move before making its own, the second mover's strategy is conditional upon the first mover's. As discussed previously, the first and second movers will each take turn to occupy one of the two distinct efficient quality levels for the ordinary and green segments. Knowing that the second mover will introduce a distinct product, how should the first mover utilize the first mover advantage to "preempt" the desirable segment? To answer this question, we will analyze the first mover's quality decisions under different exogenous and endogenous arrangements of price leadership.

As discussed in Higgins (1992) and Tyagi (1999), exogenous arrangements of price leadership in Stackelberg duopoly can be based on either *market* or *firm* characteristics. For the market-based arrangements, Deneckere et al. (1992) and Furth and Kovenock (1993) show that the firm with larger loyal segment would emerge as the price leader. The next proposition identifies the conditions that favor the introduction of a green product by the first mover under such an arrangement of price leadership.

Proposition 6 *If the firm with the larger loyal segment also acts as the price leader, the first mover in quality choice will introduce a green product under either of the two conditions*: (1) *and* (2) $n_0^g \geq 1.41$.

The proof of the proposition is given in Appendix 5. The above proposition presents an interesting result where less than 50% ($0.71/1.71 = 41.5\%$) of green customers in a market would be considered as "sufficiently large" to induce the first mover to go green. In a market where the firm with the biggest market share acts as the price leader, a phenomenon that can be observed in many industries (Higgins 1992), introducing a green product to take the relatively smaller green segment is actually a more profitable quality choice for the first mover. The reasoning is that quality choice can be used by the first mover as a vehicle to reach a favorable arrangement of price leadership under different market conditions. If going green can help the first mover secure the desirable role as a price leader or follower, it pays for the firm to preempt the green segment through introducing a green product. Additionally, since customers in the green segment have higher willingness-to-pay for a given quality set than customers in the ordinary segment, a relatively small green segment would be enough to induce the first mover to take the green quality position.

For the firm-based arrangements of price leadership, the positioning first mover can be assumed to be the exogenously specified price leader or follower due to several observable characteristics of the firm, such as reputation, historical pricing behavior, and technological capability (Tirole 1988). The next proposition describes the first mover's quality choice under such an arrangement of price leadership.

Proposition 7 *The first mover in quality choice will choose the smaller market segment if and only if it is also the price leader.*

The proof of the proposition is given in Appendix 6. The proposition leads to another result where the first mover would prefer to introduce a green product even if the green segment is still smaller than the ordinary segment. The reasoning is that, if the first mover in quality choice has to "accept" the role of price leader, taking the smaller (green) segment that is less attractive for the follower to attack (i.e., with a higher critical price) as well as with higher willingness-to-pay can be a more profitable option.

In some industries, however, there might be no apparent arrangement of price leadership, and the first mover's quality choice will depend on the endogenously determined arrangements of price leadership (Gal-Or 1985; Dowrick 1986). In the next proposition, we will use the outcomes of the role-choosing game discussed previously to analyze the first mover's quality choice based on endogenous arrangements of price leadership between the two firms.

Proposition 8 *In a market where the arrangement of price leadership is endogenously determined, the first mover will introduce a green product under either of the two conditions:* (1) $0.78 < n_0^g \leq 1.28$, *and the first and second movers act as the price follower and leader, respectively, or* (2) $n_0^g > 1.28$.

The proof of the proposition is given in Appendix 7. The proposition indicates that, when the green segment is still smaller than the ordinary segment, attaining the desirable role as the price follower is necessary to give the first mover enough incentive to introduce a green product. According to Proposition 5, not until n_o^g exceeds 0.78 would it be possible for the two firms to reach such an arrangement of price leadership as one of the two likely outcomes under Nash equilibrium through some sort of coordination mechanisms or explicit/implicit collusion. If the desirable arrangement cannot be endogenously reached, an even stronger condition, $n_o^g \geq 1.28$, is needed to induce the first mover to introduce a green product.

Figure 2.3 summarizes the results in Propositions 6, 7, and 8, which identifies the percentages of green customers in a market (from 0% to 100%) that favor the introduction of a green product by the first mover in quality choice under different arrangements of price leadership. (Note that all the values of n_o^g are converted to the percentages of green customers in the figure.) Areas A, B, and C represent the ranges of the percentage of green customers which favors the introduction of a green product by the first mover in a market where the larger firm or the first mover is the price leader, while areas D and E represent the ranges of the percentage that favors the introduction of a green product by the first mover in a market where price leadership is endogenously determined.

It should be noted, however, that the decision to go green can still be subject to a high degree of risks and uncertainties when the green segment is relatively smaller than the ordinary segment. There is no guarantee that the arrangements of price leadership which favor the introduction of a green product will be "honored" by both firms since quality choice is usually made before the pricing decision for a product (Tirole 1988). While an exogenous arrangement of price leadership may be challenged, the endogenous arrangement that favors the introduction of a green product may not be sustainable without a majority of green customers in the market

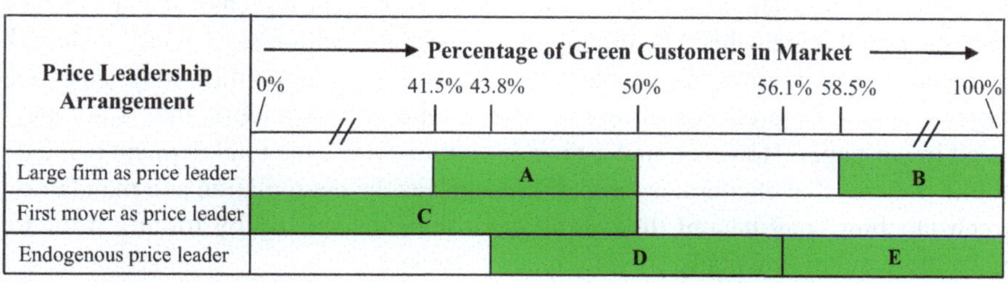

Notes: (i) Shaded areas A and B represent the ranges of the percentage of green customers which favors the introduction of a green product by the first mover in quality choice in a market where the firm with larger loyal segment acts as the price leader.
(ii) Shaded area C represents the range of the percentage of green customers which favors the introduction of a green product by the first mover in quality choice in a market where the first mover also acts as the price leader.
(iii) Shaded areas D and E represent the ranges of the percentage of green customers that favors the introduction of a green product by the first mover in quality choice in a market where the price leader is endogenously determined.

Fig. 2.3 First Mover Advantages in Green Product Development

(i.e., since it is only one of the two possible outcomes under Nash equilibrium in the role-choosing game. Nevertheless, where there is risk, there is also opportunity. The analytical results which suggest that a green firm may emerge with less than 50% of green customers in the market reveal an interesting yet important message to today's companies: The right time to go green may actually be earlier than you thought! This message is especially useful for small-to-mid sized companies which usually start with targeting a niche segment of a market. Interestingly, according to the 2019 Retail and Sustainability Survey conducted by CGS, a global provider of business applications, about 47% of consumers in American population are potential green consumers would pay more for a sustainable product. With a favorable price leadership arrangement, this could be the time for today's firms to take the first-mover advantage through introducing a green product to preempt the green segment.

2.6 Concluding Remarks

In this chapter, we develop a multi-firm, multi-segment model to analyze the simultaneous and sequential decisions of entry, quality choice, and price competition in product development for a market where customers differ in their willingness-to-pay for the environmental quality of a product. For the simultaneous game, we identify the conditions for a subgame perfect equilibrium in which a subset of the competitive firms in the market enter with distinct efficient products for multiple segments to reach the price equilibrium under the concept of price cut immune. At the equilibrium, the overall environmental quality in products sold to all the customers in the market is higher than that under monopoly since the market segmentation scheme used under monopoly (which reduces the environmental qualities of some products to prevent cannibalization) no longer works with the presence of new and potential entrants. The result that competition can benefit the environment differs from the popular view that "the monopolist is the conservationist's friend." It should be noted that policy makers should still be aware that competition may lead to a larger scale of resource extraction than does monopoly, a situation that is not analyzed in our paper. However, under the assumption where the total demand is fixed, public policies that encourage competition, such as the deregulation and mandatory green labeling, may benefit the overall environmental quality by forcing firms to make more environmentally friendly quality decisions.

For the sequential game, we analyze both the endogenous and exogenous arrangements of price leadership and the quality choice by the first mover to enter the market. According to our analyses, in a market where there exists no apparent (exogenous) arrangement of price leadership, a rule of thumb is that a firm should choose to act as the price leader when the size of its market segment is relatively small to prevent its competitor from acting as an "over protective" price leader whose low leader's price will lead to low profits for both firms. We have also analyzed how the first mover can utilize the first mover advantage by preempting the desirable market segment with a green product. Specifically, in a market where the firm with larger loyal segment acts as the price leader, the first mover may use the introduction of a green product to attain the desirable role as a price follower through preempting the relatively smaller green segment with high willingness-to-pay. Additionally, in a market where the first mover is also the price leader, introducing a green product to the relatively smaller green segment, which is less attractive for the follower to attack, would be a more profitable choice. Furthermore, in a market where the arrangements of price leadership are endogenously determined, introducing a green product would be a more profitable option if the ordinary and green firms act as the price leader and follower, respectively, given a sufficiently large green segment.

Future research is needed on the interfaces between green product design and several newly observed market, technological, and regulatory trends, such as consumers' environmentally conscious behaviors in an omni-channel environment (Chopra 2018), the new paradigm of fulfilling customers' ongoing desire to participate in green product design enabled by internet of things (IoT) (Manavalan and Jayakrishna 2019), and flexible environmental regulations (Ramanathan et al. 2018).

2.7 Appendix 1 (Proof of Proposition 1)

The proof given here is an extension of Moorthy (1984). Let p_m and $s_m^0 = \left(q_t^m, q_e^m\right)$ denote the price and quality levels of the product introduced to segment m by the monopolist. The monopolist's problem can be formulated as follows:

$$\max \sum_{m=1}^{M} n_m \left(p_m - c_t q_t^{m2} - c_e q_e^{m2}\right)$$

subject to:

$$\text{For all } m = 1,\ldots,M,\ v_t q_t^m + v_m q_e^m - p_m \geq v_t q_t^j + v_m q_e^j - p_j, \quad \text{for} \quad j = 1,\ldots,M, \tag{2.14}$$

and

$$q_t^m + q_e^m = 1. \tag{2.15}$$

Note that constraint (2.14) is the self selection constraint for each segment, while constraint (2.15) is the technology constraint. Following Moorthy (1984), only the downward self selection constraints are binding. We can rewrite the objective function as follows:

$$\max\ \pi = \sum_{m=1}^{M-1}\left[n_m\left(v_t q_t^m + v_m q_e^m - c_t q_t^{m2} - c_e q_e^{m2}\right) + A_{m+1}\left(v_{m+1} q_e^m - v_m q_e^t\right)\right] + n_M\left(v_t q_t^M + v_M q_e^M - c_t q_t^{M2} - c_e q_e^{M2}\right), \text{ where } A_m = n_m + \cdots + n_M, \tag{2.16}$$

subject to the technology constraint. The first-order conditions of the above objective function lead to the results in (2.3) and (2.4).

2.8 Appendix 2 (Technical Definitions of Price Cut Immune)

We will extend the equilibrium concept of price cut immune proposed in Davis et al. (2004) to the *N*-firm model with the following two definitions. (Note that $-k$ denotes the entire firm set that excludes firm k.)

Definition 1 (Price Cut Immune) Let $\Pi_k(p_k, p_{-k})$ denote the profit of firm k with price p_k. A pricing outcome $\{p_k^*, p_{-k}^*\}$ is *price cut immune* (PCI) if (1) for any $p_k < p_k^*$, $\Pi_k(p_k, p_{-k}^*) \le \Pi_k(p_k^*, p_{-k}^*)$, and (2) for any $p_{-k} < p_{-k}^*$, $\Pi_{-k}(p_k^*, p_{-k}) \le \Pi_{-k}(p_k^*, p_{-k}^*)$.

Definition 2 (Price Equilibrium) For any set of prices announcement by firms $-k$, p_{-k}, let $\Psi_k(p_k)$ be the set of price announcements p_k such that $\{p_k, p_{-k}\}$ is PCI. Define $\Psi_{-k}(p_{-k})$ conformably. Then $\{p_k^*, p_{-k}^*\}$ is a price equilibrium if (1) $\Pi_k(p_k^*, p_{-k}^*) = \max_{p_k \in \Psi_k(p_{-k}^*)} \Pi_k(p_k, p_{-k}^*)$ and (2) $\Pi_{-k}(p_k^*, p_{-k}^*) = \max_{p_{-k} \in \Psi_{-k}(p_k^*)} \Pi_{-k}(p_k^*, p_{-k})$.

2.9 Appendix 3 (Proof of Proposition 4)

To solve for the value of p_o^l, p_g^l, p_o^f, and p_g^f, we will first assume that the ordinary and green firms are the price leader and follower, respectively. Recall that the difference between the prices to protect and to attack is a constant K whose value is given in (2.10). Given the critical price p_o^l set by the leader, the green firm's profit from the attacking option is equal to its profit from the protecting option, that is, $(n_o + n_g)\left[(p_g^f - K) - C(s_g^*)\right] - \delta = n_g\left[p_g^f - C(s_g^*)\right] - \delta$, where the values of $C(s_o^*)$ and $C(s_g^*)$ are given in (2.9). We can then solve for the value of p_g^f as given in (2.11). To obtain the value of p_o^l, notice that the green customers are indifferent between a green product priced at p_g^f and an ordinary product priced at p_o^l (so that the green firm can set the price at p_g^f to protect its own segment), that is, $V_g(s_g^*) - p_g^f = V_g(s_o^*) - p_o^l$, where the values of $V_g(s_o^*)$ and $V_g(s_g^*)$ are given in (2.9). We can thus solve for p_o^l as given in (2.11). Similarly, we can solve for p_g^l

and p_o^f as in (2.12) if the green and ordinary firms act as the price leader and follower, respectively.

2.10 Appendix 4 (Proof of Proposition 5)

Given the values of p_o^l, p_g^l, p_o^f, and p_g^f in (2.11) and (2.12), we have

$$p_g^f - p_g^l = \frac{\left(2n_o^{g2} + n_o^g - 2\right)v_e^2}{4n_o^g\left(c_t + c_e\right)} \quad \text{and} \quad p_o^f - p_o^l = \frac{-\left(2n_o^{g2} - n_o^g - 2\right)v_e^2}{4n_o^g\left(c_t + c_e\right)}, \qquad (2.17)$$

which imply that $p_g^f \geq p_g^l$ if $n_o^g \geq 0.78$ and that $p_o^f \geq p_o^l$ if $n_o^g \geq 0.78$. It can also be verified that $p_o \leq p_o^l$, $p_o \leq p_o^f$, $p_g \leq p_g^l$, and $p_g \leq p_g^f$, where the values of p_o and p_g are given in (2.13). Therefore, if $n_o^g \geq 0.78$, then $p_g^f \leq p_g^l$ and $p_o^f \geq p_o^l$ (i.e., the green firm wants to lead, and the ordinary firm wants to follow). It follows that Outcome I on the lower left corner of Table 2.1 is a Nash equilibrium solution for $n_o^g \geq 0.78$ since neither the ordinary firm (follower) nor the green firm (leader) wants to change its role given the role chosen by its opponent. Similarly, it can be shown that Outcome II (i.e., the ordinary firm leads and the green firm follows) is the Nash equilibrium solution for $n_o^g > 1.28$ (i.e., $p_g^f > p_g^l$ and $p_o^f < p_o^l$), and that Outcomes I and II are both the Nash equilibrium solutions for $0.78 < n_o^g \leq 1.28$ (i.e., $p_g^f > p_g^l$ and $p_o^f \geq p_o^l$). Note that there is a slightly different version of Outcome III where both ordinary and green segments choose the same product offered by only one of the two firms. It can be shown, however, that the Nash equilibrium solutions of the role-choosing game described above remain unchanged.

2.11 Appendix 5 (Proof of Proposition 6)

When $n_o^g < 1$, the ordinary product has a larger loyal segment than the green product. If the first mover introduces an ordinary product (and acts as the price leader due to its larger loyal segment), its leader's profit from the ordinary segment will be $\pi_o = n_o\left[p_o^l - C\left(s_o^*\right)\right] - \delta$. Otherwise, its follower's profit from the green segment will be $\pi_g = n_g\left[p_g^f - C\left(s_g^*\right)\right] - \delta$, where the values of p_o^l, p_g^f, $C\left(s_o^*\right)$, and $C\left(s_g^*\right)$ are given in (2.9) and (2.11). We thus have $\pi_g - \pi_o = n_o\left(2n_o^{g2} - 1\right)v_e^2 / 4\left(c_t + c_e\right)$. It follows that $\pi_o \leq \pi_g$ if $n_o^g \geq 0.71$, the condition for the first mover to introduce a green product. Similarly, when $n_o^g \geq 1$, it can be shown that $\pi_g - \pi_o \geq 0$ if $n_o^g \geq 1.41$.

2.12 Appendix 6 (Proof of Proposition 7)

If the first mover acts as the exogenously specified price leader, its profit from introducing an ordinary or a green product is equal to $\pi_o = n_o\left[p_o^l - C\left(s_o^*\right)\right] - \delta$ or $\pi_g = n_g\left[p_g^l - C\left(s_g^*\right)\right] - \delta$, where the values of p_o^l, p_g^l, $C\left(s_o^*\right)$, and $C\left(s_g^*\right)$ are given in (2.9), (2.11), and (2.12). It can then be shown that $\pi_g - \pi_o = n_o\left(1 - n_o^g\right)v_e^2 / 4\left(c_t + c_e\right) \geq 0$ if $n_o^g \leq 1$. Similarly, if the first mover acts as the exogenously specified price follower, it can be shown that $\pi_g - \pi_o \geq 0$ if $n_o^g \geq 1$.

2.13 Appendix 7 (Proof of Proposition 8)

According to Proposition 5, when $n_o^g > 1.28$, the ordinary and green firms will act as the price leader and follower, respectively. The first mover's profits from introducing an ordinary and green products are $\pi_o = n_o\left[p_o^l - C\left(s_o^*\right)\right] - \delta$ and $\pi_g = n_g\left[p_g^f - C\left(s_g^*\right)\right] - \delta$, respectively, where the values of p_o^l, p_g^f, $C\left(s_o^*\right)$, and $C\left(s_g^*\right)$ are given in (2.9) and (2.11). The first mover will then prefer to introduce the green product since $\pi_g - \pi_o = n_o\left(2n_o^{g2} - 1\right)v_e^2 / 4\left(c_t + c_e\right) \geq 0$ for $n_o^g > 1.28$. When $n_o^g \leq 0.78$, first mover's profits from introducing an ordinary and green products are $\pi_o = n_o\left[p_o^f - C\left(s_o^*\right)\right] - \delta$ and $\pi_g = n_g\left[p_g^l - C\left(s_g^*\right)\right] - \delta$, respectively. The first mover will then prefer to introduce an ordinary product since $\pi_g - \pi_o = n_o\left(n_o^{g2} - 2\right)v_e^2 / 4n_o^g\left(c_t + c_e\right) < 0$ for $n_o^g \leq 0.78$. For $0.78 < n_o^g \leq 1.28$, it can be shown that the first mover's profit from introducing a green product is higher than that from introducing an ordinary product if the ordinary and green firms are the price leader and follower, respectively, which is one of the two outcomes under Nash equilibrium in the role-choosing game shown in Table 2.1.

References

Adner, R., & Zemsky, P. (2005). Disruptive technologies and the emergence of competition. *The RAND Journal of Economics 36*(2), 229–254.

Bernard, S. (2018). Multidimensional green product design. *Environmental and Resource Economics, 72*, 1–20.

Bloemhof-Ruwaard, J. M., van Beek, P., Hordijk, L., & van Wassenhove, L. N. (1995). Interactions between operational research and environmental management. *European Journal of Operational Research, 85*, 229–243.

Brandenburg, M., Govindan, K., Sarkis, J., & Seuring, S. (2014). Quantitative models for sustainable supply chain management: Developments and directions. *European Journal of Operational Research, 233*(2), 299–312.

Bouchery, Y., Corbett, C. J., Fransoo, J. C., & Tan, T. (2017). *Sustainable supply chains: A research-based textbook on operations and strategy*. Berlin/Heidelberg, Germany: Springer Science & Business Media.

Bovenberg, A. L., Goulder, H. H., & Gurney, D. J. (2005). Efficiency costs of meeting industry-distributional constraints under environmental permits and taxes. *RAND Journal of Economics, 36*(4), 951–971.

Business Week (2003). Detroit is missing the boat: Demand for hybrids is rising, but the big three are slow to get on board. Business Week (October 27, 2003).

Calcott, P., & Walls, M. (2000). Can downstream waste disposal policies encourage upstream Design for Environment? *American Economic Review, 90*(2), 233–237.

Chen, C. (2001). Design for the Environment: A quality-based model for green product development. *Management Science, 47*(2), 250–263.

Chen, C., & Liu, L. Q. (2014). Pricing and quality decisions and financial incentives for sustainable product design with recycled material content under Price leadership. *International Journal of Production Economics, 147*, 666–677.

Chen, C., & Ren, Y. (2010). Exploring the relationship between vehicle safety and fuel efficiency. *Transportation Research-Part D, 15*, 112–116.

Chen, C., & Zhang, J. (2009). The inconvenient truth about improving vehicle fuel efficiency: A multi-attribute analysis of the efficient frontier of the U.S. automobile industry. *Transportation Research-Part D, 14*, 22–31.

Chen, C., & Zhang, J. (2013). Green product design with engineering tradeoffs under technology efficient Frontiers: Analytical results and empirical tests. *IEEE Transactions on Engineering Management, 60*(2), 340–352.

Chen, C., Zhu, J., Yu, J.-Y., & Noori, H. (2012). A new methodology for evaluating sustainable product design performance with two-stage network data envelopment analysis. *European Journal of Operational Research, 221*, 348–359.

Chopra, S. (2018). The evolution of Omni-Channel retailing and its impact on supply chains. *Transportation Research Procedia, 30*, 4–13.

Christensen, C. M. (1997). *The Innovator's dilemma*. Boston, MA: Harvard Business School Press.

Christensen, C. M., & Overdorf, M. (2000). Meeting the challenge of disruptive change. *Harvard Business Review, 78*(2), 66–76.

Clark, G., Kosoris, J., Hong, L. N., & Crul, M. (2009). Design for Sustainability: Current trends in sustainable product design and development. *Sustainability, 1*(3), 409–424.

Commission of the European Communities. (2001). *Green Paper on Integrated Product Policy*, Retrieved 29 January 2020, from https://eur-lex.europa.eu/legal-content/EN/TXT/PDF/?uri=CELEX:52003DC0302&from=EN

Corbett, C. J., & DeCroix, G. A. (2001). Shared-savings contracts for indirect materials in supply chains: Channel profits and environmental impacts. *Management Science, 47*(7), 881–893.

Corbett, C. J., & Klassen, R. D. (2006). Extending the horizons: Environmental excellence as key to improving operations. *Manufacturing & Service Operations Management, 8*(1), 5–22.

Crandall, R. W., & Graham, J. D. (1989). The effect of fuel economy standards on automobiles safety. *Journal of Law and Economics, 32*(1), 97–118.

Davis, S. J., Murphy, K. M., & Topel, R. H. (2004). Entry, pricing, and product design in an initially monopolized market. *Journal of Political Economy, 112*(1), S188–S225.

de Neufville, R. (1990). *Applied systems analysis*. New York: McGraw-Hill.

Debo, L. G., Toktay, L. B., & Van Wassenhove, L. N. (2005). Market segmentation and product technology selection for remanufacturable products. *Management Science, 51*(8), 1193–1205.

Debo, L. G., Toktay, L. B., & Van Wassenhove, L. N. (2006). Joint life-cycle dynamics of new and remanufactured products. *Production and Operations Management, 15*(4), 498–513.

Dekker, R., Fleischmann, M., Inderfurth, K., & Van Wassenhove, L. N. (2004). *Reverse logistics. quantitative models for closed-loop supply chains*. Berlin: Springer-Verlag.

Deneckere, R., Kovenock, D., & Lee, R. (1992). A model of price leadership based on consumer loyalty. *Journal of Industrial Economics, XL*(2), 147–156.

Desai, P., Kekre, S., Radhakrishnan, S., & Srinivasan, K. (2001). Product differentiation and commonality in design: Balancing revenue and cost drivers. *Management Science, 47*(1), 37–51.

Dowrick, S. (1986). von Stackelberg and Cournot duopoly: Choosing roles. *The Rand Journal of Economics, 17*(2), 251–260.
Eichner, T., & Pethig, R. (2001). Product design and efficient Management of Recycling and Waste Treatment. *Journal of Environmental Economics and Management, 41*, 190–134.
Eichner, T., & Pethig, R. (2003). Corrective taxation for curbing pollution and promoting green product design and recycling. *Environmental and Resource Economics, 25*, 477–500.
Eriksson, C. (2004). Can green consumerism replace environmental regulation?—A differentiated-product example. *Resource and Energy Economics, 26*, 281–293.
Ferguson, M. E., & Toktay, L. B. (2006). The effect of competition on recovery strategies. *Production and Operations Management, 15*(3), 351–368.
Ferrer, G. (2003). Yield information and supplier Lead time in remanufacturing operations. *European Journal of Operational Research, 149*(3), 540–556.
Ferrer, G., & Swaminathan, J. M. (2006). Managing new and remanufactured products. *Management Science, 52*(1), 15–26.
Field, F. R., & Clark, J. P. (1997). A practical road to lightweight cars. *Technology Review, 100*(1), 28–35.
Fleischmann, M., Bloemhof-Ruwaard, J. M., Dekker, R., & van der Laan, E. (1997). Quantitative models for reverse logistics: A review. *European Journal of Operational Research, 103*(1), 1–17.
Fleischmann, M., Beullens, P., Bloemhof-Ruwaard, J. M., & Van Wassenhove, L. N. (2001). The impact of product recovery on logistics network design. *Production and Operations Management, 10*(2), 156–173.
Fullerton, D., & Wu, W. (1998). Policies for green design. *Journal of Environmental Economics and Management, 36*, 131–148.
Furth, D., & Kovenock, D. (1993). Price leadership in a duopoly with capacity constraints and product differentiation. *Journal of Economics, 1*(1), 1–35.
Gal-Or, E. (1985). First and second advantages. *International Economic Review, 26*, 649–653.
Geyer, R., Van Wassenhove, L. N., & Atasu, A. (2007). The economics of remanufacturing under limited component durability and finite product life cycles. *Management Science, 53*(1), 88–100.
Ginsberg, J. M. & Bloom, P. N. (2004). Choosing the right green marketing strategy. *MIT Sloan Management Review, 46*(1), 79–84.
Guide, V. D. R., & van Wassenhove, L. N. (2006). Closed-loop supply chains: An introduction to the feature issue. *Production and Operations Management, 15*(3), 345–350.
Guide, V. D. R., Souza, G. C., Van Wassenhove, L. N., & Blackburn, J. D. (2006). Time value of commercial product returns. *Management Science, 52*(8), 1200–1214.
Heese, H. S., Cattani, K., Ferrer, G., Gilland, W., & Roth, A. (2005). Competitive advantages through take-back of used products. *European Journal of Operational Research, 161*(6), 143–157.
Higgins, R. S. (1992). An economic theory of leader choice in Stackelberg models. *Journal of Economic Studies, 23*(5/6), 79–95.
Hotelling, H. (1929). The stability in competition. *Economic Journal, 39*, 41–57.
Hotelling, H. (1931). The economics of exhaustible resources. *Journal of Political Economics, 39*(2), 137–175.
Hussain, S. S. (1999). The ethics of going green: The corporate social responsibility debate. *Business Strategy and the Environment, 8*(4), 203–210.
Jayaraman, V., Patterson, R. A., & Rolland, E. (2003). The design of reverse distribution networks: Models and solution procedures. *European Journal of Operational Research, 150*(1), 128–149.
Johnson, J. P., & Myatt, D. P. (2003). Multiproduct quality competition: Fighting brands and product line pruning. *American Economic Review, 93*(3), 748–774.
Juran, J. M. (1988). *Juran's quality control handbook*. New York: McGraw Hill.
Keeney, R. L., & Raiffa, H. (1993). *Decisions with multiple objectives*. Cambridge: Cambridge University Press.
Kim, K., & Chhajed, D. (2002). Product design with multiple quality-type attributes. *Management Science, 48*(11), 1502–1511.

Krishnan, V., & Ulrich, K. T. (2001). Product development decisions: A review of the literature. *Management Science, 47*(1), 1–21.

Lilien, G. L., & Rangaswamy, A. (2003). *Marketing engineering*. Upper Saddle River, NJ: Prentice Hall.

Majumder, P., & Groenevelt, H. (2001). Competition in remanufacturing. *Production and Operations Management, 10*(2), 125–141.

Malloy, R. A. (1996). Designing with recyclate. *Plastics World, 54*, 27–28.

Manavalan, E., & Jayakrishna, K. (2019). A review of internet of things (IoT) embedded sustainable supply chain for industry 4.0 requirements. *Computers & Industrial Engineering, 127*, 925–953.

Moorthy, K. S. (1984). Market segmentation, self-selection, and product line design. *Marketing Science, 3*(4), 288–307.

Moorthy, K. S. (1985). Using game theory to model competition. *Journal of Marketing Research, 22*(3), 262–282.

Moorthy, K. S. (1988). Product and Price competition in a duopoly. *Marketing Science, 7*(2), 141–168.

Moorthy, K. S., & Png, I. P. L. (1992). Market segmentation, cannibalization, and the timing of product introductions. *Management Science, 38*(3), 345–359.

Murali, K., Lim, M. K., and Petruzzi, N. C. (2019). The effects of ecolabels and environmental regulation on green product development. *Manufacturing & Service Operations Management, 21*(3), 519–535.

Porter, M., & van der Linde, C. (1995). Green and competitive: Ending the stalemate. *Harvard Business Review, 73*(5), 120–133.

Ramanathan, R., Ramanathan, U., & Bentley, Y. (2018). The debate on flexibility of environmental regulations, innovation capabilities and financial performance—A novel use of DEA. *Omega, 75*, 131–138.

Ramdas, K. (2003). Managing product variety: An integrative review and research directions. *Production and Operations Management, 12*(1), 79–101.

Reinhardt, F. L. (1998). Environmental product differentiation: Implications for corporate strategy. *California Management Review, 40*(4), 43–73.

Sahal, D. (1981). *Patterns of technological innovation*. Reading, MA: Addison-Wesley.

Salema, M. I. G., Barbosa-Povoa, A. P., & Novais, A. Q. (2007). An optimization model for the design of a capacitated multi-product reverse logistics network with uncertainty. *European Journal of Operational Research, 179*(3), 1063–1077.

Savaskan, R. S., Bhattacharya, S., & van Wassenhove, L. N. (2004). Closed-loop supply chain models with product remanufacturing. *Management Science, 50*(2), 239–252.

Sdrolia, E., & Zarotiadis, G. (2019). A comprehensive review for green product term: From definition to evaluation. *Journal of Economic Surveys, 33*(1), 150–178.

Shaked, A., & Sutton, J. (1982). Relaxing price competition through product differentiation. *Review of Economic Studies, XLIX*, 3–13.

Smith, A. E., Ross, M. E., & Montgomery, W. D. (2002). *Implications of trading implementation design for equity-efficiency trade-offs in carbon permit allocations*. Working Paper. Charles River Associates.

Solow, R. M. (1974). The economics of resource or the resources of economics. *American Economic Review, 64*(2), 1–14.

Su, J. C., Wang, L., & Ho, J. C. (2017). The timing of green product introduction in relation to technological evolution. *Journal of Industrial and Production Engineering, 34*(3), 159–169.

Subramanian, R., Sudheer, G., & Talbot, B. (2007). Compliance strategies under permits for emissions. *Production and Operations Management, 16*(6), 763–779.

Tietenberg, T. H. (2004). *Environmental economics and policy*. Boston, MA: Pearson Addison Wesley.

Tirole, J. (1988). *The theory of industry organization*. Cambridge, MA: MIT Press.

Tyagi, R. K. (1999). Pricing patterns as outcomes of product positions. *Journal of Business, 72*(1), 135–157.

Tyagi, R. K. (2000). Sequential product positioning under differential costs. *Management Science, 46*(7), 928–940.

Urban, G. L., & Hauser, J. R. (1993). *Design and marketing of new products*. Upper Saddle River, NJ: Prentice Hall.

Wall Street Journal. (2006). Eyes on the road: Auto makers get new mantra: fuel efficiency. *Wall Street Journal*. Retrieved September 12, 2006, D.2.

White, K., Hardisty, D. J., & Habib, R. (2019). The Elusive Green Consumer. *Harvard Business Review* (July–August 2019), 3–11.

Xu, B., Xu, Q., Bo, Q., & Hu, Q. (2018). Green product development with consumer heterogeneity under horizontal competition. *Sustainability, 10*(6), 1902.

Zhu, W., & He, Y. (2017). Green product design in supply chains under competition. *European Journal of Operational Research, 258*(1), 165–180.

Chapter 3
Decision Support for Sustainable Supply Chain Management

Hendrik Reefke and David Sundaram

3.1 Introduction

Sustainability is about managing business decisions in an integrated and balanced manner with regard to economic, social, and environmental dimensions (Elkington 1998). The requirement for sustainability is acknowledged by business stakeholders and is defined as "*development that meets the needs of the present without compromising the ability of future generations to meet their own needs*" (WCED 1987).

Supply chains (SC) are crucial for supporting sustainable development due to their wide-ranging impacts and influences. Unsustainable and unaccounted SC impacts are usually not attributable to a single SC partner but are an outcome of all the interactions within the chain. SCs and sustainability requirements alike are characterised by complexities which have to be studied and integrated in practice in order to support sustainable supply chain management (SSCM). Decision makers in SCs are thus tasked with initialising strategic and operational sustainability orientations while maintaining a long-term focus.

However, SSCM suffers from a lack of guidance in the form of decision support models and understanding of relationships between SC actors and their respective activities and enabling/disabling factors. This chapter provides guidance to academics by offering an overview of the field and its development and by helping to focus their research efforts with regard to decision support accordingly. Additionally, it

H. Reefke (✉)
School of Management, Cranfield University, Cranfield, UK
e-mail: hendrik.reefke@cranfield.ac.uk

D. Sundaram
Department of Information Systems and Operations Management,
The University of Auckland Business School, Auckland, New Zealand
e-mail: d.sundaram@auckland.ac.nz

C. Chen et al. (eds.), *Pursuing Sustainability*, International Series in Operations Research & Management Science 301,
https://doi.org/10.1007/978-3-030-58023-0_3

provides advice to practitioners on how to structure their SSCM implementation efforts by outlining customised decision support mechanisms.

3.1.1 Motivation and Objectives

Corporate success depends on supply chain management (SCM) (Chen and Paulraj 2004) since significant proportions of revenues are generated through the SC (Lambert and Cooper 2000). Sustainability needs to be integrated into all SC functions in order to have the desired effect (Jayaraman et al. 2007) which is further underlined by pressures from regulations, customers, reputation, competition, and the public (Esty and Winston 2006; Lieb and Lieb 2010; Linton et al. 2007). However, economic priorities often override these requirements despite the fact that SSCM is also demanded from an economic point of view, due to for example globalisation effects, dependencies on foreign markets and imports, outsourcing, SC disruptions, or economic recessions (Lee 2010). The economic, political, social, and ethical pressures of Corporate Social Responsibility (CSR) (Garriga and Melé 2004) further motivate SSCM.

From an academic research perspective, sustainability continues to be a popular topic and the field of SSCM has grown and matured considerably (see e.g. Carter and Washispack 2018; Reefke and Sundaram 2017; Seuring and Müller 2008b; Winter and Knemeyer 2013). During this growth SSCM has been influenced by related fields including SCM, logistics, operations management, environmental management, social sciences, marketing, and strategy (Badurdeen et al. 2009; Carter and Rogers 2008). Despite these advances, fully acknowledged theories of SSCM are absent and have not been implemented in SC practice. Further targeted SSCM research is recommended in order to identify and exploit the multifaceted sustainability opportunities in SCs (Carter and Easton 2011; Colicchia et al. 2011; Dey et al. 2011; Halldórsson and Kovács 2010; Pagell and Wu 2009; Reefke and Sundaram 2017; Winter and Knemeyer 2013).

Practical implementation continues to be difficult but research has started to investigate the requirements and practices to support SSCM (Reefke and Sundaram 2017; Wagner and Svensson 2010). SC practitioners are in need of guidance in the form of decision support mechanisms in order to advance the application of sustainable SC strategies and operations. Scholars in the field have outlined the need for a solid understanding of SSCM relationships and validated decision support models (Reefke and Sundaram 2017, 2018) which would also ensure the relevance of research efforts. This chapter is intended to provide guidance to scholars in the field by addressing the following key objectives:

- Provide a synthesis of definitions and characteristics of SSCM.
- Identify research requirements regarding decision support for SSCM.
- Investigate processes that can support decisions that will ultimately transform SCs towards SSCM.

- Outline targeted decision support models in order to aid transformation and ongoing development of SSCM.

3.1.2 Article Structure

Following the outline of the research motivation and objectives for this study, the article progresses with an overview of definitions of SSCM and illustrates key characteristics of sustainable SCs. It then synthesises research recommendations from a selection of seminal articles in the field with regard to decision support in SSCM. This is followed by an account of decision models specifically targeted at outlining theoretical connections in SSCM and at providing decision support to SC practitioners. The models are further elaborated within a case context in order to illustrate their usefulness. Concluding comments reflect on how these findings contribute to the understanding of SSCM and how they can be leveraged by SC scholars.

3.2 Sustainable Supply Chain Management

This section is organised into three interconnected parts. First, the progression of SSCM as a research field is explained and a selection of academic definitions of SSCM is presented. Second, sustainability considerations are elaborated on. Third, the key characteristics of sustainable supply chains are outlined.

3.2.1 Defining Sustainable Supply Chain Management

SSCM is an area that has evolved through insights from various related fields of business research and has become an established area of academic enquiry. The growing array of structured reviews of literature on SSCM (see e.g. Carter and Washispack (2018) for a synthesis of 59 structured literature reviews) are evidence for the rapid progression of the field. Many definitions of SSCM have emerged especially in the early years of academic discourse but there is no generally acknowledged definition (Ashby et al. 2012). This may be due to the multidisciplinary nature in adding sustainability to SCM, which itself originated from fields such as purchasing, logistics, and transportation (Croom et al. 2000). Table 3.1 provides an overview of SSCM definitions proposed over the years and numbered in chronological order of publication.

From these definitions it is evident that SSCM deals with the coordination of all SC flows while the requirements of all relevant stakeholders should be met and optimised in accordance with the three sustainability dimensions. The need for an integrated and holistic focus on the triple bottom line (environmental, economic,

Table 3.1 Definitions of SSCM

Source	SSCM definition
Carter and Rogers (2008: 368)	SSCM is the strategic, transparent integration and achievement of an organization's social, environmental, and economic goals in the systemic coordination of key inter-organizational business processes for improving the long-term economic performance of the individual company and its SCs
Seuring and Müller (2008b: 1700)	SSCM is the management of material, information and capital flows as well as cooperation among companies along the SC while taking goals from all three dimensions of sustainable development, i.e. economic, environmental and social, into account, which are derived from customer and stakeholder requirements. In sustainable SCs, environmental and social criteria need to be fulfilled by the members to remain within the SC, while it is expected that competitiveness would be maintained through meeting customer needs and related economic criteria
Pagell and Wu (2009: 38)	A truly sustainable SC would at worst do no net harm to natural or social systems while still producing a profit over an extended period of time; a truly sustainable supply chain could, customers willing, continue to do business forever A sustainable SC is one that performs well on both traditional measures of profit and loss as well as on an expanded conceptualization of performance that includes social and natural dimensions. SSCM is then the specific managerial actions that are taken to make the SC more sustainable with an end goal of creating a truly sustainable chain
Baburdeen et al. (2009: 57)	SSCM involves the planning and management of sourcing, procurement, conversion and logistics activities involved during pre-manufacturing, manufacturing, use and post-use stages in the product lifecycle in closed-loop through multiple lifecycles with seamless information sharing about all product lifecycle stages between companies by explicitly considering the social and environmental implications to achieve a shared vision
Hassini et al. (2012: 70–71)	SSCM is the management of SC operations, resources, information, and funds in order to maximize the SC profitability while at the same time minimizing the environmental impacts and maximizing the social well-being. Objectives: while maximizing profits calls for reducing operations costs, minimizing the environmental impacts and maximizing the social well-being can add to the SC's operational costs. Challenges to SC managers: dealing with multiple decision makers and assessing the environmental impacts and social benefits in a multi-party SC
Ahi and Searcy (2013: 339)	The creation of coordinated SCs through the voluntary integration of economic, environmental, and social considerations with key inter-organizational business systems designed to efficiently and effectively manage the material, information, and capital flows associated with the procurement, production, and distribution of products or services in order to meet stakeholder requirements and improve the profitability, competitiveness, and resilience of the organization over the short- and long-term
Pagell and Shevchenko (2014: 45)	SSCM is the designing, organizing, coordinating, and controlling of SCs to become truly sustainable with the minimum expectation of a truly sustainable SC being to maintain economic viability, while doing no harm to social or environmental systems

and social) is pointed out frequently but many publications tend to focus on a single or two dimensions. Especially investigations into social considerations are lacking but are urgently needed in order to understand the many trade-offs in SSCM decision-making (Cruz 2009; Seuring 2013). Several academic scholars in SSCM have voiced significant critique of the prevalent triple bottom line thinking that is based on a balance of the three sustainability dimensions. A prioritisation of environmental and social considerations over economic demands is suggested as the way to achieve SC sustainability (see e.g. Markman and Krause 2016; Montabon et al. 2016).

3.2.2 *Sustainability Considerations*

SSCM has been described as an extension of green SCM with an additional focus on economic and social considerations (Ahi and Searcy 2013; Ashby et al. 2012; Sarkis 2012). Sustainability is often seen as synonymous with the triple-bottom-line (TBL) which focusses on social, environmental, and economic goals (Ahi and Searcy 2013; Seuring 2013; Winter and Knemeyer 2013). SSCM aims for the alignment of a SC in order to satisfy TBL considerations as well as concerns of SC stakeholders (Carter and Rogers 2008; Seuring and Müller 2008b). The TBL assigns equal importance to the three sustainability dimensions (Montabon et al. 2016) and has had a dominant influence on sustainability related research. Influenced by TBL thinking, SSCM has an extended focus towards improving the long-term economic performance of the SC as a whole and its individual firms while applying environmental-friendly strategies and taking social responsibilities into account (Carter and Rogers 2008; Svensson 2007). Both, inter- and intra-organisational connections need to be considered in order to support the collaborative development of SC sustainability. This integrative view of the TBL aligns with stakeholder theory, that is, capturing all SC impacts and the effects on its stakeholders (Montabon et al. 2016). The value of the TBL to achieve true sustainability in SCs has however also been questioned since the concept aims for win-win relationships and may not support trade-offs that are detrimental to a single dimension (Markman and Krause 2016; Montabon et al. 2016).

3.2.2.1 Economic Considerations

Although supposedly of equal importance, the majority of companies will firstly focus on their economic viability (Markman and Krause 2016; Montabon et al. 2016). SSCM can help companies to portray themselves as environmentally friendly and can lead to marketing advantages (Smith et al. 2010). However, a lack of profit would likely result in the termination or reduction of practices associated exclusively with environmental or social benefits. Hassini et al. (2012) acknowledge the difficulty of achieving multiple TBL goals and regard SC profitability as vital while environmental and social impacts should be managed to one's best abilities. They put emphasis on the reduction of operational costs in order to maximise profits and point towards potential trade-offs between the profitability of a SC as a whole and that of individual companies. Seuring and Müller (2008b) and Font et al. (2008) draw attention to the need to focus on customer requirements, that is, that competitiveness is contingent on satisfying the needs of their customers by employing market focussed resources and meeting economic performance criteria.

Researchers have identified several economic motivators for SSCM, including potential increases in shareholder value, customer demand, market share, employee retention, business opportunities, innovation, reputation, as well as risk reductions (Esty and Winston 2006; Seuring and Müller 2008b). Other economically relevant attributes include improved SC relationships, increased levels of collaboration and information sharing, reduction of opportunistic behavior, and the development of sustainability characteristics which are difficult to replicate (Carter and Rogers 2008; Kleindorfer et al. 2005). Based on the economic considerations mentioned, connections can thus be drawn between applying SSCM, the development of unique SC and company specific attributes, and higher profitability coupled with competitive advantages.

3.2.2.2 Environmental Considerations

The term "sustainability" still carries different meanings and it is frequently associated with economic and environmental dimensions only (Montabon et al. 2016; Sarkis 2012). Sustainability initiatives, and the introduction of sustainability into SCs, started with environmental considerations and also researchers have predominantly focussed on assessing the resulting economic implications (Ahi and Searcy 2013; Ashby et al. 2012; Berger et al. 2001; Font et al. 2008). This focus has become known as "green SCM" and is concerned with environmental innovations, product life cycle considerations, and the selection of members for the SC based on environmental practices and strategies (Berger et al. 2001; Srivastava 2007). Such an environmental SC focus has been pushed forward by various factors including regulations, customer perceptions, risk management, efficiency gains, and cost.

Generally SCs should reduce environmental impacts and impose no harm on the environment. The reduction of waste material and especially hazardous waste is crucial (Linton et al. 2007; Nunen et al. 2005). This should be coupled with more efficient usage of natural resources, materials, and energy (Nunen et al. 2005). Unintended by-products from manufacturing, product usage, and disposal need to be avoided which likely requires the inclusion of sustainability considerations into product design, the forward SC and reverse logistics in order to manage product recovery processes. Environmental performance of a SC is thus directly connected to the design and lifecycle of a product, which determine whether production and usage follow environmental considerations and whether resources can be effectively recovered at the end of a product's life (Badurdeen et al. 2009; Linton et al. 2007).

3.2.2.3 Social Considerations

The inclusion of social aspects into sustainable operations and SC research is less explored (Carbone et al. 2012; Sarkis 2012; Seuring 2013; Walker et al. 2014). Social measures are harder to quantify than economic and environmental aspects (Goldschmidt et al. 2013) and environmental issues often supersede social considerations in shaping organisational sustainability (Carbone et al. 2012). Some researchers see only few synergies between these dimensions at corporate and SC levels, requiring the development of distinct capabilities for improvement in either dimension (Carbone et al. 2012). However, social and environmental dimensions have also been described as co-dependent (Orlitzky et al. 2003) with predominantly win-win relationships (Seuring and Müller 2008a).

Social sustainability is linked to CSR, which is driven by ethical values and codes of conduct in order to balance the interests of corporate stakeholders (Ashby et al. 2012). Increasing employee motivation is driving CSR since employees generally prefer to work for "responsible" companies. Focussing on corporate sustainability is positively correlated with financial performance, mainly due to reputation gains (Orlitzky et al. 2003) but it is not certain if this relationship applies in SCs to the same extent (Seuring and Müller 2008a). Other potential benefits triggered by a social SC orientation include employee health and safety, fair treatment and social equity, fostering employee skills and abilities, and adding human capital to society (Ashby et al. 2012).

Hassini et al. (2012) point out that social well-being can increase operational costs. Especially low-cost producers are likely to neglect social considerations in their SCs which often operate in socially sensitive industries (Hoejmose et al. 2013). CSR is also not equivalent with true sustainability since companies and SCs can engage in socially praiseworthy actions without addressing the root causes of their unsustainable behavior (Markman and Krause 2016). A social SC focus ultimately requires an extension beyond organisational and SC considerations towards improving quality of life and equity in the communities affected by the SC operations (Vachon and Mao 2008).

3.2.3 Sustainable Supply Chain Characteristics

3.2.3.1 Sustainable Supply Chain Focus

Significant proportions of business revenues are generated through SCs which are frequently networks of globally dispersed companies faced with different regulatory restrictions, market conditions, and customer requirements (Cooper et al. 1997; Lambert and Cooper 2000). SC strategies are often based on the availability of cheap transport (Halldórsson and Kovács 2010) as well as production costs, quality, delivery terms, and flexibility considerations. SCs often rely on global sourcing and neglect energy efficiencies in order to save on labour or storage costs (Halldórsson and Kovács 2010). The resulting global networks do not necessarily lend themselves to quick operational shifts or strategic reorientations due to their complexity, dependence on infrastructure, economic importance, and the predominately cost based sourcing considerations. Further, SCs are often characterised by only limited visibility and control between individual companies. There are various systemic interactions that decision makers need to take into consideration when making changes to a SC (Mentzer et al. 2001). The introduction of sustainability principles is challenging in any organisational environment but associated difficulties naturally increase with the number of companies and stakeholders involved. Tackling sustainability challenges requires a holistic SC response (Carter and Easton 2011) and sustainability considerations need to be integral to SC decisions in order to address the wide-reaching impacts of business conduct (Jayaraman et al. 2007; Kleindorfer et al. 2005). Linton et al. (2007) even describe SCs as a catalyst for the broader adoption of sustainability. With this shift in focus, the sustainability debate needs to evolve as well towards a strategic focus on SSCM (Carter and Rogers 2008; Jayaraman et al. 2007; Seuring and Müller 2008b). Decision makers in SCs find themselves in a crucial position (Sarkis 1998) since their decisions have significant negative as well as positive sustainability impacts (Carter and Easton 2011; Murphy and Poist 2003). Svensson (2007) adds that decision makers need to take into account their connections across multiple SCs, which requires visibility of all interfaces.

3.2.3.2 Building Blocks for a Sustainable Supply Chain

In order to foster SSCM, Carter and Rogers (2008) emphasise the need to focus on systemic relationships between inter-organisational business processes in order to integrate and achieve holistic sustainability goals. Other authors reflect especially on the dynamics between multiple decision makers across the SC members (Hassini et al. 2012). Challenges are likely to arise when assessing and quantifying sustainability impacts and benefits that a SC causes. In turn, decisions regarding potential countermeasures, monetary investments, resource allocations, or the equal distribution of potential benefits pose challenges. Individual members in a SC may try to gain benefits while avoiding necessary investments, that is, members may aim for local optimisation instead of a collaborative sustainability effort (Carter and Rogers 2008; Shepherd and Günter 2006).

According to Christopher (1998) SCM relies "on cooperation and trust and the recognition that properly managed the whole can be greater than the sum of its parts". Mentzer et al. (2001) emphasise that a true SC orientation is characterised by the visibility of tactical implementations and SC strategy across the chain. Applying these insights, a building block for SSCM is a shared or common vision which is a key component for any SC transformation (Tan et al. 1998) and has been described as essential for success in SCM (Mentzer et al. 2001). In another step towards SSCM, Seuring and Müller (2008b) suggest that SC wide sustainability goals should be based on a balanced review of all SC stakeholder requirements, while competitiveness is simultaneously ensured by meeting customer needs. When determining sustainability goals, the need for a full SC perspective is generally emphasised that encompasses all SC activities such as procurement, production, distribution, pre-manufacturing, manufacturing, use, and post-use stages (Ahi and Searcy 2013; Badurdeen et al. 2009). Visibility and control of the various SC flows have to be maintained and fostered, including capital flows, information and decision flows, logistics and material flows, as well as product and service flows (Ahi and Searcy 2013; Hassini et al. 2012; Seuring and Müller 2008a). Hence, the integration of inter-organisational business systems (Ahi and Searcy 2013) as well as seamless information sharing (Badurdeen et al. 2009) are described as essential. Following this argument, Carter and Rogers (2008) identify transparency as a key component of SSCM based on active stakeholder engagement and visibility of supplier operations. Horizontal and vertical SC collaboration and strategic relationships have been associated with success in sustainability initiatives and competitive advantage (Harris et al. 2010; Lee 2008). They can improve relations and aid in developing trust and long-term relationships between suppliers, that is, characteristics which are not easily duplicated by competitors (Carter and Rogers 2008). In terms of environmental improvements, close SC relationships can lead to the elimination of waste through quality management, as well as coordinated pollution reduction, control, and prevention (Bansal and McKnight 2009). Lastly, SC practitioners should be more proactive about communicating their SSCM efforts as well as concerns in order to further develop their reputation, increase SC coordination, and lower costs (Carter and Rogers 2008; Krause et al. 2009).

3.3 Research Requirements on Decision Support

The field of SSCM has expanded quickly and authors have outlined research directions respectively. In order to provide an informed overview of research requirements on decision support for SSCM, we summarised research recommendations from literature that investigate the field from different angles. Reefke and Sundaram (2017) utilised a Delphi approach in order to extract research suggestions from domain experts. Seuring (2013) explored the application of modelling approaches whereas Winter and Knemeyer (2013) targeted their article specifically at suggesting avenues of research. Ashby et al. (2012) performed a structured review with a focus on social and environmental aspects while Hassini et al. (2012) focussed on performance aspects. Carter and Easton (2011) finally derive future research directions from a selection of SC journals.

It is evident from the research requirements in Table 3.2 that decision support is yet to be developed in support of the various aspects of SSCM. Despite the fact that the field has seen many useful contributions over the years, practical decision support and prescriptive methods or tools are largely absent. The recommendations include the investigation of biases and trade-off decisions (Carter and Easton 2011), as well as a focus on cost allocations and unaccounted SC impacts (Reefke and

Table 3.2 The need for decision support in SSCM (adapted from Reefke and Sundaram 2017)

Source	Decision support—research requirements
Reefke and Sundaram (2017)	• Actual costs of supply chain operations, for example unaccounted environmental and social impacts
	• Future of supply chains, for example long-term outlook and restructuring needs
	• Investments into sustainability and their justifications
	• Implementation hurdles of sustainability initiatives, for example time and cost requirements
	• Long supply chains and resulting special requirements
	• Transportation modes, for example which mode works best for each commodity
	• Cost allocations, for example for sustainability efforts and unaccounted supply chain impacts
	• Missing theory development to guide practice, for example lack of strategic models and applicable frameworks
Seuring (2013)	• How can the social dimension be integrated into respective models?
	• Interrelation among all three dimensions of sustainability and models thereof
	• How does environmental and social performance impact supply chain performance?
	• Establish the links to the literature on strategic supply chain design, supply chain performance and collaboration
Winter and Knemeyer (2013)	• Research should look at the connection between managerial components and sustainability efforts, in an effort to better understand how managerial practices can influence the success or failure of sustainability initiatives
	• Companies need specific guidelines and an explicit toolbox that supports their efforts to reach their sustainability objectives, for example structural management components and adequate control mechanisms
	• The development and validation of appropriate metrics and scorecards in support of SSCM
	• The development of estimation tools and techniques to provide financial justifications
	• Investigate how suppliers can engage their customers on sustainability initiatives or to better understand how sustainable supply chain initiatives can be used to enhance a company's brand and/or marketing efforts
Ashby et al. (2012)	• The role of supply chain relationships in achieving sustainability
	• Life cycle analysis and the concept of closed loop supply chains could provide a more connected view of sustainability in supply chains
	• An integrated, holistic, and relational viewpoint is vital for progressing SSCM from "greening" to a "virtuous circle" that addresses sustainability interactions as well at all stages
	• Translating SSCM theory developed through more focussed approaches into actual supply chain practice should be a key priority

(continued)

Table 3.2 (continued)

Source	Decision support—research requirements
Hassini et al. (2012)	• Pricing, as part of the value proposition to the customer, should be more strongly emphasised
	• Address inventory management within sustainable supply chains since traditional inventory models focus on economic aspects
	• How should SMEs and large firms approach investment in and adoption of sustainable practices?
	• Research into performance assessments of sustainable supply chain, for example metrics, composite indicators, compatibility with existing theory
Carter and Easton (2011)	• Research to dig deeper into individual industries as sampling frames to identify specific types of sustainability activities and assess the applicability of specific theories
	• Investigate the relationship between company environmental and social performance versus economic performance
	• Examine how bounded rationality and perceptions of opportunism within the context of SSCM impact the decision to source domestically or even locally, as opposed to internationally, and how supply chain governance structures are affected
	• Examination of the biases that can enter the individual decision-making process, and how these biases can impact the efficacy of SSCM initiatives
	• Investigation of how individual managers can influence and gain the commitment of key internal stakeholders to bring SSCM projects to fruition

Sundaram 2017). Prominent decision areas in need of support furthermore include the various facets of risk management (Hassini et al. 2012; Winter and Knemeyer 2013). More targeted decision support extends to the development of performance management tools such as indicators, metrics, and scorecards (Hassini et al. 2012; Winter and Knemeyer 2013). Developing an understanding of human resources in support of SSCM are called for by investigating managerial components and practices (Winter and Knemeyer 2013) as well as extended supply chain relationships and the role of individual decision makers (Carter and Easton 2011).

It can be summarised that the field is lacking practical insights and advice on how to implement and progress a strategic orientation towards sustainability supported by sustainable supply chain operations (see e.g. Ashby et al. 2012; Carter and Easton 2011; Hassini et al. 2012; Reefke and Sundaram 2017; Seuring and Müller 2008b; Winter and Knemeyer 2013). SC managers need guidelines and prescriptive support in order to guide long term SC planning and daily operations. Thus, decision-making tools need to be able to deal with this multifaceted nature of SSCM, the many systemic interconnections between SC actors, and the trade-offs involved in decision-making processes. The following sections outline several models targeted at providing such support mechanisms.

3.4 Decision Models

This section introduces two models that have been developed particularly for addressing the need for practical decision support in SSCM. These models are the outcome of a rigorous research study (Reefke and Sundaram 2018) and may prove to be instrumental for the understanding of SSCM relationships and SSCM application. Both models are based on seminal management approaches aimed at (1) transforming SC towards SSCM and (2) developing SSCM towards higher levels of maturity.

3.4.1 SSCM Transformation

Decision makers in SCs are tasked with introducing sustainability principles into their daily tasks and operations. Such shifts in operating principles take time to plan, implement, and control. Hence, procedural support is required that provides guidance in this endeavour. The SSCM transformation model put forward in this section builds on existing transformation approaches and adopts high level transformation steps. However, it is then customised for SSCM through the definition of requirements and support mechanisms for each step.

3.4.1.1 Models of Transformation

The ability to adapt and transform to changing requirements is vital for successful businesses and SCs (Beamon 1999; Lee 2004). The ability to transform and continuously improve SC processes from strategic and operational perspectives are thus also central for sustainable development. A selection of relevant transformation approaches is therefore introduced here and their underlying structures are synthesised in Table 3.3.

A variety of transformation models can be found in literature, both generic as well as specialised. Transformation models have a practical orientation by nature but are frequently lacking comprehensive academic evaluations regarding for example ease of implementation, speed of transformation, and the retention of success. In support of the approach put forward for SSCM, several useful scholarly evaluations of transformation models outlining their positive characteristics alongside their limitations can be pointed out. Furthermore, such models are commonplace in management consulting and have been successfully applied throughout various industries (see e.g. de Mast and Lokkerbol 2012; Schroeder et al. 2008). The transformation approaches shown are based on foundational decision-making models such as Simon (1977), Mintzberg et al. (1976), Hage (1980), and Langley et al. (1995).

It is apparent from the synthesis provided in Table 3.3 that the transformation models reviewed share common elements, phases, and structures. They generally

Table 3.3 Synthesis of transformation models

Authors	Objectives	Phases/steps/levels
Deming (1986)	PDCA—Continuous process improvement	(1) Plan, (2) Do, (3) Check, and (4) Act
Heskett et al. (1994)	Service profit chain—Increase service quality and profitability	(1) Internal service quality, (2) Employee satisfaction, (3) Employee retention/productivity, (4) External service value, (5) Customer satisfaction, (6) Customer loyalty, and (7) Revenue growth/profitability
Scheer et al. (2003)	ARIS—Enterprise modelling and business process management	(1) Strategise, (2) Design, (3) Implement, and (4) Control
Gilmour et al. (2004)	IBM Consulting—Value creation and growth cycle	(1) Increase sales and market share, (2) Earn more margin, (3) Reinvest in processes and technology, (4) Drive greater productivity, (5) Invest in differentiators, and (6) Deliver greater value to customers
Schroeder et al. (2008)	DMAIC—Process analysis and control	(1) Define, (2) Measure, (3) Analyse, (4) Improve, and (5) Control
The Natural Step (2009)	The Natural Step—Sustainability implementation	(1) System, (2) Success, (3) Strategic, (4) Actions, and (5) Tools

start with an exploratory discovery/learning phase, followed by a planning/designing phase with a subsequent implementation/monitoring/control phase. Most approaches emphasise targeted, key performance indicator (KPI) driven analysis and continuous performance control in order to prevent shifting back to previous patterns. Realising the potential benefits requires that sustainable processes become institutionalised as routines in the organisation (Schroeder et al. 2008) which would need to be adapted and extended across the SC. Due to the wide spread usage and acceptance of structured transformation approaches, it is rational to leverage the underlying structures as a basis for SC transformations towards sustainability. A targeted decision support model in this regard is introduced next.

3.4.1.2 SSCM Transformation Model

The high level SSCM transformation model, shown in Fig. 3.1, is influenced and informed by the transformation approaches outlined in Table 3.3. The bottom-up development of its detailed elements makes it unique to SSCM, as further elaborated in Reefke and Sundaram (2018).

Transformation approaches frequently start with "discovering" and "learning" in order to guide process and, if applicable, sustainability transformations. Discovery is about the evaluation of external and internal sustainability requirements, while learning is about assessing internal capabilities and support mechanisms. "Strategising" deals with the development of an SSCM strategy and respective SC processes. Other transformation approaches include similar steps (e.g. Scheer et al. 2003; The Natural Step 2009) since strategic decisions have long-term implications

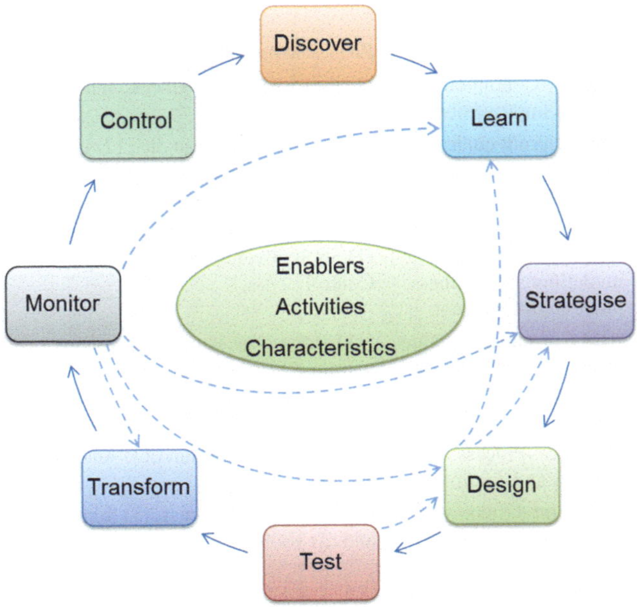

Fig. 3.1 SSCM transformation model (adapted from Reefke and Sundaram 2018)

for all SC stakeholders. In this step it is recommended to define transformation goals and balance resulting trade-offs accordingly. The "design" phase translates the strategy into implementable activities and methodical procedures first. This is followed by "testing", offering validation before the actual implementation takes place. Unfavourable test results are accounted for through feedback loops linking to previous steps so that remedial actions can be taken and implemented into preceding steps. The final design can then be implemented through the transformation phase which puts the strategy into action. "Monitor" facilitates an assessment regarding the success of the transformation in order to inform SC stakeholders accordingly. Feedback loops again ensure that issues can be addressed from the most appropriate phase. The "control" phase finally focusses on the success of individual process transformations and the transformation towards the end goal of becoming a sustainable SC as a whole. Accordingly, full and partial cycles are recommended in order to support sustainability improvements at process and strategic levels.

The SSCM transformation model is designed to provide targeted decision support and also serves the purpose of managing sustainability risks at strategical and operational levels. The SSCM transformation model puts forward a selection of cyclical steps, activities, and requirements. The model elements are informed by following a top-down approach for establishing the high level structure and a bottom-up approach for customising it towards the specific application of SSCM. As elaborated in Reefke and Sundaram (2018), a ranking proposes that certain elements are particularly supportive, or even essential, for each step. While no model or framework can be truly comprehensive, it offers a wide-ranging overview that can guide transformation efforts by supporting prioritisation and resource allocation. The intention is to establish a common methodology across SCs in order to guide transformation efforts, align sustainability goals, and enable continuous improvements through its cyclical nature. This SSCM decision model may thus be instrumental for implementing a sustainable SC strategy and the resulting transformation of underlying structures, processes, and systems. Furthermore, it can be the starting point for establishing a SC-wide sustainability culture.

The succeeding sections establish processes on how these outcomes can be developed or matured further.

3.4.2 SSCM Maturity

The SSCM maturity model introduced in this section puts forward the notion that SCs exhibit different levels of maturity with regard to their sustainability integration and application. It is informed by the high-level structures of existing maturity models and, as with the SSCM transformation model, introduces customised building blocks that are unique to SSCM. The idea that higher levels of SSCM proficiency can be reached over time and fostered through the application of targeted developments underlies this model. As such, the SSCM maturity model is meant to be utilised as a decision support tool in order to make existing SCs more sustainable but is also intended to guide the development of truly sustainable SCs.

3.4.2.1 Overview of Maturity Models

Maturity models are designed to aid businesses in creating an overview of their own processes. They establish a structured approach based on a common vision and language. Furthermore, the maturity concept allows for the prioritisation of goals and activities and guides respective performance management and benchmarking activities (Carnegie Mellon—Software Engineering Institute 2002; Lockamy and McCormack 2004; McCormack 2001). The maturity concept assumes that business characteristics can be categorised into levels of development describing associated behavioural, regulative, and performance standards. A maturity level can be described as an evolutionary plateau of process improvement (Carnegie Mellon—Software Engineering Institute 2002) which is further based on the assumption that processes can be organised into stages of development (McCormack et al. 2008). Such staggered developmental stages are often central to successful SC initiatives, transformations, and a long-term development strategy (Stevens 1993). Maturity models generally encompass elements that relate to definition, measurement, management, and business process control. Scholars associate higher maturity with better managerial control, improved forecasting, lower costs, effective goal attainment, and successful continuous improvement methodologies (Lockamy and McCormack 2004; McCormack et al. 2008).

In summary, maturity development represents a logical approach to progression, especially in bigger systems like SCs, since process changes and improvements are best implemented successively (Carnegie Mellon—Software Engineering Institute 2002; Lockamy and McCormack 2004). Table 3.4 synthesises the objectives, aims, and model structures of selected maturity models, i.e. the definition of levels according to distinct developmental stages. While aimed at describing different parts of business or SC developments, similarities between the level descriptions are evident. Furthermore, developmental goals and requirements are generally defined for each level in order to guide decision makers in their efforts and in the allocation of resources.

3.4.2.2 SSCM Maturity Model

Companies and SCs generally aim to continuously improve, or mature, their processes, structures, policies, and capabilities in order to increase competitiveness. As illustrated in the previous section, maturity models define distinct levels of development and offer a structured approach for improvement. The maturity concept is well established in literature and scholars accept the notion that maturity models can guide managerial decision makers (Carnegie Mellon—Software Engineering Institute 2002; Lockamy and McCormack 2004). Thus, it is advisable to adopt the general structural elements and logic of established seminal work but to adapt these aspects according to the purpose at hand. Accepting this approach, a maturity model for SSCM should provide and detail the following aspects:

- outline the purpose of the transformation,
- provide a common language by setting goals, objectives and guidelines,
- determine responsibilities,
- establish a clear direction and shared vision,
- help users to communicate and evaluate their decisions,
- outline a progression strategy between the current state and the long-term strategy.

Based on these requirements, an SSCM maturity model was developed as shown in Fig. 3.2. Associated descriptions of each level alongside specific goals and requirements can be seen in Table 3.5. The top level structure has been informed by the maturity models reviewed and summarised in Table 3.4. Their insights and essential building blocks were leveraged towards a design targeted at SSCM. The proposed maturity progression is organised in six levels ranging from "un-aware and non-compliant" at the lowest level (1) towards "extended and sustainability leadership" (level 6). As shown in Table 3.5, the levels correspond to specific stages of SSCM maturity and provide directions and a vision for further development. Goals and requirements are identified at each level, thereby establishing an overall SSCM vision as well as a development strategy. The model maintains a neutral,

Table 3.4 Synthesis of maturity models

Authors	Objectives	Levels
Stevens (1993)	SC integration	1—Baseline 2—Functional integration 3—Internal integration 4—External integration
Starik and Rands (1995)	Sustainability levels and systems	1—Individual 2—Organisational 3—Political-economic 4—Social-cultural 5—Ecological
Beamon (1999)	Environmental management in SCs	1—Problem solving 2—Managing for compliance 3—Managing for assurance 4—Managing for eco-efficiency 5—Fully integrated
Veleva and Ellenbecker (2001)	Sustainability performance indicators	1—Facility compliance/conformance 2—Facility material use and performance 3—Facility effect 4—SC and product life cycle 5—Sustainable systems
Carnegie Mellon—Software Engineering Institute (2002)	Process maturity	1—Ad hoc/individual competencies 2—Increased visibility 3—Standardised processes 4—Performance management 5—Continuous improvement
Yusuf et al. (2004)	SC agility	Three maturity stages of SC agility for the intertwined customer interaction, asset configuration, and knowledge leverage dimensions
Lockamy and McCormack (2004)	SCM and process maturity	1—Ad hoc and undefined 2—Defined 3—Linked 4—Integrated 5—Extended
Marshall and Toffel (2005)	Sustainability requirements	1—Endanger human life 2—Reduce life expectancy 3—Species extinction and human rights violation 4—Subjective values of the quality of life
OMG (2008)	BPMM—Process maturity	1—Initial 2—Managed 3—Standardised 4—Predictable 5—Innovating
Boone et al. (2009)	GAIA—SSCM maturity	1—Genesis: Compliance focus 2—Advancing: Develop strategic objectives 3—Innovating: Coordinated strategy 4—Accelerating: Vision, goals and proactive behaviour

(continued)

Table 3.4 (continued)

Authors	Objectives	Levels
GCIO and OMG (2009)	Green business maturity	1—Ad hoc 2—Common understanding 3—Governance structure 4—Internal optimisation and extension 5—Economic returns from green initiatives

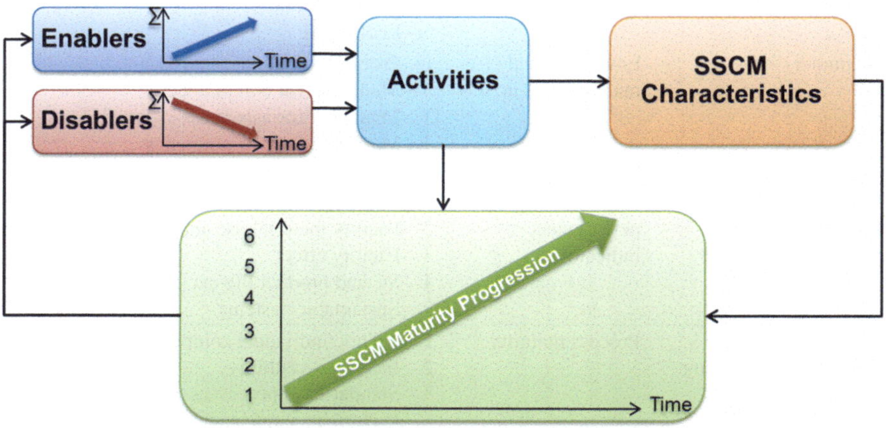

Fig. 3.2 SSCM maturity model (adapted from Reefke and Sundaram 2018)

Table 3.5 SSCM maturity levels (Reefke and Sundaram 2018)

Level	Description	Goals and requirements
6	Processes are systematically managed through continuous improvement. Full SC collaboration embracing sustainability leadership position	*Continue* to optimise processes and ensure future leadership position
5	Sustainability has become a fully integrated concept and SC has moved towards proactive measures	*Propagate* strategic concepts and move towards leadership position
4	SC is linked and includes a comprehensive sustainability performance measurement system	*Develop* from compliance level towards proactive sustainability efforts
3	Sustainability goals/standards have been defined and SC members are compliant with regulations	*Establish* key indices to measure sustainability performance within SC
2	Sustainability measures are disconnected from strategic direction. Compliance on a basic level	*Align* sustainability goals and efforts with defined processes. Establish consistency
1	SC is unaware and non-compliant to any regulations and undertakes no sustainability efforts	*Create* sustainability awareness. Introduce sustainability initiatives

generalisable approach in order to remain customisable towards specific SC requirements and decision challenges.

The SSCM maturity model, shown in Fig. 3.2, maps the dynamic relationships between maturity and SSCM factors. The logic of the decision model can be summarised as follows:

> *The existence of certain enabling factors helps a SC to perform activities that support SSCM, whereas disablers prevent a SC from doing so. In combination these result in certain characteristics that the SC possesses. As a SC develops such characteristics, it reaches higher levels of maturity. Along with higher maturity levels the amount of disabling factors and/or their effects decrease and the amount of enabling factors and/or their effects increase, which allow for more activities directed at sustainability in SCs.*

Reefke and Sundaram (2018) identified a total of 96 SSCM factors which are further separated into 26 enablers, 21 disabler, 23 activities, and 26 characteristics. As shown in Fig. 3.2, the factors are logically interconnected and were furthermore evaluated with regard to their comparative importance. The relationships put forward by this decision model can be illustrated through a selection of SSCM factors relating to performance measurement. The activity "definition and measurement of clear key performance indicators" may only be feasibly performed if "performance measurement tools for consistent and accurate measurement" (enabler) exist. On the flipside, it can be hindered by disabling factors such as a "focus on short term financial performance" or a "misguided focus in the sustainability movement" which may potentially lead to performance targets that are not reflective of a balanced sustainability orientation. Accurate and timely performance measurement will support the characteristics "alignment and synchronisation of SC and sustainability initiatives and goals" as well as "true cost allocation". Hence, this logic implies that the development of SSCM characteristics necessitates performing certain activities which, in turn, require respective enabling factors and the absence or appropriate managerial control of disablers.

3.5 Illustration of the SSCM Models

In order to demonstrate the applicability and usefulness of the SSCM transformation and maturity models, this section firstly introduces the concept of SSCM progression. This is followed by a well-publicised case scenario of the Mattel SC that illustrates how the models may be utilised.

3.5.1 SSCM Progression

As evident from the review of the literature, process transformation and maturity development are tightly linked concepts. Based on this realisation, the stepwise SSCM transformation approach can be integrated with the dynamics in sustainable SCs put forward by the SSCM maturity model.

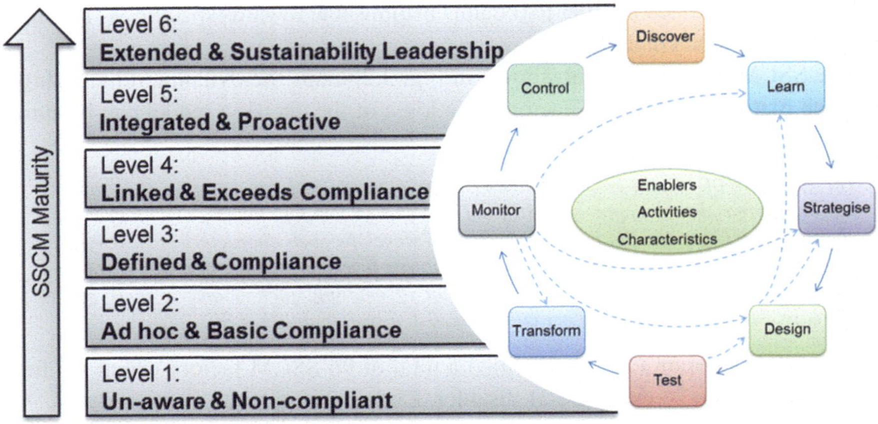

Fig. 3.3 SSCM maturity progression

As shown in Fig. 3.3, this maturity progression proposes a cyclical multi-step approach, commonly used for transformation methodologies (Scheer et al. 2003; The Natural Step 2009) as well as iterative, convergent and linked decision-making processes (Simon 1977; Mintzberg et al. 1976; Langley et al. 1995). The progression between the levels of SSCM maturity is supported through an iterative transformation process consisting of defined phases which can be performed on a continuous basis. The time dimension is addressed specifically by outlining the strategic progression between the current state of a SC, its long-term vision, and transitional states of development. This approach follows the recommendations of seminal work in the field and addresses the need for a virtuous SSCM improvement methodology that is grounded in a deeper understanding of relationships and interactions in SSCM (Ashby et al. 2012).

The SSCM factors, i.e. enablers, disablers, activities, and characteristics, constitute essential building blocks for SCs aiming to become more sustainable. The categorisation of these factors into distinct but interconnected categories and their respective evaluations guided the design of the SSCM transformation and maturity models. These research artefacts may address the uncertainties and fill in the unknowns that decision makers tasked with SSCM are confronted with. The presentation of higher-level relationships allows for an easier grasp of their applicability and implications by providing a context and by illustrating the causal relationships between the factor groups. It needs to be acknowledged that empirical data can be interpreted in multiple ways (Ketokivi and Mantere 2010). Hence, other categorisations of the factors may also be coherent. Despite this potential shortcoming, SC scholars can use the SSCM factors and decision models in order to develop a better theoretical understanding of the logical relationships present in sustainable SCs. Furthermore, the SSCM models provide ample scope for related research endeavours, for example testing the relationships put forward empirically through case studies or survey instruments. Particular importance may be allocated to testing the presence/strengths of specific factors (or groupings of factors) in (un)successful SSCM projects. Based on the research insights, SC practitioners can utilise the adaptable nature of the decision models alongside the importance evaluations of the factors. SSCM developments can be customised according to unique SC requirements and the different developmental stages, that is, levels of maturity. In summary, the research artefacts presented can benefit SC stakeholders by allowing them to assess, implement, transform, continuously develop and advance SC sustainability.

3.5.2 *Case Illustration*

In this section, the models and their details presented are utilised for analysing the Mattel SC (as described in e.g. Gilbert and Wisner 2010; Hoyt et al. 2008; Schmidt 2008).

Mattel is an American multinational company that designs, manufactures, and markets toys worldwide. It had to recall millions of toys due to two keys problems (a) they contained magnets which came loose and were swallowed by children and (b) they were coated with paints that contained lead leading to poisoning (Gilbert and Wisner 2010). At the heart of this recall was a supply chain that was not sustainable in many different dimensions. At least seven of the SSCM disablers identified in Reefke and Sundaram (2018) were apparent in this SC:

(a) "focus on short term financial results"—CEOs often focus on short term results to keep shareholders happy. But organisations and CEOs will need to orient to a longer-term view of performance and health for the survival of their organisations and supply chains. This could often lead to short-term pain.
(b) "price war battles"—this is yet another area where organisations and supply chains fail by competing purely on price at the expense of quality. Here again such price wars may be used to increase revenue in the short term or could be used in the longer term to gain customers.
(c) "competition forces cost reductions"—high competition often leads organisations and SCs to try to reduce costs. This often leads organisation such as Mattel to shift their manufacturing to low labour cost countries.
(d) "collaboration"—many organisations like Mattel shift their manufacturing to low labour cost countries which are different linguistically, legally, socially, as well as culturally. This often leads to a lack of collaboration or an impedance mismatch in collaboration further leading to misunderstanding particularly around requirements, governance and compliance. This problem is even more accentuated the further we move along the tiers to second and third tier suppliers.
(e) The above decisions also lead to "long distances to import/export goods" which in turn lead to low visibility in the SC. The lack of visibility could be due to the reasons mentioned above in (d) as well as due to different information systems or even a lack of robust/formal systems further up the supply chain.

(f) One of the problems of long distances is "unverified claims of sustainable practices" of suppliers. And the flip side is also lack of visibility into unsustainable practices of suppliers and along the supply chain as a whole.
(g) Furthermore, the long distances usually result in a "dependence on fossil energy" for transportation. This dependence also extends to manufacturing.

In contrast more than seven of the SSCM Enablers (Reefke and Sundaram 2018) need to be strengthened or enhanced in the context of Mattel in order to improve the sustainability of their SC (Hoyt et al. 2008). Chief among these are:

(a) "sustainable material inputs"—this was one of the key problems with Mattel. The paint was contaminated with lead. Quality control measures need to be undertaken and strengthened both at the supplier end as well as inward goods of Mattel.
(b) "collaboration with suppliers"—collaboration with suppliers or lack thereof was at the root of the disaster. It is vital to improve visibility up and down the supply chain. Distances, time zone, language, and cultural differences, exacerbates problems of collaboration. There is a vital need to formalise collaboration through sound processes underpinned by groupware, workflow, and knowledge management systems.
(c) "performance measurement" of various vital KPI's related to quality, research and development, and production need to be instituted, monitored, benchmarked, and controlled. Tight quality control procedures in the supply chain including certification of suppliers, testing of raw materials, work-in-process, and final products.
(d) They also need a "realization of benefits through sustainability efforts", "awareness and acceptance of necessary time and cost investments", and "documentation of the impacts of SC". In Mattel's case it was literally "adapt or die". They had to invest time and money into sustainability efforts in order to survive. In fact, they created the post of senior vice president of corporate social responsibility to address the fallout.
(e) There is an apparent need for application of "models, frameworks, roadmaps to support the transformation towards SSCM".
(f) Furthermore an "efficient information/communication technology to increase sharing and updates" would have allowed Mattel to be aware of the problem early on and address it in a proactive manner rather than reactively. As mentioned earlier cross-national inter-organisational knowledge, workflow, and knowledge management system are vital for sustainable supply chains.

Ameliorating the disablers and enhancing the enablers identified and introducing key SSCM support activities would have enabled Mattel to progress on their sustainability journey. Almost every one of the activities identified in Reefke and Sundaram (2018) is relevant for Mattel, even extending towards "reverse logistics" considering the product recall. These activities would help the Mattel SC to develop the identified SSCM characteristics. And as these characteristics develop the hope is that the "un-aware and non-compliant" parts of the Mattel SC would be able to

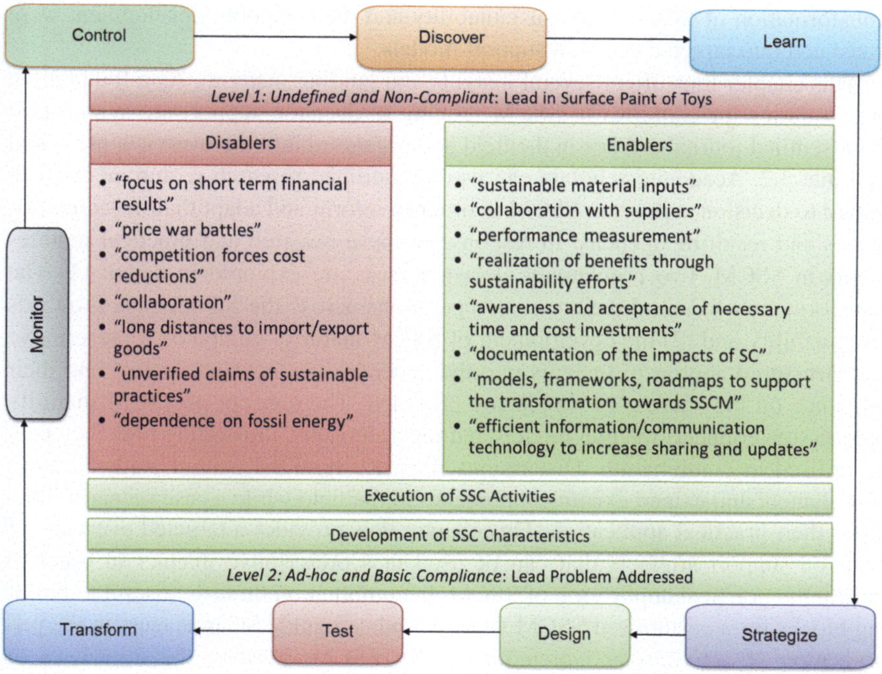

Fig. 3.4 Maturity progression illustration (adapted from Reefke and Sundaram 2018)

move towards higher maturity levels and potentially (if desired) to "extended and sustainability leadership" (Fig. 3.2).

The disablers, enablers, activities and characteristics are intertwined and have causal relationships. But how does a SC go about ameliorating the disablers, enhancing the enablers, and executing the activities? It is in this context that the SSCM transformation model illustrated in Fig. 3.1 and its detailed elements come to the fore. Applying this model systematically over a period of time could enable the SC to mature in terms of sustainability (Fig. 3.2). The connection between the two models is shown in Fig. 3.3 and further illustrated through the Mattel case in Fig. 3.4.

3.6 Conclusion

This chapter aims to illuminate the multiple facets of practices that can support or hinder SSCM, thereby providing a more solid theoretical foundation of SSCM relationships for scholars in the field. Targeting the objectives outlined in the introduction, definitional constructs of SSCM and building blocks of sustainable SCs were deconstructed and explored. Further, decision-making processes for the transformation of SCs towards sustainability and their ongoing development were translated into targeted decision support models.

This chapter thus offers several insights and guiding material: As a foundation, opportunities for enquiries in SSCM relating to decision support are synthesised from seminal journal articles in the field and organised into the overview presented in Table 3.2. Academic scholars can use the outlined research requirements with regard to decision support in SSCM in order to inform and adapt their research priorities and resulting agendas. In response to these research and practical requirements in SSCM, two problematic decision areas are explored in detail: (1) The transformation SCs and their various operations across the SC members towards sustainability and (2) the development of SSCM maturity guided by a prescriptive and structured approach. Insights into the dependencies between factors and their influence on the success of SSCM are provided. The concept of SSCM maturity progression connects the two models and their elements, illustrating how they may be utilised in combination. Discussions illustrate how the overall study findings complement and extend existing approaches in the field while a case scenario illustrates their practical application. This chapter thus provides a targeted overview of decision support artefacts that can be used in a prescriptive manner in order to inform the practical application of SSCM. In combination these are useful as building blocks for a customised SSCM strategy and can guide SC managers in the prioritisation of activities and prerequisites for SSCM. Further, the overviews of definitions and research requirements coupled with the decision models inform and guide scholars in the field in structuring and targeting future research avenues.

References

Ahi, P., & Searcy, C. (2013). A comparative literature analysis of definitions for green and sustainable supply chain management. *Journal of Cleaner Production, 52*, 329–341. https://doi.org/10.1016/j.jclepro.2013.02.018.

Ashby, A., Leat, M., & Hudson-Smith, M. (2012). Making connections: A review of supply chain management and sustainability literature. *Supply Chain Management: An International Journal, 17*(5), 497–516. https://doi.org/10.1108/13598541211258573.

Badurdeen, F., Metta, H., & Gupta, S. (2009). Taxonomy of research directions for sustainable supply chain management. In *IIE Annual Conference Proceedings* (p. 1256).

Bansal, P., & McKnight, B. (2009). Looking forward, pushing back and peering sideways: Analyzing the sustainability of industrial symbiosis. *Journal of Supply Chain Management, 45*(4), 26–37.

Beamon, B. M. (1999). Measuring supply chain performance. *International Journal of Operations & Production Management, 19*(3), 275–292. https://doi.org/10.1108/01443579910249714.

Berger, G., Flynn, A., Hines, F., & Johns, R. (2001). Ecological modernization as a basis for environmental policy: Current environmental discourse and policy and the implications on environmental supply chain management. *Innovation: The European Journal of Social Science Research, 14*(1), 55–72.

Boone, L. M., Colaianni, A. J., Hardison, J. R., Shafer, J. J., Shepherd, N. J., Ramaswamy, M. S., & Wilkerson, T. H. (2009). The GAIA supply chain sustainability maturity model: LMI Research Institute.

Carbone, V., Moatti, V., & Vinzi, V. E. (2012). Mapping corporate responsibility and sustainable supply chains: An exploratory perspective. *Business Strategy and the Environment, 21*(7), 475–494. https://doi.org/10.1002/bse.1736.

Carnegie Mellon—Software Engineering Institute. (2002). Capability Maturity Model Integration (CMMI), version 1.1. In *CMMISM for systems engineering, software engineering, integrated product and process development, and supplier sourcing (CMMI-SE/SW/IPPD/SS, v1.1). Version 1.1* (pp. 1–729). Pittsburgh, PA: Carnegie Mellon—Software Engineering Institute.

Carter, C. R., & Easton, P. L. (2011). Sustainable supply chain management: Evolution and future directions. *International Journal of Physical Distribution and Logistics Management, 41*(1), 46–62. https://doi.org/10.1108/09600031111101420.

Carter, C. R., & Rogers, D. S. (2008). A framework of sustainable supply chain management: Moving toward new theory. *International Journal of Physical Distribution and Logistics Management, 38*(5), 360–387.

Carter, C. R., & Washispack, S. (2018). Mapping the path forward for sustainable supply chain management. *Journal of Business Logistics, 39*(4), 242–247.

Chen, I. J., & Paulraj, A. (2004). Towards a theory of supply chain management: The constructs and measurements. *Journal of Operations Management, 22*(2), 119–150.

Christopher, M. (1998). *Logistics and supply chain management: Strategies for reducing cost and improving service*. London: Financial Times/Pitman.

Colicchia, C., Melacini, M., & Perotti, S. (2011). Benchmarking supply chain sustainability: Insights from a field study. *Benchmarking: An International Journal, 18*(5), 705–732. https://doi.org/10.1108/14635771111166839.

Cooper, M. C., Ellram, L. M., Gardner, J. T., & Hanks, A. M. (1997). Meshing multiple alliances. *Journal of Business Logistics, 18*(1), 67–90.

Croom, S., Romano, P., & Giannakis, M. (2000). Supply chain management: An analytical framework for critical literature review. *European Journal of Purchasing & Supply Management, 6*(1), 67–83.

Cruz, J. M. (2009). The impact of corporate social responsibility in supply chain management: Multicriteria decision-making approach. *Decision Support Systems, 48*(1), 224–236. https://doi.org/10.1016/j.dss.2009.07.013.

de Mast, J., & Lokkerbol, J. (2012). An analysis of the six sigma DMAIC method from the perspective of problem solving. *International Journal of Production Economics, 139*, 604–614. https://doi.org/10.1016/j.ijpe.2012.05.035.

Deming, W. E. (1986). Out of the crisis. Cambridge, MA: MIT Press.

Dey, A., LaGuardia, P., & Srinivasan, M. (2011). Building sustainability in logistics operations: A research agenda. *Management Research Review, 34*(11), 1237–1259.

Elkington, J. (1998). *Cannibals with foerks: The triple bottom line of 21st century business*. Gabriola Island, BC; Stony Creek, CT: New Society Publishers.

Esty, D. C., & Winston, A. S. (2006). *Green to gold: How smart companies use environmental strategy to innovate, create value, and build competitive advantage*. New Haven, CT: Yale University Press.

Font, X., Tapper, R., Schwartz, K., & Kornilaki, M. (2008). Sustainable supply chain management in tourism. *Business Strategy and the Environment, 17*(4), 260–271. https://doi.org/10.1002/bse.527.

Garriga, E., & Melé, D. (2004). Corporate social responsibility theories: Mapping the territory. *Journal of Business Ethics, 53*(1–2), 51–71. https://doi.org/10.1023/b:busi.0000039399.90587.34.

GCIO, & OMG. (2009). Green Business Maturity Model: Green Computing Impact Organization (GCIO) & Object Management Group (OMG).

Gilbert, J., & Wisner, J. (2010). Mattel, Lead paint, and magnets: Ethics and supply chain management. *Ethics & Behavior, 20*(1), 33–46. https://doi.org/10.1080/10508420903482491.

Goldschmidt, K., Harrison, T., Holtry, M., & Reeh, J. (2013). Sustainable procurement: Integrating classroom learning with university sustainability programs. *Decision Sciences Journal of Innovative Education, 11*(3), 279–294. https://doi.org/10.1111/dsji.12007.

Gilmour, B., Maine, D. and Chu, J. (2004). Consumer Products 2010-executing to lead in a world of extremes. IBM Institute for Business Value, 26.

Hage, J. (1980). *Theories of organizations: Form, process and transformation*. New York: Wiley.

Halldórsson, Á., & Kovács, G. (2010). The sustainable agenda and energy efficiency: Logistics solutions and supply chains in times of climate change. *International Journal of Physical Distribution and Logistics Management, 40*(1/2), 5–13. https://doi.org/10.1108/09600031011018019.

Harris, I., Rodrigues, V. S., Naim, M., & Mumford, C. (2010). Restructuring of logistics systems and supply chains. In A. McKinnon, S. Cullinane, M. Browne, & A. Whiteing (Eds.), *Green logistics: Improving the environmental sustainability of logistics* (pp. 101–123). London: Kogan Page.

Hassini, E., Surti, C., & Searcy, C. (2012). A literature review and a case study of sustainable supply chains with a focus on metrics. *International Journal of Production Economics, 140*(1), 69–82. https://doi.org/10.1016/j.ijpe.2012.01.042.

Heskett, J., Jones, T., Loveman, G., Sasser, E., & Schlesinger, L. (1994). Putting the Service-Profit Chain to work. *Harvard Business Review*, March-April, 164–174.

Hoejmose, S., Brammer, S., & Millington, A. (2013). An empirical examination of the relationship between business strategy and socially responsible supply chain management. *International Journal of Operations & Production Management, 33*(5), 589–621. https://doi.org/10.1108/01443571311322733.

Hoyt, D., Lee, H. L., & Tseng, M. (2008). *Unsafe for children: Mattel's toy recalls and supply chain management Reference no. GS63* (p. 20). Stanford, CA: Stanford Business School.

Jayaraman, V., Klassen, R., & Linton, J. D. (2007). Supply chain management in a sustainable environment. *Journal of Operations Management, 25*(6), 1071–1074.

Ketokivi, M., & Mantere, S. (2010). Two strategies for inductive reasoning in organizational research. *Academy of Management Review, 35*(2), 315–333.

Kleindorfer, P. R., Singhal, K., & Van Wassenhove, L. N. (2005). Sustainable operations management. *Production and Operations Management, 14*(4), 482–492.

Krause, D. R., Vachon, S., & Klassen, R. D. (2009). Special topic forum on sustainable supply chain management: Introduction and reflections on the role of purchasing management. *Journal of Supply Chain Management, 45*(4), 18–25. https://doi.org/10.1111/j.1745-493X.2009.03173.x.

Lambert, D. M., & Cooper, M. C. (2000). Issues in supply chain management. *Industrial Marketing Management, 29*(1), 65–83.

Langley, A., Mintzberg, H., Pitcher, P., Posada, E., & SaintMacary, J. (1995). Opening up decision making: The view from the black stool. *Organization Science, 6*(3), 260–279.

Lee, H. L. (2004). The triple-a supply chain. *Harvard Business Review, 82*(10), 102–112. 157.

Lee, H. L. (2008). Embedding sustainability: Lessons from the front line. *International Commerce Review: ECR Journal, 8*(1), 10–20.

Lee, H. L. (2010). Don't tweak your supply chain—Rethink it end to end. *Harvard Business Review, 88*(10), 62–69.

Lieb, K. J., & Lieb, R. C. (2010). Environmental sustainability in the third-party logistics (3PL) industry. *International Journal of Physical Distribution and Logistics Management, 40*(7), 524–533. https://doi.org/10.1108/09600031011071984.

Linton, J. D., Klassen, R., & Jayaraman, V. (2007). Sustainable supply chains: An introduction. *Journal of Operations Management, 25*(6), 1075–1082.

Lockamy, A., & McCormack, K. (2004). The development of a supply chain management process maturity model using the concepts of business process orientation. *Supply Chain Management: An International Journal, 9*(4), 272–278.

Markman, G. D., & Krause, D. (2016). Theory building surrounding sustainable supply chain management: Assessing what we know, exploring where to go. *Journal of Supply Chain Management, 52*(2), 3–10. https://doi.org/10.1111/jscm.12105.

Marshall, J. D., & Toffel, M. W. (2005). Framing the elusive concept of sustainability: A sustainability hierarchy. *Environmental Science & Technology, 39*(3), 673–682. https://doi.org/10.1021/es040394k

McCormack, K. (2001). Supply chain maturity assessment: A roadmap for building the extended supply chain. *Supply Chain Practice, 3*(4), 4–21.

McCormack, K., Ladeira, M. B., & de Oliveira, M. P. V. (2008). Supply chain maturity and performance in Brazil. *Supply Chain Management: An International Journal, 13*(4), 272–282. https://doi.org/10.1108/13598540810882161.

Mentzer, J. T., DeWitt, W., Keebler, J. S., Min, S., Nix, N. W., Smith, C. D., & Zacharia, Z. G. (2001). Defining supply chain management. *Journal of Business Logistics, 22*(2), 1–25.

Mintzberg, H., Raisinghani, D., & Theoret, A. (1976). The structure of unstructured decision processes. *Administrative Science Quarterly, 21*, 246–275.

Montabon, F., Pagell, M., & Wu, Z. (2016). Making sustainability sustainable. *Journal of Supply Chain Management, 52*(2), 11–27. https://doi.org/10.1111/jscm.12103.

Murphy, P. R., & Poist, R. F. (2003). Green perspectives and practices: A "comparative logistics" study. *Supply Chain Management: An International Journal, 8*(2), 122–131. https://doi.org/10.1108/13598540310468724.

Nunen, J., Zuidwijk, R., & Moonen, H. (2005). Smart and sustainable supply chains. In P. Vervest, E. van Heck, K. Preiss, & L.-F. Pau (Eds.), *Smart business networks* (pp. 159–167). Berlin: Springer.

OMG. (2008). *Business Process Maturity Model (BPMM)*. Object management group. Retrieved from http://www.omg.org/spec/BPMM/1.0/PDF.

Orlitzky, M., Schmidt, F. L., & Rynes, S. L. (2003). Corporate social and financial performance: A meta-analysis. *Organization Studies, 24*(3), 403–441. https://doi.org/10.1177/0170840603024003910.

Pagell, M., & Wu, Z. (2009). Building a more complete theory of sustainable supply chain management using case studies of 10 exemplars. *Journal of Supply Chain Management, 45*(2), 37–56. https://doi.org/10.1111/j.1745-493X.2009.03162.x.

Pagell, M., & Shevchenko, A. (2014). Why research in sustainable supply chain management should have no future. *Journal of Supply Chain Management, 50*(1), 44-55. https://doi.org/10.1111/jscm.12037

Reefke, H., & Sundaram, D. (2017). Key themes and research opportunities in sustainable supply chain management—Identification and evaluation. *Omega, 66*(part B), 195–211. https://doi.org/10.1016/j.omega.2016.02.003.

Reefke, H., & Sundaram, D. (2018). Sustainable supply chain management: Decision models for transformation and maturity. *Decision Support Systems, 113*, 56–72. https://doi.org/10.1016/j.dss.2018.07.002.

Starik, M., & Rands, G. P. (1995). Weaving an integrated web: Multilevel and multisystem perspectives of ecologically sustainability organizations. *Academy of Management Review, 20*(4), 908–935.

Sarkis, J. (1998). Evaluating environmentally conscious business practices. *European Journal of Operational Research, 107*(1), 159–174. https://doi.org/10.1016/s0377-2217(97)00160-4.

Sarkis, J. (2012). A boundaries and flows perspective of green supply chain management. *Supply Chain Management: An International Journal, 17*(2), 202–216. https://doi.org/10.1108/13598541211212924.

Scheer, A.-W., Abolhassan, F., Jost, W., & Kirchmer, M. (2003). *Business process change management: ARIS in practice*. Berlin: Springer.

Schmidt, C. W. (2008). Face to face with toy safety: Understanding an unexpected threat. *Environmental Health Perspectives, 116*(2), 71–76.

Schroeder, R. G., Linderman, K., Liedtke, C., & Choo, A. S. (2008). Six sigma: Definition and underlying theory. *Journal of Operations Management, 26*, 536–554. https://doi.org/10.1016/j.jom.2007.06.007.

Seuring, S. (2013). A review of modeling approaches for sustainable supply chain management. *Decision Support Systems, 54*(4), 1513–1520. https://doi.org/10.1016/j.dss.2012.05.053.

Seuring, S., & Müller, M. (2008a). Core issues in sustainable supply chain management—a Delphi study. *Business Strategy and the Environment, 17*(8), 455–466. https://doi.org/10.1002/bse.607.

Seuring, S., & Müller, M. (2008b). From a literature review to a conceptual framework for sustainable supply chain management. *Journal of Cleaner Production, 16*(15), 1699–1710. https://doi.org/10.1016/j.jclepro.2008.04.020.

Shepherd, C., & Günter, H. (2006). Measuring supply chain performance: Current research and future directions. *International Journal of Productivity and Performance Management, 55*(3/4), 242–258.

Simon, H. A. (1977). *The new science of managerial decision* (3rd ed.). Englewood Cliffs, NJ: Prentice-Hall.

Smith, N. C., Palazzo, G., & Bhattacharya, C. (2010). Marketing's consequences: Stakeholder marketing and supply chain CSR issues. INSEAD working paper no. 2010/17/INSEAD Social Innovation Centre.

Srivastava, S. K. (2007). Green supply-chain management: A state-of-the-art literature review. *International Journal of Management Reviews, 9*(1), 53–80.

Stevens, G. C. (1993). Integrating the supply chain. *International Journal of Physical Distribution and Logistics Management, 19*(8), 3–8. https://doi.org/10.1108/EUM0000000000329.

Svensson, G. (2007). Aspects of sustainable supply chain management (SSCM): Conceptual framework and empirical example. *Supply Chain Management: An International Journal, 12*(4), 262–266.

Tan, K. C., Kannan, V. R., & Handfield, R. (1998). Supply chain management: Supplier performance and firm performance. *International Journal of Purchasing and Materials Management, 34*(3), 2–9.

The Natural Step. (2009). *The natural step—Our story*. Retrieved from http://www.naturalstep.org/en/our-story.

Vachon, S., & Mao, Z. (2008). Linking supply chain strength to sustainable development: A country-level analysis. *Journal of Cleaner Production, 16*(15), 1552–1560. https://doi.org/10.1016/j.jclepro.2008.04.012.

Veleva, V., & Ellenbecker, M. (2001). Indicators of sustainable production: Framework and methodology. *Journal of Cleaner Production, 9*(6), 519–549.

Wagner, B., & Svensson, G. (2010). Sustainable supply chain practices: Research propositions for the future. *International Journal of Logistics Economics and Globalisation, 2*(2), 176–186. https://doi.org/10.1504/IJLEG.2010.032432.

Walker, H., Seuring, S., Sarkis, J., & Klassen, R. (2014). Sustainable operations management: Recent trends and future directions. *International Journal of Operations & Production Management, 34*(5). https://doi.org/10.1108/IJOPM-12-2013-0557.

WCED. (1987). *Our common future*. Retrieved from New York, NY.

Winter, M., & Knemeyer, A. M. (2013). Exploring the integration of sustainability and supply chain management: Current state and opportunities for future inquiry. *International Journal of Physical Distribution and Logistics Management, 43*(1), 18–38. https://doi.org/10.1108/09600031311293237.

Yusuf, Y. Y., Gunasekaran, A., Adeleye, E. O., & Sivayoganathan, K. (2004). Agile supply chain capabilities: Determinants of competitive objectives. European Journal of Operational Research, 159(2), 379–392. https://doi.org/10.1016/j.ejor.2003.08.022.

Chapter 4
DEA Application in Sustainability 1996–2019: The Origins, Development, and Future Directions

Haibo Zhou, Yi Yang, Yao Chen, Joe Zhu, and Yu Shi

Abstract Sustainable development and sustainability assessment have been of great interest to both academe and practitioners in the past decades. In this study, we review the literature on data envelopment analysis (DEA) applications in sustainability using citation-based approaches. A directional network is constructed based on citation relationships among DEA papers published in journals indexed by the Web of Science database from 1996 to 2019. We first draw the citation chronological graph to present a complete picture of literature development trajectory since 1996. Then we identify the local main DEA development paths in sustainability research by assigning an importance index, namely search path count (SPC), to each link in the citation network. The local main path suggests that the current key route of DEA applications in sustainability focus on the environmental sustainability. Through the Kamada–Kawai layout algorithm, we find four research clusters in the literature including corporate sustainability assessment, regional sustainability assessment, sustainability composite indicator construction, and sustainability performance analysis. For each of the clusters, we further identify the key articles based on citation network and local citation scores, demonstrate the developmental trajectory of the literature, and suggest future research directions.

Keywords Data envelopment analysis (DEA) · Sustainability · Literature survey
Citation analysis

H. Zhou
School of Economics and Management, Southeast University,
Nanjing, Jiangsu, People's Republic of China

Y. Yang · Y. Chen
Manning School of Business, University of Massachusetts at Lowell, Lowell, MA, USA
e-mail: Yi_Yang@uml.edu; Yao_Chen@uml.edu

J. Zhu (✉) · Y. Shi
Foisie School of Business, Worcester Polytechnic Institute, Worcester, MA, USA
e-mail: jzhu@wpi.edu; yshi2@wpi.edu

C. Chen et al. (eds.), *Pursuing Sustainability*, International Series
in Operations Research & Management Science 301,
https://doi.org/10.1007/978-3-030-58023-0_4

4.1 Introduction

The concept of sustainability stems from ecology and describes the use of a regenerative natural system in such a way that this system retains its essential properties and its population can naturally be replenished. In more general terms, sustainability is the endurance of systems and processes. The organizing principle for sustainability is sustainable development, which finds its way into the economics and management areas in 1987 when sustainable development was first initiated as an environmentally friendly, economically feasible and socially acceptable growth pattern in the Brundtland Commission (formally named as the World Commission on Environment and Development (WCED)). Since then, thousands of initiatives have been taken at the local, national, and global levels in an attempt to address different aspects of the sustainability challenges (Mebratu 1998).

Since the early 2000s, firms have been pressured to pay attention to the triple bottom line of sustainability—profit, people, and planet (Elkington 2002) because of the increasing demand for natural resources (clean water, crude oil, woods, metals, etc.) whose supply continues to diminish, raised concerns about various unethical corporate practices, and the development of the emerging markets with supply-chain constraints (Tang and Zhou 2012). As a result, the need for measuring sustainable development is widely recognized (e.g., Tyteca 1998). So far, sustainability assessment has served four major purposes: decision making and management, advocacy, participation and consensus building, and research and analysis (Parris and Kates 2003), and it has been applied at different levels: national (e.g., Munksgaard et al. 2005; Coli et al. 2011), regional or urban community (e.g., Hu et al. 2005; Munda and Saisana 2011), industry sectorial (e.g., Zofio and Prieto 2001; Peres-Neto et al. 2006), and corporate (e.g., Figge and Hahn 2004; Kuosmanen and Kuosmanen 2009). In the beginning, sustainability assessment mainly focused on environmental sustainability problems covering only economic and environmental dimensions. More recently, this line of research has started to focus on prospects for lasting net gains and the acceptability of trade-off rules among the environmental, economic, and social dimensions (Gibson 2006; Pope et al. 2004; Winfield et al. 2010).

Accordingly, three categories of indicators and methods for sustainability evaluation have emerged in the literature. *System analysis* is an approach that takes into consideration of both relationships among the internal components of the system, and relationships between the internal components and the external environment (e.g., Ulanowicz 2009; Goerner et al. 2009; Antonio et al. 2012). *Flow analysis* evaluates system sustainability through resource utilization efficiency that only considers the relationship between internal components and the external environment (e.g., Balocco et al. 2004; Paoli et al. 2008; Campbell and Garmestani 2012). Finally, *indicator enumeration* mainly chooses indicators from environmental, economic, social, and institutional aspects to evaluate the system sustainability without considering either of the relationships mentioned above (e.g., Yli-Viikari 1999; Ness et al. 2007; Ou and Liu 2010).

Data Envelopment Analysis (DEA) (Charnes et al. 1978) is a method for evaluating performance of peer decision-making units (DMUs) with multiple performance measures that are termed as inputs and outputs. DEA first establishes an "efficient frontier" formed by a set of DMUs that exhibit best practices and then assigns the efficiency level to other non-frontier units according to their distances to the efficient frontier. Over the years, DEA has been enriched and modified. Numerous DEA models have been developed and used in various applications, including sustainability research. In general, there are three approaches to employing DEA models in sustainability literature (Choi and Zhang 2011): traditional DEA models with simple translation of data (Lovell et al. 1995; Yeh et al. 2010), traditional DEA models treating undesirable outcomes as inputs (Hu and Wang 2006; Zhang et al. 2008), and DEA models employing the concept of weak disposability technology (Färe and Grosskopf 2004; Zhou et al. 2008a). Researchers have applied DEA models to address corporate, regional and national sustainability issues as well as those related to supply chain.

Although DEA has been extensively applied in sustainability, few surveys to the best of our knowledge besides Dakpo et al. (2016) have been conducted to systematically review the current status of the literature and discuss the future research direction. Although Dakpo et al. (2016) make a critical review on methods integrating environmental aspects into productive efficiency, their study focuses on only environmental factors, especially the undesirable outputs in production technology modeling, and does not include social factors, another important part of sustainability. In addition, their review is based on subjective and qualitative analyses rather than objective quantitative analysis methods. To fill the gap, our study collects 320 relevant papers published from 1996 to March 2016 and analyzes the research status of DEA applications in sustainability through citation analysis of bibliometrics. Using the citation analysis software HistCite, we conduct a visual analysis and construct a citation chronological graph to identify the main development route and key publications of DEA application in sustainability. Then with the help of Pajek software, we discover the major research clusters as well as the local main paths, and further identify future research directions in each research cluster. Moreover, our review highlights the importance of reliable sustainability measures and introduces current major DEA approaches in sustainability evaluation.

This paper is organized as follows. In the next section, we describe the data and methods used in this study. Section 4.3 discusses the basic statistics for DEA applications in sustainability. Section 4.4 presents the major findings through citation chronological graph and main path analysis. Section 4.5 identifies the major research clusters and draws the development trajectories of each cluster, presenting the most-cited works in each research area. The last section draws conclusions including implications and insights from the analysis results.

4.2 Review Methods

4.2.1 Data Source and Collection

To facilitate a coherent review, we use systematic searches and formal summaries of the literature to integrate major studies in the area. ISI Web of Science (WOS) is used as the data source to collect relevant scholarly work. WOS is the world's leading citation database with a multidisciplinary coverage of over 10,000 high impact journals in science and social science as well as proceedings of over 120,000 international conferences. Specifically, we select the databases within WOS including Science Citation Index Expanded (SCIE), Social Sciences Citation Index (SSCI), Conference Proceedings Index–Science (CPI-S), and Conference Proceedings Index–Social Science and Humanities (CPI-SSH).

We start with an exhaustive search in the databases using combinations of sustainability related keywords (i.e., sustainability, sustainable, green, and social responsibility) with the term of DEA or data envelopment analysis in the fields of title, abstract, author keyword, or Keywords Plus®. A sample of 475 articles is retrieved from the databases after the initial search. We then read through these papers' abstracts to assess whether they dealt with DEA applications in sustainability. When we are unsure, we download and read the full publications. Non-DEA or non-sustainability papers are manually examined and excluded from the dataset. In the course of manual checking and screening, we find out that some papers, although listed as DEA or sustainability in the Keywords Plus® field, contain limited contents about DEA or sustainability. For these cases, we conduct a partition analysis on the citation network to find out the outliers and then remove them from the dataset. After the manual checking and screening, the final sample consists of 320 articles published from 1996 to March 2016 in various subject areas including corporate sustainability assessment, sustainability composite indicators construction, sustainability performance analysis, and regional sustainability development assessment. Out of the 320 articles, 120 were published in journals with a 2015 ABS (Chartered Association of Business Schools) journal ranking of 3 or above. We then developed a detailed summary for each article in the final sample.

4.2.2 Citation-Based Review Methods

To further probe the origins and current state of the literature, we employ two citation-based methods: the main path analysis and Kamada–Kawai algorithm. We believe these citation-based methods can complement the traditional qualitative review methods by bringing a level of objectivity and quantification. In recent years, the citation-based methods have been applied increasingly across a variety of research fields such as literature research (Liu et al. 2013), journal evaluation (Garfield 1972), and scholar assessment (Schildt et al. 2006).

The main path analysis, introduced by Hummon and Dereian (1989), is a well-known method that traces the main knowledge flow in a scientific discipline through citation data. This network-based method treats scientific publications as nodes of a network. Then, citation information is used to establish links among nodes, and a link's direction points from the cited document to the citing one. The first step in finding the main path is to identify the importance of each citation link in the network. This is achieved by counting the times a citation link has been traversed. In this study, we choose to use the search path count (SPC) recommended by Batagelj (2003) to do the counts. The SPC for each link is defined as the total number of times a link is traversed assuming that one exhausts all efforts in searching out all paths from all the *sources* (nodes that are cited, but cite no other nodes) to all the *sinks* (nodes that cite other nodes, but are not cited). After SPCs for all the links are calculated, we start the main path search from all the sources by applying the "priority first search" algorithm as Hummon and Dereian (1989) suggested. That is, at any node, one always chooses the next link in the path with the highest SPC as the outgoing link. By applying the choice rule repeatedly until hitting a sink, a main path is constructed.

Another citation-based method used in this study is Kamada–Kawai algorithm, an automatic graph drawing algorithm based on the idea that graph can be considered as a dynamic system of *springs*. In that system, every pair of nodes, connected by edge, is connected by spring. The optimization procedure tries to minimize the total energy of the system using the two-dimensional Newton–Raphson method, which is known in the multidimensional scaling (MDS) community as *the stress function*. The strength of a spring between two vertices is inversely proportional to the square of the shortest distance (in graph terms) between those two vertices. Essentially, vertices that are closer in the graph-theoretic sense (i.e., by following edges) will have stronger springs and therefore be placed closer together. An advantage of this method is that it can be applied straightforwardly to drawing edge-weighted graphs (Harel and Koren 2002).

4.3 Literature Overview

4.3.1 Publications Over Time

Figure 4.1 demonstrates the number of publications per year from 1996 to 2015. The literature of DEA application in sustainability can be traced back to the year of 1996 when Färe et al. first introduced input-oriented DEA methods containing "bad output" pollution variables to obtain environmental performance indicators for US fossil fuel-fired electric utilities. Since then, DEA has found its way into a broad spectrum of applications in the sustainability area such as energy and environment efficiency (Zhang et al. 2008; Sueyoshi and Goto 2014a, b), and corporate social responsibility (Chen and Delmas 2011). The number of publications

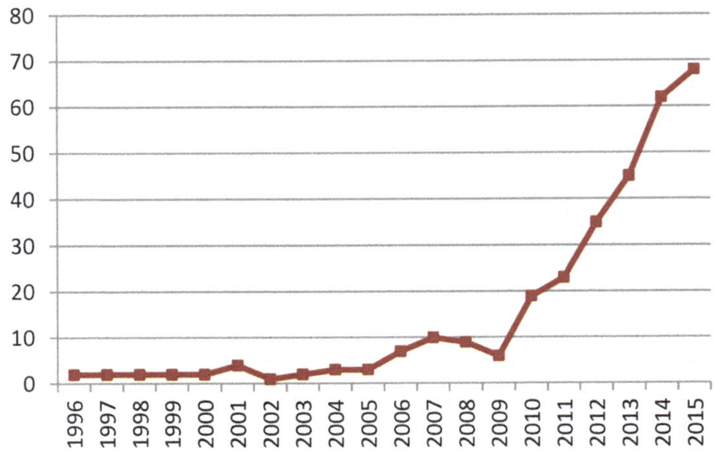

Fig. 4.1 Distribution of publications over time

has flourished in the past 5 years. The recent increased interest in this area may be due to the 2009 United Nations Climate Change conference, commonly known as the Copenhagen Summit, which raised climate change policy to the highest political level, and therefore drew attention of the academe to the issue of sustainability.

4.3.2 Publication Outlets and Scholarly Community

Approximately half (48%) of the reviewed articles were published in 20 journals (see Fig. 4.2) and these journals present a wide range of research scope from specialized journals in energy and environment journals to general operational management journals. Among the 20 journals, seven journals published at least 10 articles, including Energy Economics, Journal of Cleaner Production, Energy Policy, Sustainability, Ecological Economics, Applied Energy, and European Journal of Operational Research.

Table 4.1 lists the top 10 DEA authors in order according to the total local citation scores of their published papers in sustainability area. Local citation score (LCS) is based on total citations of paper A received from local collection, and shows the citation frequency within the collection, indicating the relative importance of certain paper. In this study, we calculate a researcher's total local citation score (TLCS) as the sum of LCS of his/her all papers. The TLCS can reflect the relative importance of the researcher in certain area. As can be seen from the table, Zhou P, Ang BW, Sueyoshi T, Färe R, and Grosskopf S are the top five DEA researchers in sustainability area based on the TLCS.

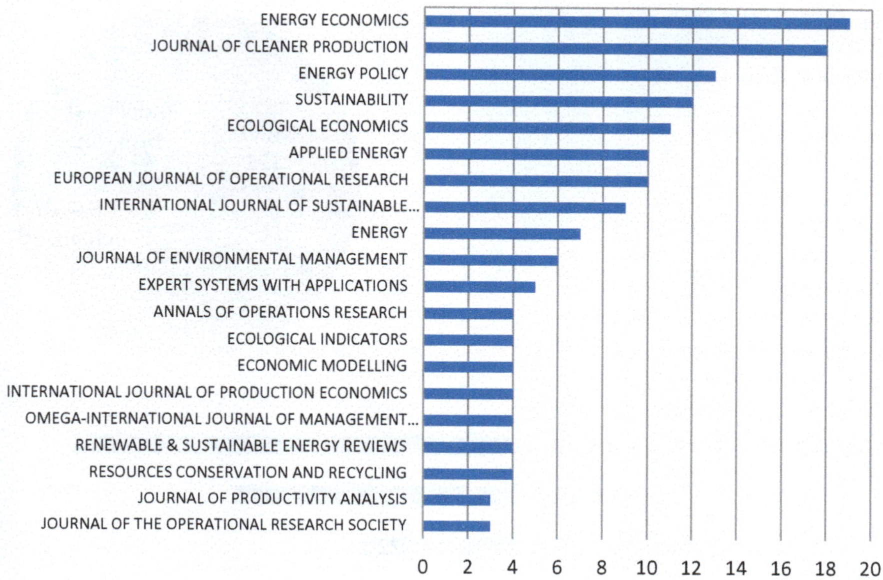

Fig. 4.2 Distribution of reviewed papers among top 20 journals

Table 4.1 Top 10 DEA researchers in sustainability area according to their total LCS

Ranking	Authors	Total LCS	Total number of papers
1	Zhou P	117	7
2	Ang BW	100	5
3	Sueyoshi T	91	14
4	Färe R	66	4
5	Grosskopf S	66	4
6	Poh KL	63	2
7	Goto M	54	7
8	Tyteca D	49	3
9	Chung YH	37	1
10	Zhang B	36	2

4.3.3 Research Analysis Unit and Application Area

When analyzing the content of the 320 articles, we divide the publications into four categories based on the unit of analysis: individual firm, supply chain, industry, and region. Figure 4.3 is the breakdown pie chart of the number of papers in each category. We find that a majority (56%) of the reviewed studies focus on corporate sustainability and industrial sustainability using firms as DMUs. On the other hand, a growing attention has been given to sustainability issues in different stages along the supply chain. There are also 121 papers studying regional sustainability from the macro-level perspective. The region here can be a city, a province, or a country.

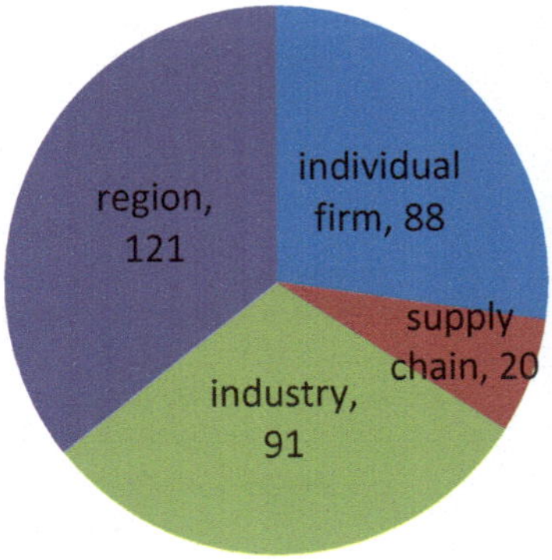

Fig. 4.3 Distribution of reviewed papers using different analysis units

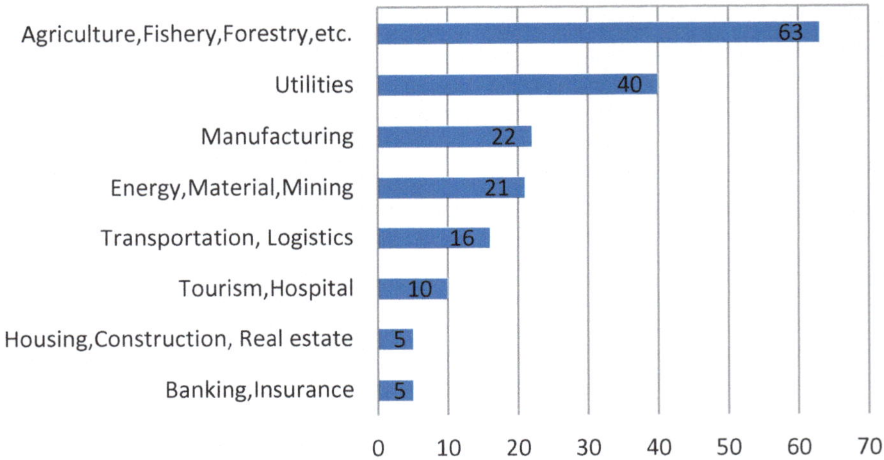

Fig. 4.4 Distribution of reviewed papers by industry sector

In Fig. 4.4, we list various industry sectors studied by at least 5 reviewed papers. The top five most frequently studied industries include Agriculture, Utilities, Manufacturing, Energy, and Transportation and Logistics. It is not surprising because these industries have significant impacts on environmental and social sustainability.

4.3.4 DEA Methodologies Employed

The majority of the reviewed papers focus on analytical models, and a variety of DEA methods have been applied in developing these models. Based on Liu et al. (2013, 2016) and Zhou et al. (2008b), we classify DEA methods used in the sustainability research into six main groups: (1) Traditional DEA models including CCR and BCC models; (2) Slack-based models (SBM) and intertemporal DEA models,

Table 4.2 Distribution of reviewed papers in different DEA method groups

Research methodology	Number of papers
Traditional DEA	92
SBM and dynamic DEA	61
Extending models	51
Two-stage contextual factor evaluation framework	40
Special data	13
Two-stage network DEA	14
Others	62

especially DEA-based Malmquist productivity index; (3) Extending models including assurance region, dual factor, cross-efficiency, and super-efficiency; (4) Two-stage contextual factor evaluation framework that first obtains efficiency scores through DEA analysis and then correlates these scores with various contextual factors either by ordinary least squares regressions (OLS), Tobit regressions, or maximum likelihood estimation (MLE). (5) Models handling special types of data such as fuzzy, ordinal, qualitative, and negative data; (6) Two-stage network DEA. Table 4.2 presents the number of studies in each method category.

As Table 4.2 shows, the most frequently used DEA methods in sustainability study are Traditional DEA models (e.g., CCR and BCC models). Typically, traditional DEA models use the radial measure and calculate efficiency based on the input excesses and output shortfalls. Recently, more and more advanced DEA models (those in Categories 2–6) have been developed and applied in sustainability research in response to the growing demand for analysis accuracy and data complexity.

Among those advanced methods, SBM and intertemporal DEA (Category 2) are the most popular ones used in a total of 61 studies. SBM introduced by Tone (2001) has been frequently used to evaluate the sustainability of DMUs at both the micro and macro levels by including undesirable outputs in the model (Goto et al. 2014; Chen and Xie 2015). Recently, new SBM models are also developed, for example, SBM based on directional distance functions (Färe and Grosskopf 2010), sequential slack-based efficiency measure (SSBM) (Zhang and Kim 2014). Intertemporal DEA, on the other hand, employs the Malmquist model to handle dynamic time series data, and has been widely applied in evaluating the intertemporal sustainability performance of both corporations (Graham 2009) and regions (Lei et al. 2013).

Extending DEA models (Category 3) are the second popular advanced DEA methods used in sustainability study. As an extension to conventional DEA models, this group methods include assurance region on multipliers (Wey 2015), dual factor handling the case where factors simultaneously play both input and output roles (Kumar et al. 2014; Mirhedayatian et al. 2014), cross-efficiency DEA for peer evaluation (Lee and Saen 2012), and super-efficiency DEA for further ranking the efficient DMUs (Li and Lin 2015b).

Two-stage contextual factor evaluation framework (Category 4), introduced by Simar and Wilson (2007), is also widely used in sustainability analysis. This framework usually obtains the efficiency scores through DEA methods first, especially using bootstrapping methods to construct a base for statistical inference, and then conducts contextual factor analysis through a series of statistical regression analyses. The purpose of this line research is to find the determinants and influential factors of sustainability, or to portray the relationships between sustainability and environmental factors (Assaf et al. 2012; Chen et al. 2014; Gadanakis et al. 2015; Picazo-Tadeo et al. 2011).

DEA methods in Categories 5 and 6 have not been extensively employed in sustainability studies. But recently, more and more attention has been given to models handling special types of data in sustainability research, including fuzzy data (Azadi et al. 2015), ordinal data (Chen and Delmas 2011), qualitative data (Zeydan et al. 2011), negative data (DiMaria 2014), and so on. The Two-stage network DEA, which takes into account the inner operational mechanism of the subsystems in each DMU under evaluation (Chen et al. 2013), is also gaining popularity.

In addition to the six major categories of DEA methodologies, several new trends are observed in the literature. First, our literature review reveals that two or more DEA methods could be used simultaneously in one study. Second, in sustainability evaluation, the concept of material balance is being incorporated into the production model (Coelli et al. 2007). Last, recent research has been combining DEA methods with more non-DEA methods such as analytic hierarchy process (AHP) (Wey 2015), principal component analysis (PCA) (Dong et al. 2015), analysis of variance (ANOVA) (Kim et al. 2011), artificial neural network (ANN) (Chuang et al. 2011), and hierarchical clustering method (Xie et al. 2016).

4.4 Citation Chronological Graph and Local Main Path Analysis

In this section, we present our major findings through the citation-based methods. Among the 320 reviewed articles, we first identify the top 50 papers with the highest local citation scores (LCS). Then, based on the top 50 papers, we draw a citation chronological graph that provides the foundation for the local main path analysis on the development trajectory of the research contents.

4.4.1 Findings from Citation Chronological Graph

The citation chronological graph indicates all important papers during the development of a discipline and their relationships based on citations during the active years. Using Histcite software, we draw the citation chronological graph of the top

50 DEA publications in sustainability (see Fig. 4.5). In the figure, a circle represents a paper with its serial number inside and the circle size reflects its LCS value. The bigger the circle is, the higher the paper's LCS is, and the more significant it is. The arrow indicates the direction of citation, from a citing paper to a cited paper. The ordinate is a timeline from 1996 to 2015.

In Fig. 4.5, there are a total of 141 links between the 50 nodes (major papers). As we can see, the DEA application in sustainability research begins with Färe et al. (1996). But, the number of citations is small in the early years. Papers with more citations have been published since 2008, showing that the DEA application research in sustainability area has become more and more popular recently. The review paper by Zhou et al. (2008b) has the maximum LCS value of 52, which provides a good summary on previous DEA applications in energy and environment efficiency assessment as well as a reference for future sustainability research. The paper with the second highest LCS value of 37 is Chung et al. (1997) that proposes a new index called the Malmquist–Luenberger productivity index which overcomes the shortcomings of the original Malmquist index. The Malmquist–Luenberger index readily allows for inclusion of undesirable outputs without requiring information on shadow prices. Zhang et al. (2008) is another highly cited paper with a LCS value of 34, which analyzes eco-efficiency of industrial system using DEA, introduces the concept of industrial system eco-efficiency, and invites many researchers into the regional sustainable development assessment area. It is noteworthy that there are some nodes isolated or with fewer links away from the main stream, indicating some interesting research areas that have not been well developed. To further refine the research topic and the development process, we next employ the local main path analysis.

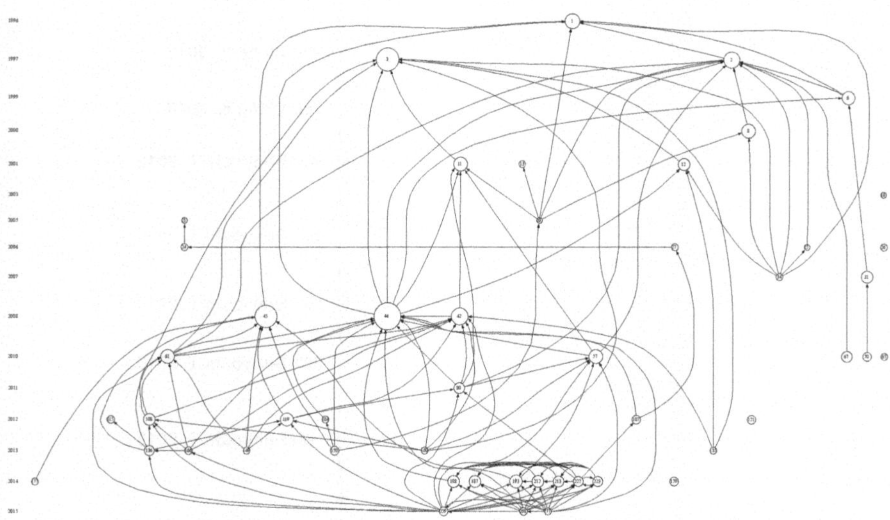

Fig. 4.5 Citation chronological graph of DEA papers in sustainability (LCS-count50)

4.4.2 Findings from Local Main Path Analysis

The local main path indicates the most significant knowledge route at each juncture of knowledge dissemination for a scientific discipline (Liu et al. 2013). Using Pajek software, we identify the local main path in the literature, and present it in Fig. 4.6. In the figure, the arrow indicates the direction of knowledge flow from the cited publication to the citing one, and the line thickness reflects the search path count (SPC) value. The thicker the line is, the more significant the route is.

The local main path consists of 16 papers, which constitute the backbone of the whole network and play an important role in knowledge flow of the field. Table 4.3 lists the 16 papers in detail.

Fig. 4.6 Local main path of DEA application research in sustainability

Table 4.3 Detailed information of the studies in the local main path

ID	Title	Authors	Journal	Year
1	An activity analysis model of the environmental performance of firms—application to fossil-fuel-fired electric utilities	Färe R, Grosskopf S, Tyteca D	Ecological Economics	1996
2	Linear programming models for the measurement of environmental performance of firms—concepts and empirical results	Tyteca D.	Journal of Productivity Analysis	1997
6	Towards indicators of sustainable development for firms: a productive efficiency perspective	Callens I, Tyteca D.	Ecological Economics	1999
44	A survey of data envelopment analysis in energy and environmental studies	Zhou P, Ang B W, Poh K L.	European Journal of Operational Research	2008
42	Linear programming models for measuring economy-wide energy efficiency performance	Zhou P, Ang B W	Energy Policy	2008
57	Total factor carbon emission performance: a Malmquist index analysis	Zhou P, Ang B W, Han J Y.	Energy Economics	2010
80	Evaluation of potential reductions in carbon emissions in Chinese provinces based on environmental DEA	Guo X D, Zhu L, Fan Y, et al.	Energy Policy	2011
119	Efficiency and abatement costs of energy-related CO_2 emissions in China: a slacks-based efficiency measure	Choi Y, Zhang N, Zhou P	Applied Energy	2012
136	Energy and emissions efficiency patterns of Chinese regions: a multi-directional efficiency analysis	Wang K, Wei Y M, Zhang X	Applied Energy	2013
166	China's regional energy and environmental efficiency: a range-adjusted measure-based analysis	Wang K, Lu B, Wei Y M	Applied Energy	2013
259	China's regional sustainability and diversified resource allocation: DEA environmental assessment on economic development and air pollution	Sueyoshi T, Yuan Y	Energy Economics	2015a
268	Environmental assessment on coal-fired power plants in US north-east region by DEA non-radial measurement	Sueyoshi T, Goto M	Energy Economics	2015a
277	DEA environmental assessment in time horizon: radial approach for Malmquist index measurement on petroleum companies	Sueyoshi T, Goto M	Energy Economics	2015b
291	Japanese fuel mix strategy after disaster of Fukushima Daiichi nuclear power plant: Lessons from international comparison among industrial nations measured by DEA environmental assessment in time horizon	Sueyoshi T, Goto M	Energy Economics	2015c
290	Comparison among US industrial sectors by DEA environmental assessment: Equipped with analytical capability to handle zero or negative in production factors	Sueyoshi T, Yuan Y	Energy Economics	2015b
305	Marginal rate of transformation and rate of substitution measured by DEA environmental assessment: Comparison among European and north American nations	Sueyoshi T, Yuan Y	Energy Economics	2016

The origin of the local main path starts with Färe et al. (1996), in which the environmental performance concept is first proposed. It uses input-oriented DEA methods containing "bad output" pollution variables to evaluate environmental performance in a similar manner to the earlier hyperbolic DEA methods used by Färe et al. (1989), and introduces the weak disposability concept to account for the fact that the bad outputs (pollution) cannot be freely disposed, thereby laying out the foundation of DEA environmental sustainability evaluation for future research. Following Färe et al. (1996) and Tyteca (1997) uses three different DEA models—undesirable output-oriented DEA model, both inputs and undesirable outputs-oriented DEA model, and output-oriented DEA model to calculate environmental performance indicators, and illustrates the approaches by using the same data set as Färe et al. (1996). Their results show that different decision makers (e.g., a public decision maker vs. company's manager) should choose different DEA models in accordance with their purposes when evaluating environmental performance. The third paper on the path Callens and Tyteca (1999) first propose the evaluation of corporate sustainability using DEA methods. In comparison with prior studies on environmental performance evaluation that only focus on economic and environmental dimensions, their study points out three-way efficiency (economic, social and environmental) as a necessary (but not sufficient) step towards sustainability. Zhou et al. (2008b) make a literature review on the application of DEA in energy and environmental (E&E) issues over the period of 1983 to 2006.

The next two papers on the local main path are still in the research stream of energy and environmental efficiency. Zhou and Ang (2008b) fill the gap by evaluating energy efficiency within a joint production framework of both desirable and undesirable outputs. Zhou et al. (2010) extend CO_2 emission performance research from cross-sectional to time-series analysis by introducing a Malmquist CO_2 emission performance index (MCPI). They further propose bootstrapping MCPI for sensitivity analysis and statistical inferences, and make a multiple regression analysis, which invites more and more studies to use the two-stage contextual factor evaluation framework. Guo et al. (2011) further use the similar DEA methods to compute potential carbon emission reductions for energy conservation technology (ECT) and energy structural adjustment (ESA).

Most of the studies prior to 2012 use the radial efficiency measures that cannot capture all the technical inefficiency. To fill the gap in the literature, a group of different DEA models have been developed recently. For example, Choi et al. (2012) employ a slacks-based efficiency measure, which measures all the slack variables of inputs and outputs and can find out all the sources of inefficiency. Wang et al. (2013b) utilize the multi-directional efficiency analysis (MEA) approach proposed by Bogetoft and Hougaard (1999). MEA selects benchmarks such that the input contractions or output expansions are proportional to the potential improvement in each input or output variable separately, so that not just the efficiency status but also the efficiency patterns can be detected. In the next paper, Wang et al. (2013c) apply the Range-Adjusted Measure based DEA (RAM-DEA) models proposed by Cooper et al. (1999) and combine both energy performance and environmental performance for each DMU under a unified treatment in order to measure China's regional integrated energy and environmental efficiency.

The more recent references on local main path, conducted by Sueyoshi and his group, are mainly about sustainability evaluation in the context of natural and managerial disposability. Sueyoshi and Yuan (2015a) are the first to discuss the radial measurement of scale efficiency under the natural and managerial disposability concepts. They also change the research direction to China's environmental pollution measured by PM2.5 and PM10 instead of CO_2. Afterwards, Sueyoshi and Goto (2015a) measure the scale efficiency by two non-radial models and examine the influence of the small number ε on unified and scale efficiencies for the first time. After Sueyoshi and Yuan (2015a) and Sueyoshi and Goto (2015a, b, c) are a series of studies integrating time horizon into the analysis. By incorporating Malmquist index into the proposed DEA environmental assessment and linking the index to natural and managerial disposability, these two papers examine the occurrence of frontier shifts over time. Sueyoshi and Goto (2015b) employ a radial approach, while Sueyoshi and Goto (2015c) use non-radial models to handle zero and negative data points. Furthermore, Sueyoshi and Yuan (2015b) extend the work of Sueyoshi and Goto (2015c) by proposing a new DEA approach to handle zero and negative values for both the radial and non-radial measurements, discomposing the negative desirable output into positive and negative parts. Sueyoshi and Yuan (2016a) discuss a new use of DEA environmental assessment to measure Marginal Rate of Transformation (MRT) and Rate of Substitution (RS) among production factors. Traditionally, MRT & RS measurements are subject to the problem of instability caused by multipliers or dual variables. To overcome the problem that, this study, for the first time, applies a new multiplier restriction method into the assessment.

The local main path analysis shows that the current key route of DEA application in sustainability focuses on environmental sustainability: Namely, how to minimize the negative influence on the environment while maximizing economic outputs. Along the local main path, research topics have evolved from sustainability evaluation to sustainability improvement; research methods have extended from radial approaches (hyperbolic DEA) to non-radial ones; research perspectives have changed from static to intertemporal analysis; and research scenarios have extended from the weak disposability to the natural and managerial disposability. More recent development of the literatures starts to evaluate scale efficiency in the sustainability context with some special models aimed at handling special types of data. Multiplier restriction is also adapted into sustainability evaluation. In addition, the significance of specific parameters used in DEA models has been discussed including the non-Archimedean small number and the type of Damages to Return (DTR), Returns to Scale (RTS), and Damages to Scale (DTS).

Although many topics and methods have been examined, all of the studies along the local main path focus on the environmental aspect of sustainability. The other two major components of sustainability triple bottom line—*economic* and *social* sustainability seem under investigated in the literature. In particular, there is a lack of studies evaluating interactive impacts between the three components (i.e., *social-environmental*, *environmental-economic*, and *social-economic*) on sustainability issues. In addition, several methodological challenges remain and call for further exploration such as time lag between inputs and their effects, structural differences among industries when evaluating regional sustainability, self-sufficiency rates in energy supply of nations, occurrence of multiple solutions, and future uncertainty.

Because the local main path selects the route with the highest SPC value at every branching point, some important works may be missing from this path. In the following section we adopt a different citation-based method—Kamada–Kawai algorithm to draw the whole picture of the citation network and identify research clusters accordingly.

4.5 Research Topic Clusters

Using Kamada–Kawai algorithm in Pajek software, we construct the citation network and then identify research clusters of DEA applications in sustainability. We first delete those vertices with an input degree of 1 or less[1] to select the significant papers. After this operation, we get a new network with 65 significant papers. Then we use Kamada–Kawai algorithm to draw the layout of their citation network (see Fig. 4.7). The arrows indicate the direction of citations, and the solid circles represent published papers with a serial number, author name(s), and published year beside it. The citation network is divided into four research topic clusters by Kamada–Kawai algorithm as listed in Table 4.4: (1) Corporate sustainability assessment, (2) Sustainability composite indicators construction, (3) Sustainability performance analysis, and (4) Regional sustainability development assessment.

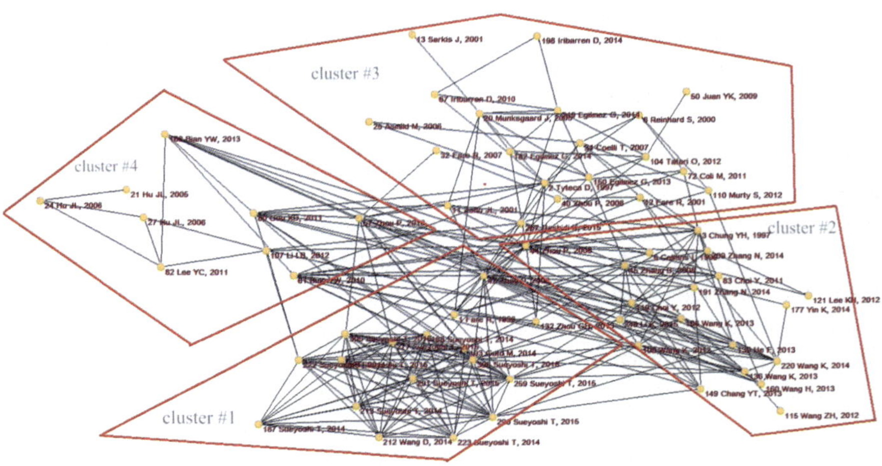

Fig. 4.7 Citation network graph

[1] The input degree of a vertex means the number of edges pointing to this vertex, which represents the vertex paper is cited by how many other papers.

Table 4.4 Research topic clusters

Cluster	Research topic	Main research content
1	Corporate sustainability assessment	Measure corporate eco-efficiency and social performance using different DEA methodologies
2	Sustainability composite indicators construction	Methodology development of DEA in sustainability composite indicator, mainly in macro level
3	Sustainability performance analysis	Analyzing the impact factors of sustainability, and the relationships of different sustainability dimensions
4	Regional sustainability development assessment	Evaluate the sustainability of regional development using various DEA models based on energy usage and resource usage

Before moving to each specific research topic cluster, it is worthy of pointing out that knowledge exchange and connections exist between clusters. Much of the knowledge exchange occurs between sustainability composite indicator construction (Cluster 2) and the application of such indictors in sustainability assessment. For example, Zhang et al. (2008) connects Clusters 1 and 2 by making the DEA methods for constructing sustainability composite indicators be applied to corporate sustainability assessment. Egilmez et al. (2013) is an important node publication connecting Clusters 1 and 3 by extending the sustainability assessment to the impact factors analysis of sustainability. Some seminal works could connect all four research clusters such as Färe et al. (1996) and Zhou et al. (2008b).

In the successive sections, we will summarize the major findings in each of the research topic clusters as well as make suggestions on future research directions.

4.5.1 *Corporate Sustainability Assessment*

4.5.1.1 Current Status of the Literature

The corporate sustainability assessment is perhaps the most popular DEA application in sustainability research. This is evident by the number of corporate sustainability assessment papers published. In this research topic, scholars use various DEA models to measure the eco-efficiency or corporate social responsibility (CSR) to support the decision-making process of corporations. Table 4.5 lists the indicators usually used as DEA inputs and outputs for firm DMU evaluation.

From Table 4.5, we can find that the commonly used indicators for corporate economic inputs include assets, capital and various cost-related indicators at the firm level. The most frequently used environmental input is energy consumption, then comes to land use and investment for environment protection. Carbon footprint, as a measure of the total amount of carbon dioxide, nitrogen oxides, and methane emissions from fossil fuel combustion is a newly adopted indicator by researchers. In social dimension, the most frequently used input indicator is human labor. There are also some other social input indicators such as investment in

Table 4.5 Indicators used in corporate sustainability assessment

	Input indicator	Output indicator
Economic	Assets, capital, materials and machinery, R&D cost, administrative expenses, marketing expense, operating cost, transportation cost, staff cost, technical risk, commercial risk	Output yield, revenue or net income, sales, profit, return on assets, value-added, market share, Tobin's q and market value, intangible assets
Environmental	Energy consumption, land use, investment in CO_2 abatement, investment for environment protection, carbon footprint, energy footprint, water footprint, waste generated, waste treatment cost, pesticide risk, erosion	Hydrocarbon emissions, carbon monoxide emissions, carbon dioxide emissions, nitrogen oxide emissions, SO_2 emission, pollution prevention and treatment, waste recycled, Agri-environmental payments, environmental certification, estimated CO_2 saving, environmental costs savings initiatives, climate change, environmental management and innovation, environmental strength
Social	Cost of work safety, labor health, human labor, investment in customer relationship management, delivery punctuality and accuracy, supplier rejection rate, qualitative control	Quality, flexibility, service and customer satisfaction, human rights, delivery punctuality and accuracy, capacity and safety, community, diversity, social contribution, corporate transparency, cooperation

customer relationship management (Akdeniz et al. 2010), qualitative control (Kuo and Lin 2012), and investment in work safety (Ødegaard and Roos 2014). As for the outputs, output yield, revenue, and sales are three most popular indicators in economic output. The most frequently used environmental output indicators are all kinds of pollution emissions such as carbon monoxide emissions, carbon dioxide emissions, nitrogen oxide emissions, and SO_2 emission. The social output indicators include service and customer satisfaction (Kim et al. 2011), human rights (Lee and Saen 2012), and social contribution (Chen and Delmas 2011).

Färe et al.'s study (1996) is the first to use DEA methods to obtain corporate environmental performance by taking the weak disposability of bad outputs into account. Specifically, they decompose overall productive efficiency into input efficiency and environmental efficiency. Tyteca (1997) proposes the consideration of three different DEA models—undesirable output-oriented index model, both inputs and undesirable outputs-oriented index model, and normalized undesirable output oriented index model, and uses them to measure the environmental performance of US. fossil fuel-fired electric utilities. A number of subsequent studies have used similar approaches in other industrial applications (e.g., Ball et al. 1994; Weber 1996; Piot-Lepetit and Vermersch 1998; Sharma et al. 1999; Zofio and Prieto 2001; Zhou et al. 2013). De Koeijer et al. (2002) first propose the concept of sustainability efficiency, and integrate DEA estimates of environmental and economic efficiency in a sustainability index. After the review article by Zhou et al. (2008b), more and more advanced DEA methods have been applied to address special problems in corporate sustainability assessment, including dual-role factors model (Lee and Saen 2012), directional distance functions (DDF) for different facets of eco-efficiency (Picazo-Tadeo et al. 2012), Economic Input–Output Life Cycle Assessment (EIO-LCA) for assessing continuing impact of economic activities (Egilmez et al. 2013), Em + DEA method for energy analysis (Iribarren et al. 2014), and network DEA for studying the internal process of corporate production (Zhu et al. 2014). In the research stream of environmental efficiency evaluation, undesirable outputs are typically included in DEA models mostly as either inputs, some form of translated outputs or weakly disposable outputs. However, the efficiency findings from different studies can be hardly comparable because they are contingent upon the specific definition of undesirable outputs (Scheel 2001). To overcome this problem, Asmild and Hougaard (2006) propose a two-step directional DEA approach to disaggregate undesirable nutrient surpluses into nutrient flows, which does not need to make the specific definition of undesirable outputs and makes the outputs (nutrient removal) desirable. Another way to avoid the problems associated with defining and dealing with undesirable outputs is proposed by Coelli et al. (2007). Specifically, they incorporate the materials balance concept into the production model and produce a new environmental efficiency measure that can be decomposed into technical and allocative components.

Another stream of research in the area focuses on how to improve corporate sustainability through technology innovation using different DEA models, such as non-radial approach (Sueyoshi and Goto 2014a), radial and non-radial integrated approach (Sueyoshi and Wang 2014), and radial measurement with subcomponent measures (Sueyoshi and Goto 2014b). Sueyoshi and his group also measure the scale efficiency and the influence of the small number ε for the first time (Sueyoshi and Goto 2015a). Afterwards, they extend the corporate sustainability research to the intertemporal dimension by using DEA Malmquist index (Sueyoshi and Goto 2015b, c). They also propose some new DEA approaches to handle zero and negative values for both the radial and non-radial measurements (Sueyoshi and Goto 2015c; Sueyoshi and Yuan 2015b). Sueyoshi and Yuan (2016a) discuss the new use of DEA environmental assessment with multiplier restriction to measure Marginal Rate of Transformation (MRT) and Rate of Substitution (RS) among production factors. This research route is part of the key route of DEA application in sustainability as discussed before.

A third research theme emerged in this area is related to corporate social responsibility (CSR). CSR refers to a company's positive impact on society and the environment, through its operations, products, or services and through its interaction with key stakeholders such as employees, customers, investors, communities, and suppliers. Belu (2009) first proposes a cross-sectional DEA output-oriented model for analyzing the relationship between corporate economic and financial performance and social responsibility. Chen and Delmas (2011) utilize the DEA model for ordinal data (Cook and Zhu 2006) to create a single CSP efficiency index and make a comparison between current aggregation approaches to justify the DEA approach's advantages. Chen and Delmas (2012) then develop a new eco-inefficiency frontier model that uses the additive inefficiency index and allow firms to select their own directions for improvement to reach the efficiency frontier. But the existing DEA papers for CSR are still relatively rare compared to the increasing attention to CSR in both business practices and academic research.

In summary, current DEA applications in corporate sustainability assessment mainly include three research topics: evaluating corporate eco-efficiency, improving corporate environmental efficiency, and measuring corporate social performance. A wide range of DEA methods have been developed and applied into this research area from the classical DEA model, to hyperbolic efficiency measures, to models handling dual-role factors and cross-efficiency, DDF, LCA + DEA, Em + DEA, network DEA, multiplier restriction, the Malmquist index, and so on. The incorporation of the materials balance concept into DEA research is also an important development.

4.5.1.2 Future Directions

There are several directions for future DEA application in corporate sustainability. First, DEA methods such as network DEA and integrated DEA can be combined with other data analysis methods such as AHP, PCA, ANOVA, Sensitivity Analysis, and econometrics. The combination of DEA and other data analysis methods can provide a more accurate way for indicator selection, an objective evaluation of corporate sustainability and a proper explanation for corporate unsustainability. For example, network DEA could be further developed in corporate sustainability by considering the stage property of production process to reduce evaluation errors and find the deep-seated reasons for corporate unsustainability.

In addition, the problem associated with defining and dealing with undesirable outputs in DEA sustainability evaluation is getting more and more attention in the literature. Although two approaches, including the directional network DEA approach and the incorporation of the materials balance concept into the production model, have been proposed to mitigate such problems, future research should further develop these two methods as well as other more effective models for the handling of undesirable output problems.

Finally, existing literature mainly focuses on the environmental aspect of sustainability, and only a few articles study the social aspect. Thus, more research is needed to apply DEA models in the CSR area from CSR modeling to empirical investigations on the relationship between CSR and corporate sustainability. Furthermore, a synthesized consideration of all the triple bottom lines of sustainability (*economic*, *environmental*, and *social*) is needed in future research on corporate sustainability assessment.

4.5.2 Sustainability Composite Indicators Construction

4.5.2.1 Current Status of the Literature

A composite indicator (CI) is a mathematical aggregation of a set of individual indicators that measure multi-dimensional concepts but usually have no common units of measurement (Nardo et al. 2005). It has been widely accepted as a tool for performance monitoring, benchmarking, policy analysis and public communication in the sustainability field. DEA has been extensively applied to CI construction.

The use of DEA in CI construction can be divided into two groups. One group follows the tradition of DEA by first identifying inputs and outputs and then constructing an aggregated index using the common DEA procedure. The first paper in this group, Chung et al. (1997) introduce a directional distance function and use it as a component in a new productivity index—Malmquist–Luenberger productivity index that readily models joint production of goods and bads. Later, their index method is applied to a series of sustainability studies (e.g., Lin et al. 2011; He et al. 2013). Since then, researchers have developed various DEA models to construct sustainability CIs. Callens and Tyteca (1999) build indicators based on the concepts of cost–benefit analysis and the principles of productive efficiency. By allowing for the assessment of business participation into sustainable development, they demonstrate that economic, social and environmental efficiency is a necessary (but not sufficient) step towards sustainability. Reinhard et al. (2000) compare DEA and SFA methods for the calculation of efficiency. Tsolas and Manoliadis (2003) then apply DEA methods to establish a sustainability index including environmental impact for thermal electrical power production in Greece. Zhou and Ang (2008b) present several DEA-type linear programming models to measure economy-wide energy efficiency performance considering undesirable outputs. Zhang et al. (2008) envision the undesirable outputs as inputs and use a CCR DEA model to establish the index of industrial eco-efficiency. Their work includes the environmental impacts related to both resource use and pollution emissions. To better select the important input and output indicators, Azad and Ancev (2010) use DEA methods to compute the component distance functions in order to construct an environmental performance index (EPI) (Färe et al. 2004). Pérez et al. (2013) combine principal component analysis (PCA) and DEA to address some objections related to the aggregation procedure. Murty et al. (2012) first criticize the modeling of pollutants as inputs or weakly disposable outputs and propose a new model of pollution-generating technologies as an intersection of an intended-production technology of the firm and nature's residual-generation set, which they call *by-production technology*. They then use the intersection of two other types of distance function rather than the directional distance functions or hyperbolic distance functions, in the intended-production technology and a residual-generation technology to obtain the efficiency indices.

In this line of research, new DEA models have been proposed to better reflect the real situation, such as Slacks-based efficiency Measure (Choi et al. 2012), Multi-directional Efficiency Analysis (MEA) approach (Wang et al. 2013c), and Range-Adjusted Measure based DEA (RAM-DEA) models (Wang et al. 2013d). Kumar et al. (2014) propose Green DEA (GDEA), which is built on an existing DEA model with weight restrictions, providing a common framework for future research in terms of a green supplier selection strategy and other multi criteria decision making problems. Dong et al. (2015) evaluate a composite indicator that builds on a combination of non-negative PCA and common-weight DEA. In order to avoid "discriminating power problem" and "technical regress" occurred in previous DEA models, Li and Lin (2015a) establish a new environmental production possibility set by combining the super-efficiency and sequential DEA models, and construct the CI of energy efficiency performance using a meta-frontier framework with the improved directional distance function (DDF).

The other group of research in CI construction first transforms the sub-indicators into the same type of variables (benefit or cost type) and then aggregates them into a CI by DEA models. Zhou et al. (2007) first adopt this method to construct a CI for modeling the sustainable energy development of 18 APEC economies in 2002. The proposed approach uses two sets of weights that are most and least favorable for each entity, thereby providing a more reasonable and encompassing CI. Similarly, Zhou et al. (2010) propose another DEA-like model considering the weighted product (WP) method instead of weighted additive one for data aggregation. Hatefi and Torabi (2010) modify the DEA-like model of Zhou's to introduce a common weight MCDA–DEA model to construct CIs. Blancard and Hoarau (2013) follow the same approach and apply the multiplicative optimization approach of Zhou et al. (2010) to construct CIs. Giambona and Vassallo (2014) aggregate CIs via a DEA-BoD (benefit-of-doubt) approach with weights determined endogenously by imposing proportion constraints. Wang (2015) extends existing approaches in MCDA-DEA field by establishing a generalized framework to construct a CI. He introduces the slacks-based CI combining with the Malmquist index for both static and dynamic analysis.

4.5.2.2 Future Directions

In this research cluster, the applications of DEA mainly focus on the macro level by constructing the CI for regions or industries. Our review reveals two key research routes in the CI construction: One is identifying inputs and outputs, and the other is firstly transforming all the sub-indicators into the same type of variables (benefit or cost type).

In future study, DEA methods can be used to directly construct the CIs, or to indirectly compute the key variables of a CI, like the component distance function. Since the DEA methods in CI construction has been well established, the future research should pay more attention to the combination of DEA methods with other statistics approaches such as PCA and AHP in CI construction. Weight setting is another area that requires most attention, especially when transforming the sub-indicators with either weight restrictions subjectively or determined endogenously. What is also notable is that the material-balance condition is a valuable direction for by-production technology in building sustainability composite indicators. In addition, current CIs mainly contain economic and environmental indicators but have limited social indicators. Thus, future CI construction should pay more attention to social indicators so that it could better reflect the social sustainability performance.

4.5.3 Sustainability Performance Analysis

4.5.3.1 Current Status of the Literature

Research in sustainability performance analysis mainly focuses on evaluating the relationships between different aspects of sustainability such as economic sustainability and environmental sustainability and identifying the impact factors of sustainability typically through the two-stage approach. The leading paper on the development trajectory, Zheng et al. (1998), is the first to use the two-stage approach (DEA methods in the first step and Tobit regression second) to analyze determinants of technical efficiency. Following them, Sarkis and Cordeiro (2001) use the two-stage approach to test the corporate environmentalism-financial performance linkage. After these pioneer works, research is diverged into two streams according to the methodologies used. One stream is mainly based on the two-stage approach, including static and dynamic analysis. Ylvinger (2003) points out the necessity of unified-assessment in sustainability performance evaluation including economic, environmental and social aspects. He then uses seven always-solvable DEA models in one-time period to measure product performance and policy performance to identify their impact-factors. This line of research is extended by Sarkis (2006) to a time series context using DEA models in the first stage and non-parametric statistics (the Mann-Whitney *U*-test) in the second to investigate the relationship between environmental performance and adoption of environmental and risk management practices. His work invites a series of subsequent two-stage studies with the integration of various DEA models and other methods such as variance analysis (Lopez-Cabrales et al. 2006), simulation models (Speelman et al. 2009), correlation analysis (Nuti et al. 2011), Tobit regressions (Shieh 2012), and sensitivity analysis (West 2015). In addition, Bi et al. (2016) utilize a combination of factor analysis and a DEA-Tobit two-stage method to assay the low-carbon technological innovation performance of China's manufacturing industry under global value chain.

Graham (2009) is the first to use an environmentally sensitive DEA Malmquist productivity index to see how environmental services influence measured productivity in the long run. The ability to incorporate environmental impacts without price data into a Malmquist productivity index makes the index attractive for the present study. Lundgren and Zhou (2017) take one step further by applying DEA firstly to calculate the Malmquist firm performance indexes and then using a panel vector auto-regression methodology to examine the casual relationship between three dimensions of firm performance productivity and environmental investment. Ødegaard and Roos (2014) combine Malmquist index and bootstrap DEA to analyze the contribution of labor quality attributes toward firm productivity. Then Arabi et al. (2014) introduces a new slacks-based model for Malmquist–Luenberger (ML) Index incorporating bad outputs. Recently, the equality of the efficiency score distributions is also getting scholar's attention. For example, Kenjegalieva et al. (2009) utilize the bootstrap-based Li test adapted by Simar and Zelenyuk to estimate and statistically compare the distributions of estimated efficiency scores.

The other research stream mainly includes the papers using the input–output approach in a life-cycle context as a complement to DEA methods in sustainability analysis. Munksgaard et al. (2005) introduce the input–output approach operating in a life cycle context into establishing general measures of environmental quality as a complement to DEA methods. EIO-LCA is employed to quantify the environmental impacts associated with the economic activities (Egilmez and Park 2014). Egilmez et al. (2014) apply the above EIO-LCA and DEA methods to supply chain sustainability analysis and measure the sensitivity of environmental impact indicators using a sensitivity analysis.

In addition to these two main research streams discussed above, some other methods have been applied to find out the impact factors of sustainability. For example, Färe et al. (2007a, b) investigate the association between pollution abatement activities and traditional productivity by comparing productivity when bad output production is regulated vs. not regulated. Wang and Wei (2016) develop a new decomposition method to examine the contributions of each individual energy input and undesirable output toward productivity change. Other methodological development in this area include two-stage environmental network DEA model (Li et al. 2012), variable coefficient test and fuzzy theory (Song et al. 2015), counterfactual experiment (Shi et al. 2015), strategy map (Sánchez 2015) and three-stage DEA model (Li and Lin 2016).

4.5.3.2 Future Directions

Currently the two main methodologies used for sustainability analysis are the two-stage approach, and integrated methodology of DEA and EIO-LCA. Although the two-stage research stream has evolved from static analysis to intertemporal analysis, most of the DEA models still treat the production process as a "black-box" when analyzing the impact factors of the overall efficiency. However, a production process could have multiple stages, and therefore evaluating the efficiency of every single stage separately would be necessary and useful to diagnose and improve the overall efficiency of production activities. So in future research, more network DEA models should be used in combination with the two-stage contextual factor analysis. What is also notable is that recently some papers have used the Li test adapted by

Simar and Zelenyuk (2006) to test the differences among distributions of performance scores. This nonparametric test is particularly suitable for DEA analyses and could be applied to future two-stage analyses on the determinants of sustainability.

On the other hand, the research stream with integration of DEA and EIO-LCA has not been extended to the dynamic dimension yet. Environmental impacts of economic activities often have time lag and thus cross-sectional data usually cannot accurately test economic activities' impacts over time. In the future, longitudinal data should be collected and employed in the EIO-LCA with dynamic DEA models to assess the dynamic impact of environmental factors. In addition, the current input output life cycle assessment mainly focuses on the economic activities' impacts on environment rather than their social influences. To investigate the whole picture of sustainability issues, more advanced DEA models are needed to incorporate social aspects in the sustainability strategic decision-making processes.

4.5.4 Regional Sustainable Development Assessment

4.5.4.1 Current Status of the Literature

DEA is frequently used to evaluate the sustainability of regional development in support of the formulation of regional sustainable development policy. Table 4.6 lists the indicators usually used by DEA for region DMUs evaluation.

As shown in Table 4.6, the most popular economic input is capital including both material and financial capital. Energy consumption is the most used environmental input. In some cases, CO_2 emissions are also used as environmental input for regional sustainability measurement (Zhou et al. 2007). Labor and population are the commonly used social input indicators along with other social input indicators such as Gini indicator (Bosetti and Buchner 2009), human capital and degrees of

Table 4.6 Indicators used in regional sustainability assessment

	Input indicator	Output indicator
Economic	Capital (material and financial), consumption expense, budget, transportation costs, R&D expenditure, patent applications	GDP, gross regional product (GRP), value added of industries, total revenue, output yield, ANS (adjusted net saving)
Environmental	Energy consumption, CO_2 emissions, pollution investment, increase in temperature, land use, precipitation average, total local (PM10, NOx), soil loss and nitrogen loss	Sulfur dioxide emission, PM10, NO_2, soot, industrial dust, solid waste, CO_2 emission, investment in waste water collection
Social	Human labor, population, inequity indicator, Gini indicator, human capital, unemployment, degrees of market openness	Total number of visitors, employment, crude birth rate, City satisfaction scores, Gini indicator, number of hospital beds and doctors

market openness (Lei et al. 2013). In output dimension, the economic and environmental indicators for a region are similar to those at the corporate level including GDP and CO_2 emission. The adopted social outputs for a region include employment, crude birth rate, number of hospital beds and doctors (Munda and Saisana 2011), city satisfaction scores (Akyol and Koster 2013), and Gini indicator (Zhang et al. 2014).

Our review reveals two different streams in the literature to apply DEA models to the regional sustainable development assessment. One stream is about static sustainability research based on sectional data. This line of research begins with Lu and Lo (2007a) who create a cross-efficiency measure (CEM) based on Seiford and Zhu (2002)'s data translation approach to deals with undesirable outputs and false positive index (FPI) and then to further analyze the regional development of China. In their next paper in 2007, they sequentially integrate CEM and cluster analysis to construct a benchmark-learning roadmap for inefficient regions. Akyol and Koster (2013) extend these CEM models by making the tradeoffs between different objectives in economic, environmental and social development. To mitigate the issue of incomparability of different DEA models, Bian and Yang (2010) extend Shannon-DEA procedure (Soleimani-Damaneh and Zarepisheh 2009) by integrating different resource and environment performance models when measuring DMUs' performance. Munda and Saisana (2011) present a methodological framework based on a non-linear/non-compensatory multicriteria approach (Munda 2005; Munda and Nardo 2009) and DEA, and combine with sensitivity analysis for assessing regional sustainability. Houshyar et al. (2012) use a combination of fuzzy logic and DEA models to evaluate the energy use sustainability for better estimation. Chang et al. (2013) propose a non-radial DEA model with the slacks-based measure (SBM) to analyze the environmental efficiency.

The other research stream mainly focuses on intertemporal sustainability assessment. Many researchers have applied traditional DEA models and Malmquist index to analyze regional dynamic relative macroeconomic performance, energy efficiency, and water efficiency in China (e.g., Hu et al. 2005, 2006; Hu and Wang 2006; Zhou et al. 2010; Guo et al. 2011; Li and Hu 2012; Chen et al. 2015). In order to use panel data to estimate the production frontier, Zhou and Ang (2008a) present a production-theoretical decomposition analysis (PDA) approach based on the Shephard input distance function and the environmental DEA technology concepts. Different from the traditional DEA models, Lee et al. (2011) create an input-saving index by comparing the actual energy inputs to target energy inputs, and compute the energy-saving targets for 27 regions in China during the period 2000–2003. DEA window analysis technique and corresponding rank sum test are also combined with energy and emission performance evaluation models in order to give a dynamic evaluation of regional sustainability (Wang et al. 2012, 2013d). To deal with the modeling of the undesirable outputs, Wang and Wei (2014) introduce a hybrid model to evaluate the regional energy and emissions efficiency. Their model ensures the economically meaningful jointness of good and bad outputs while constraining shadow prices of bad outputs to their expected sign as in Leleu (2013).

Alfonso Piña and Pardo Martínez (2016) use DEA models to measure urban sustainability, incorporating indicators related to social performance for the first time.

Recently, some new DEA models have been applied in regional sustainable development assessment, such as non-radial directional distance function (Wang et al. 2013a; Xie et al. 2018; Lee and Choi 2018), sequential generalized directional distance approach (Zhang et al. 2014), non-oriented DDF model (Chang 2015), driving forces-pressure-state-impact-policy (DPSIP) model (Kuo and Tsou 2015), and bounded adjusted measure (BAM) (Rashidi and Saen 2015). These new DEA models can be applied in both static and dynamic assessment for regional sustainability. The DEA methods are used in combination with other methods in regional sustainability evaluation, for example, Shannon's entropy index (Lo Storto 2016).

4.5.4.2 Future Directions

Many DEA models have found their applications in regional sustainability development assessment from static analysis to dynamic analysis in combination with many other approaches such as sensitivity analysis and fuzzy logic. Although the unified assessment with integration of multi-DEA methods and other methods is becoming a trend in regional sustainable development, there are still some limitations to be solved. First, the current regional research does not consider difference industrial structures among different regions. It may not be accurate to treat all regions as homogeneous DMUs. In future research, we can try to use non-homogeneous DEA models to evaluate regional sustainability. Second, the existing literature mostly considers the region as a "black-box" and future research could use network DEA models to break the box for further study. Third, as to the research content, the previous regional studies are mainly based on issues related to environmental sustainability such as energy use. Social sustainability so far has been under investigated and therefore future studies should consider using indicators of social welfare.

4.6 Conclusion

Our study fills a gap in the DEA literature by conducting a systematic survey on DEA applications in sustainability. This survey covers DEA papers listed in the Web of Science database from 1996 through March 2016. With the assistance of three citation analysis methods—citation chronological graph, main path method and Kamada–Kawai algorithm, we identify citation networks, significant paths, important papers, and research clusters in DEA application in sustainability. The review results show that the current key route of DEA application in sustainability focuses on the eco-efficiency measurement that maximizes economic outputs while minimizing negative environmental influence. One of the key challenges in DEA sustainability research is how to deal with undesirable or bad outputs. Our review reveals three major approaches: the weak disposability approach, the natural and managerial disposability, and the materials balance approach. Four major research clusters are identified: corporate sustainability assessment, sustainability composite indicators construction, sustainability performance analysis and regional sustainability development assessment. More recent research endeavors have also dabbled in these four dimensions comprehensively. From March 2016 to the end of 2018, 21 papers regarding sustainability and DEA, with a Chartered Association of Business Schools ranking of 3 or above, were published. In 2016, Zimmer, Fröhling and Schultmann assessed the status of scientific literature on sustainable supplier management (SSM), put forward an SSM framework and conduct a comprehensive content analysis. Intriguingly, they found that a large increase in the number of papers using DEA occurred in 2014, with 9 out of 16 papers on evaluating the efficiency of supply chain metrics. In terms of regional sustainability development assessment, research predominantly encompasses ecological, urban economical and agricultural sectors. Specifically, Ji et al. (2016) constructed a DEA model for sustainable transportation in China, to help stakeholders reach transportation objectives with less consumption and emission. In addition, Fei and Lin (2016) employed DEA to measure China's agricultural energy efficiency and revealed inhomogeneity of performance between eastern and western regions. Their observations greatly contributed to the sustainable development and energy conservation in China's agricultural sector. With regard to sustainability performance analysis, Huang and Coelho (2017) developed new efficiency models based on DEA, inefficiency models based on reverse DEA (RDEA, Huang and Kao 2012) and relative overall performance models based on Inefficiency Countervailed DEA (ICDEA, Huang and Kao 2011), to evaluate sustainability performance of coral reef protection in the Coral Triangle region. Moreover, Tajbakhsh and Hassini (2018) presented a two-stage DEA model to compute the efficiency of power generation facilities, and used the proposed method to construct a comprehensive sustainability performance measure for the fossil-fuel power stations. As for corporate sustainability assessment, one preeminent demand of corporate sustainability achievement is the employment of multiple renewable energy resources. Khanjarpanah et al. (2018) proposed a novel algorithm using double frontier network data envelopment analysis with both single-period and multi-period programming models to measure the efficiency of potential hybrid power plant locations, in order to determine the best candidate locations for hybrid power plant establishment.

Five interesting phenomena are observed from the development trajectories analysis on the four focal clusters. First, there is a pattern of technology-adoption process by DEA researchers. Early adopters start with the classical DEA models and cautiously suggest the usefulness of the methodology. After DEA is accepted in the field, researchers tend to adopt the newly developed approaches and models once they are available. Second, a pattern of research perspective development can be found. The DEA application in sustainability usually begins with static measurement, and then extends to dynamic measurement. As to the evaluated objectives, previous research often starts with external evaluation treating DMUs as black-box, and then turns to internal research considering different internal structures and production processes. Typically, Wu et al. (2017) proposed a modified two-stage network DEA model to incorporate internal conflicts and cooperation among multiple manufacturing stages, by including both desirable and undesirable outputs from the sustainable manufacturing process and applying the model to a group of Chinese iron and steel makers' wastewater recycling and reusing system. They found the proposed model more effective in calculating efficiency and identifying sources of inefficiency in both stages.

Third, more and more sustainability evaluation studies start to make comprehensive assessments on objectives, from traditional economic assessment to early environmental sustainability, and then to recent social sustainability. Fourth, there is a significant trend emphasizing on the combination of DEA with other analytical methods such as AHP, game theory, LCA, and regression analysis. For more accurate and reliable assessment results. Fifth, the two-stage process model, a simple form of the network DEA model, has drawn much attention lately across applications. The typical two-stage process model breaks the limitations of traditional black-box model with a subdivision production process and can evaluate the sustainability of different process stages or supply chain aspects. Furthermore, Izadikhah and Saen (2016) went beyond the existing two-stage process models involving positive inputs and outputs, by taking into account negative input–intermediate–output data. The method was tested in quantitative examples and a case study of 29 Iranian supply chains to prove its practicality and competency. Izadikhah et al. (2018) continued employing DEA models dealing with negative data, to elect the most sustainable supplier according to economic, environmental and social standards, in supply chain management. Additionally, Izadikhah and Saen (2018) extended their previous initiative on sustainable supply chain by modeling a new stochastic two-stage DEA after pointing out the possible presence of stochastic data and undesirable data. Similarly, Moheb-Alizadeh and Handfield (2018) considered economic, environmental, and social dimensions in sustainable supplier selection and order allocation with stochastic demand. On the other hand, Zhu et al. (2018) accentuated the existence of non-comparability among decision making units, and established a cross-like DEA efficiency model of non-homogeneous DMUs to evaluate efficiency and compare a DMU with other units in the absence of homogeneity.

Several gaps in the literature are also revealed. First, the extant DEA application research on sustainability in OR/MS still mainly focuses on the economic and environmental measures (emissions, remanufacturing, waste reduction, etc.) while models that examine the social measures are lacking in the literature. The lack of social measures may be due to the unavailability of micro and/or macro data related to social indicators (e.g., social contribution, service and customer satisfaction) and methodological challenges to incorporate social measures in DEA models (e.g., how to deal with the relationship between social indicators and economic/environmental indicators). New DEA methods should be developed to incorporate the interaction of social, economic, and environmental measures, and to well characterize the complexity of social network. In other words, when evaluating the DMUs' social sustainability, we should consider their social network relationship and the relationship with the external environment. Second, there are three dimensions of the synthesis research of sustainability: content, process, and context. The extant papers mainly focus on the content, but the analysis on the process and context of sustainability is underinvestigated, for example, different industrial structures when evaluating regional sustainability and self-sufficiency rates in energy supply of nations. Third, the current research on sustainability evaluation does not contain the institutional dimension that could have a significant impact on sustainability. Future research could synthesize the institutional perspective with the triple bottom line of sustainability. In addition to the gaps mentioned above, other methodological challenges, such as the time lag between inputs and output, the occurrence of multiple solutions, future cost, and energy uncertainty, are still waiting to be explored.

The interpretation of these results should, nevertheless, take the following limitations into consideration. First, the data are collected from the WOS database and do not include all DEA papers published in journals that are not included in the WOS. Second, albeit much effort has been made to select correct DEA sustainability papers, we may still miss some DEA papers, we believe that these papers take up a very small percentage of the total papers and do not alter the major analysis results. Third, although highly cited publications, to some extent, represent the core content and evolution of the research, when compared with the whole number of articles, the limited amount of data can cause inevitable one-sidedness, which is also a limitation and drawback of chronological diagram research. Fourth, the results of the local main path analysis are subject to citation noise, a general limitation of the citation analysis. Citation noise occurs when "remote" citation occurs occasionally when a paper cites others, not because of a close connection with the main subject, but merely because of a connection in a broad sense such as the same application area, the same general method, or even just because of applying DEA methodology (Liu et al. 2013). The tail portions of the main path are especially sensitive to these noises as the number of citation count becomes fewer there. Thus, one should be cautious when interpreting the results close to the tail. Finally, there has been a debate on whether the DEA-based approach is applicable to sustainability analysis in the literature. Some scholars (e.g., Callens and Tyteca 1999; Huppes and Ishikawa 2005) argue that efficiency in DEA models is a relative concept that may not be suitable to the evaluation of sustainability performance, an absolute concept that has to do with absolute magnitudes concerning the absorption capacity of ecosystems. An efficient DMU does not imply sustainability, as efficiency is only a necessary condition (or intermediate step) for sustainability. But in our opinion, first of all, it is debatable whether sustainability performance is only an absolute concept. An assessment on relative sustainability levels could provide a benchmark system that helps companies find the most cost-effective way to achieve a reduction in environmental degradation and policymakers adopt policies aimed at achieving improvements rather than simply restricting economic activities (Gómez-Limón et al. 2012; Kuosmanen and Kortelainen 2005). For those reasons, we believe DEA is still a valuable tool for sustainability performance evaluation.

In conclusion, sustainability is an area that is gaining interest, and DEA has been proved to be an appropriate evaluation method for sustainability in the literature. The development on DEA methodologies and applications in sustainability should continue to flourish. We hope that this review can be a useful and inspirational source for further DEA research on sustainability.

Acknowledgments The current chapter is based upon Zhou, H., Yang, Yi., Chen, Y., and Zhu, J., Data envelopment analysis application in sustainability: The origins, development and future directions, European Journal of Operational Research, Vol. 264, Issue 1 (2018), 1–16, with permission from Elsevier.

References

Akdeniz, M. B., Gonzalez-Padron, T., & Calantone, R. J. (2010). An integrated marketing capability benchmarking approach to dealer performance through parametric and nonparametric analyses. *Industrial Marketing Management, 39*(1), 150–160.

Akyol, D. E., & Koster, R. (2013). Non-dominated time-window policies in city distribution. *Production and Operations Management, 22*(3), 739–751.

Alfonso Piña, W. H., & Pardo Martínez, C. I. (2016). Development and urban sustainability: An analysis of efficiency using data envelopment analysis. *Sustainability, 8*(2), 148.

Antonio, B., Cristina, B., & Stefano, A. (2012). Cities as ecosystems: Growth, development and implications for sustainability. *Ecological Modelling, 245*, 185–198.

Arabi, B., Munisamy, S., Emrouznejad, A., & Shadman, F. (2014). Power industry restructuring and eco-efficiency changes: A new slacks-based model in Malmquist–Luenberger index measurement. *Energy Policy, 68*, 132–145.

Asmild, M., & Hougaard, J. L. (2006). Economic versus environmental improvement potentials of Danish pig farms. *Agricultural Economics, 35*(2), 171–181.

Assaf, A. G., Josiassen, A., & Cvelbar, L. K. (2012). Does triple bottom line reporting improve hotel performance? *International Journal of Hospitality Management, 31*(2), 596–600.

Azad, M. A., & Ancev, T. (2010). Using ecological indices to measure economic and environmental performance of irrigated agriculture. *Ecological Economics, 69*(8), 1731–1739.

Azadi, M., Jafarian, M., Saen, R. F., & Mirhedayatian, S. M. (2015). A new fuzzy DEA model for evaluation of efficiency and effectiveness of suppliers in sustainable supply chain management context. *Computers & Operations Research, 54*, 274–285.

Ball, V. E., Lovell, C. A., Nehring, R. F., & Somwaru, A. (1994). *Incorporating indesirable outputs into models of production: An application to US agriculture*. Cahiers d'économie et Sociologie Rurales, (31).

Balocco, C., Papeschi, S., Grazzini, G., & Basosi, R. (2004). Using exergy to analyze the sustainability of an urban area. *Ecological Economics, 48*(2), 231–244.

Batagelj, V. (2003). *Efficient algorithms for citation network analysis*. arXiv preprint cs/0309023.

Belu, C. (2009). Ranking corporations based on sustainable and socially responsible practices. A data envelopment analysis (DEA) approach. *Sustainable Development, 17*(4), 257.

Bi, K. X., Huang, P., & Wang, X. X. (2016). Innovation performance and influencing factors of low-carbon technological innovation under the global value chain: A case of Chinese manufacturing industry. *Technological Forecasting and Social Change, 111*, 275–284.

Bian, Y., & Yang, F. (2010). Resource and environment efficiency analysis of provinces in China: A DEA approach based on Shannon's entropy. *Energy Policy, 38*(4), 1909–1917.

Blancard, S., & Hoarau, J. F. (2013). A new sustainable human development indicator for small island developing states: A reappraisal from data envelopment analysis. *Economic Modelling, 30*, 623–635.

Bogetoft, P., & Hougaard, J. L. (1999). Efficiency Evaluations Based on Potential (Non-Proportional) Improvements. *Journal of Productivity Analysis, 12*(3), 233–247. https://doi.org/10.1023/A:1007848222681

Bosetti, V., & Buchner, B. (2009). Data envelopment analysis of different climate policy scenarios. *Ecological Economics, 68*(5), 1340–1354.

Callens, I., & Tyteca, D. (1999). Towards indicators of sustainable development for firms: A productive efficiency perspective. *Ecological Economics, 28*(1), 41–53. https://doi.org/10.1016/S0921-8009(98)00035-4

Campbell, D. E., & Garmestani, A. S. (2012). An energy systems view of sustainability: Emergy valuation of the San Luis Basin, Colorado. *Journal of Environmental Management, 95*(1), 72–97.

Chang, M. C. (2015). Room for improvement in low carbon economies of G7 and BRICS countries based on the analysis of energy efficiency and environmental Kuznets curves. *Journal of Cleaner Production, 99*, 140–151.

Chang, Y. T., Zhang, N., Danao, D., & Zhang, N. (2013). Environmental efficiency analysis of transportation system in China: A non-radial DEA approach. *Energy Policy, 58*, 277–283.

Charnes, A., Cooper, W. W., & Rhodes, E. (1978). Measuring the efficiency of decision making units [J]. *European Journal of Operational Research, 2*(6), 429–444.

Chen, Y., Cook, W.D., Kao, C., and Zhu, Joe, Network DEA pitfalls: Divisional efficiency and frontier projection under general network structures, *European Journal of Operational Research*, Vol. 226 (2013), 507–515.

Chen, C. M., & Delmas, M. (2011). Measuring corporate social performance: An efficiency perspective. *Production and Operations Management, 20*(6), 789–804.

Chen, C. M., & Delmas, M. A. (2012). Measuring eco-inefficiency: A new frontier approach. *Operations Research, 60*(5), 1064–1079.

Chen, S., & Xie, Z. (2015). Is China's e-governance sustainable? Testing Solow IT productivity paradox in China's context. *Technological Forecasting and Social Change, 96*, 51–61.

Chen, Y., Cook, W. D., Kao, C., & Zhu, J. (2014). *Network DEA pitfalls: Divisional efficiency and frontier projection. Data envelopment analysis* (pp. 31–54). New York: Springer.

Chen, J., Song, M., & Xu, L. (2015). Evaluation of environmental efficiency in China using data envelopment analysis. *Ecological Indicators, 52*, 577–583.

Choi, Y., & Zhang, N. (2011). Assessing the sustainable performance of Chinese industrial sector. *African Journal of Business Management, 5*(13), 5261–5270.

Choi, Y., Zhang, N., & Zhou, P. (2012). Efficiency and abatement costs of energy-related CO2 emissions in China: A slacks-based efficiency measure. *Applied Energy, 98*, 198–208. https://doi.org/10.1016/j.apenergy.2012.03.024

Chuang, C. L., Chang, P. C., & Lin, R. H. (2011). An efficiency data envelopment analysis model reinforced by classification and regression tree for hospital performance evaluation. *Journal of Medical Systems, 35*(5), 1075–1083.

Chung, Y. H., Färe, R., & Grosskopf, S. (1997). Productivity and undesirable outputs: A directional distance function approach. *Journal of Environmental Management, 51*(3), 229–240.

Coelli, T., Lauwers, L., & Van Huylenbroeck, G. (2007). Environmental efficiency measurement and the materials balance condition. *Journal of Productivity Analysis, 28*(1–2), 3–12.

Coli, M., Nissi, E., & Rapposelli, A. (2011). Monitoring environmental efficiency: An application to Italian provinces. *Environmental Modelling & Software, 26*(1), 38–43.

Cook, W. D., & Zhu, J. (2006). Rank order data in DEA: A general framework. European Journal of Operational Research, 174(2), 1021–1038. https://doi.org/10.1016/j.ejor.2005.01.063

Cooper, W. W., Park, K. S., & Pastor, J. T. (1999). RAM: A Range Adjusted Measure of Inefficiency for Use with Additive Models, and Relations to Other Models and Measures in DEA. Journal of Productivity Analysis, 11(1), 5–42. https://doi.org/10.1023/A:1007701304281.

Dakpo, K. H., Jeanneaux, P., & Latruffe, L. (2016). Modelling pollution-generating technologies in performance benchmarking: Recent developments, limits and future prospects in the nonparametric framework. *European Journal of Operational Research*, Vol. 250, pp. 347–359. https://doi.org/10.1016/j.ejor.2015.07.024

De Koeijer, T. J., Wossink, G. A. A., Struik, P. C., & Renkema, J. A. (2002). Measuring agricultural sustainability in terms of efficiency: The case of Dutch sugar beet growers. *Journal of Environmental Management, 66*(1), 9–17.

DiMaria, C. H. (2014). Sustainability matter. *Quality & Quantity, 48*(3), 1257–1269.
Dong, F., Mitchell, P. D., Knuteson, D., Wyman, J., Bussan, A. J., & Conley, S. (2015). Assessing sustainability and improvements in US Midwestern soybean production systems using a PCA–DEA approach. *Renewable Agriculture and Food Systems, 31*, 1–16.
Egilmez, G., & Park, Y. S. (2014). Transportation related carbon, energy and water footprint analysis of US manufacturing: An eco-efficiency assessment. *Transportation Research Part D: Transport and Environment, 2014*(32), 143–159.
Egilmez, G., Kucukvar, M., & Tatari, O. (2013). Sustainability assessment of US manufacturing sectors: An economic input output-based frontier approach. *Journal of Cleaner Production, 53*, 91–102.
Egilmez, G., Kucukvar, M., Tatari, O., & Bhutta, M. K. S. (2014). Supply chain sustainability assessment of the US food manufacturing sectors: A life cycle-based frontier approach. *Resources, Conservation and Recycling, 82*, 8–20.
Elkington, J. (2002). *The triple bottom line of the 21st century*. Oxford: Oxford Press.
Färe, R., & Grosskopf, S. (2004). Modeling undesirable factors in efficiency evaluation: Comment. *European Journal of Operational Research, 157*(1), 242–245.
Färe, R., & Grosskopf, S. (2010). Directional distance functions and slacks-based measures of efficiency. *European Journal of Operational Research, 200*(1), 320–322.
Färe, R., Grosskopf, S., Lovell, C. K., & Pasurka, C. (1989). Multilateral productivity comparisons when some outputs are undesirable: a nonparametric approach. *The Review of Economics and Statistics, 71*, 90–98.
Färe, R., Grosskopf, S., & Tyteca, D. (1996). An activity analysis model of the environmental performance of firms-application to fossil-fuel-fired electric utilities. *Ecological Economics, 18*(2), 161–175.
Färe, R., Grosskopf, S., & Pasurka, C. A., Jr. (2001). Accounting for air pollution emissions in measures of state manufacturing productivity growth. *Journal of Regional Science, 41*(3), 381–409.
Färe, R., Grosskopf, S., & Hernandez-Sancho, F. (2004). Environmental performance: An index number approach. *Resource and Energy Economics, 26*(4), 343–352.
Färe, R., Grosskopf, S., & Pasurka, C. A. (2007a). Pollution abatement activities and traditional productivity. *Ecological Economics, 62*(3), 673–682.
Färe, R., Grosskopf, S., & Whittaker, G. (2007b). Network DEA. In *Modeling data irregularities and structural complexities in data envelopment analysis* (pp. 209–240). New York: Springer.
Fei, R., & Lin, B. (2016). Energy efficiency and production technology heterogeneity in China's agricultural sector: A meta-frontier approach. *Technological Forecasting and Social Change, 109*, 25–34.
Figge, F., & Hahn, T. (2004). Sustainable value added—Measuring corporate contributions to sustainability beyond eco-efficiency. *Ecological Economics, 48*(2), 173–187.
Gadanakis, Y., Bennett, R., Park, J., & Areal, F. J. (2015). Evaluating the sustainable intensification of arable farms. *Journal of Environmental Management, 150*, 288–298.
Garfield, E. (1972). *Citation analysis as a tool in journal evaluation*. Washington, DC: American Association for the Advancement of Science.
Giambona, F., & Vassallo, E. (2014). Composite indicator of social inclusion for European countries. *Social Indicators Research, 116*(1), 269–293.
Gibson, R. B. (2006). Sustainability assessment: Basic components of a practical approach. *Impact Assessment and Project Appraisal, 24*(3), 170–182.
Goerner, S. J., Lietaer, B., & Ulanowicz, R. E. (2009). Quantifying economic sustainability: Implications for free-enterprise theory, policy and practice. *Ecological Economics, 69*(1), 76–81.
Gómez-Limón, J. A., Picazo-Tadeo, A. J., & Reig-Martínez, E. (2012). Eco-efficiency assessment of olive farms in Andalusia. *Land Use Policy, 29*(2), 395–406.
Goto, M., Otsuka, A., & Sueyoshi, T. (2014). DEA (data envelopment analysis) assessment of operational and environmental efficiencies on Japanese regional industries. *Energy, 66*, 535–549.

Graham, M. (2009). Developing a social perspective to farm performance analysis. *Ecological Economics, 68*(8), 2390–2398.

Guo, X. D., Zhu, L., Fan, Y., & Xie, B. C. (2011). Evaluation of potential reductions in carbon emissions in Chinese provinces based on environmental DEA. *Energy Policy, 39*(5), 2352–2360.

Harel, D., & Koren, Y. (2002). Graph drawing by high-dimensional embedding. In *International Symposium on Graph Drawing* (pp. 207–219). Berlin: Springer.

Hatefi, S. M., & Torabi, S. A. (2010). A common weight MCDA–DEA approach to construct composite indicators. *Ecological Economics, 70*(1), 114–120.

He, F., Zhang, Q., Lei, J., Fu, W., & Xu, X. (2013). Energy efficiency and productivity change of China's iron and steel industry: Accounting for undesirable outputs. *Energy Policy, 54*, 204–213.

Houshyar, E., Azadi, H., Almassi, M., Davoodi, M. J. S., & Witlox, F. (2012). Sustainable and efficient energy consumption of corn production in Southwest Iran: Combination of multi-fuzzy and DEA modeling. *Energy, 44*(1), 672–681.

Hu, J. L., & Wang, S. C. (2006). Total-factor energy efficiency of regions in China. *Energy Policy, 34*(17), 3206–3217.

Hu, J. L., Sheu, H. J., & Lo, S. F. (2005). Under the shadow of Asian Brown clouds: Unbalanced regional productivities in China and environmental concerns. *The International Journal of Sustainable Development & World Ecology, 12*(4), 429–442.

Hu, J. L., Wang, S. C., & Yeh, F. Y. (2006). Total-factor water efficiency of regions in China. *Resources Policy, 31*(4), 217–230.

Huang, Y., & Coelho, V. R. (2017). Sustainability performance assessment focusing on coral reef protection by the tourism industry in the coral triangle region. *Tourism Management, 59*, 510–527.

Huang, Y.-T., & Kao, J.-J. (2012). Inefficiency countervailed DEA (IC-DEA) method for assessing corporate environmental performance. *Journal of the Operational Research Society, 63*(4), 470–477. https://doi.org/10.1057/jors.2011.56

Hummon, N. P., & Dereian, P. (1989). Connectivity in a citation network: The development of DNA theory. *Social Networks, 11*(1), 39–63.

Huppes, G., & Ishikawa, M. (2005). A framework for quantified eco-efficiency analysis. *Journal of Industrial Ecology, 9*(4), 25–41. https://doi.org/10.1162/108819805775247882

Iribarren, D., Vázquez-Rowe, I., Rugani, B., & Benetto, E. (2014). On the feasibility of using energy analysis as a source of benchmarking criteria through data envelopment analysis: A case study for wind energy. *Energy, 67*, 527–537.

Izadikhah, M., & Saen, R. F. (2016). Evaluating sustainability of supply chains by two-stage range directional measure in the presence of negative data. *Transportation Research Part D: Transport and Environment, 49*, 110–126.

Izadikhah, M., & Saen, R. F. (2018). Assessing sustainability of supply chains by chance-constrained two-stage DEA model in the presence of undesirable factors. *Computers & Operations Research, 100*, 343–367.

Izadikhah, M., Saen, R. F., & Roostaee, R. (2018). How to assess sustainability of suppliers in the presence of volume discount and negative data in data envelopment analysis? *Annals of Operations Research, 269*(1–2), 241–267.

Ji, X., Wu, J., & Zhu, Q. Y. (2016). Eco-design of transportation in sustainable supply chain management: A DEA-like method. *Transportation Research Part D: Transport and Environment, 48*, 451–459.

Kenjegalieva, K., Simper, R., Weyman-Jones, T., & Zelenyuk, V. (2009). Comparative analysis of banking production frameworks in eastern european financial markets. *European Journal of Operational Research, 198*(1), 326–340. https://doi.org/10.1016/j.ejor.2008.09.002

Khanjarpanah, H., Jabbarzadeh, A., & Seyedhosseini, S. M. (2018). A novel multi-period double frontier network DEA to sustainable location optimization of hybrid wind-photovoltaic power plant with real application. *Energy Conversion and Management, 159*, 175–188.

Kim, H. G., Choi, C. Y., Woo, J. W., Choi, Y., Kim, K., & Wu, D. D. (2011). Efficiency of the modal shift and environmental policy on the Korean railroad. *Stochastic Environmental Research and Risk Assessment, 25*(3), 305–322.

Kumar, A., Jain, V., & Kumar, S. (2014). A comprehensive environment friendly approach for supplier selection. *Omega, 42*(1), 109–123.

Kuo, R. J., & Lin, Y. J. (2012). Supplier selection using analytic network process and data envelopment analysis. *International Journal of Production Research, 50*(11), 2852–2863.

Kuo, H. F., & Tsou, K. W. (2015). Application of environmental change efficiency to the sustainability of urban development at the neighborhood level. *Sustainability, 7*(8), 10479–10498.

Kuosmanen, T., & Kortelainen, M. (2005). Measuring eco-efficiency of production with data envelopment analysis. *Journal of Industrial Ecology, 9*(4), 59–72.

Kuosmanen, T., & Kuosmanen, N. (2009). How not to measure sustainable value (and how one might). *Ecological Economics, 69*(2), 235–243.

Lee, H., & Choi, Y. (2018). Greenhouse gas performance of Korean local governments based on non-radial DDF. *Technological Forecasting and Social Change, 135*, 13–21.

Lee, K. H., & Saen, R. F. (2012). Measuring corporate sustainability management: A data envelopment analysis approach. *International Journal of Production Economics, 140*(1), 219–226.

Lee, Y. C., Hu, J. L., & Kao, C. H. (2011). Efficient saving targets of electricity and energy for regions in China. *International Journal of Electrical Power & Energy Systems, 33*(6), 1211–1219.

Lei, M., Zhao, X., Deng, H., & Tan, K. C. (2013). DEA analysis of FDI attractiveness for sustainable development: Evidence from Chinese provinces. *Decision Support Systems, 56*, 406–418.

Leleu, H. (2013). Shadow pricing of undesirable outputs in nonparametric analysis. European *Journal of Operational Research, 231*(2), 474–480. https://doi.org/10.1016/j.ejor.2013.05.028

Li, L. B., & Hu, J. L. (2012). Ecological total-factor energy efficiency of regions in China. *Energy Policy, 46*, 216–224.

Li, K., & Lin, B. (2015a). Metafroniter energy efficiency with CO2 emissions and its convergence analysis for China. *Energy Economics, 48*, 230–241.

Li, K., & Lin, B. (2015b). The improvement gap in energy intensity: Analysis of China's thirty provincial regions using the improved DEA (data envelopment analysis) model. *Energy, 84*, 589–599.

Li, K., & Lin, B. (2016). Impact of energy conservation policies on the green productivity in China's manufacturing sector: Evidence from a three-stage DEA model. *Applied Energy, 168*, 351–363.

Li, Y., Chen, Y., Liang, L., & Xie, J. (2012). DEA models for extended two-stage network structures. *Omega, 40*(5), 611–618.

Lin, W., Yang, J., & Chen, B. (2011). Temporal and spatial analysis of integrated energy and environment efficiency in China based on a green GDP index. *Energies, 4*(9), 1376–1390.

Liu, J. S., Lu, L. Y., Lu, W. M., & Lin, B. J. (2013). A survey of DEA applications. *Omega, 41*(5), 893–902.

Liu, J. S., Lu, L. Y., & Lu, W. M. (2016). Research fronts in data envelopment analysis. *Omega, 58*, 33–45.

Lo Storto, C. (2016). Ecological efficiency based ranking of cities: A combined DEA cross-efficiency and Shannon's entropy method. *Sustainability, 8*(2), 124.

Lopez-Cabrales, A., Valle, R., & Herrero, I. (2006). The contribution of core employees to organizational capabilities and efficiency. *Human Resource Management, 45*(1), 81–109.

Lovell, C. K., Pastor, J. T., & Turner, J. A. (1995). Measuring macroeconomic performance in the OECD: A comparison of European and non-European countries. *European Journal of Operational Research, 87*(3), 507–518.

Lu, W. M., & Lo, S. F. (2007a). A benchmark-learning roadmap for regional sustainable development in China. *Journal of the Operational Research Society, 58*(7), 841–849.

Lundgren, T., & Zhou, W. (2017). Firm performance and the role of environmental management. *Journal of Environmental Management, 203*, 330–341.

Mirhedayatian, S. M., Azadi, M., & Saen, R. F. (2014). A novel network data envelopment analysis model for evaluating green supply chain management. *International Journal of Production Economics, 147*, 544–554.

Mebratu, D. (1998). Sustainability and sustainable development: Historical and conceptual review. *Environmental Impact Assessment Review, 18*(6), 493–520. https://doi.org/10.1016/S0195-9255(98)00019-5

Moheb-Alizadeh, H., & Handfield, R. (2018). An integrated chance-constrained stochastic model for efficient and sustainable supplier selection and order allocation. *International Journal of Production Research, 56*(21), 6890.

Munda, G. (2005). "Measuring sustainability": A multi-criterion framework. *Environment, Development and Sustainability, 7*(1), 117–134.

Munda, G., & Nardo, M. (2009). Noncompensatory/nonlinear composite indicators for ranking countries: A defensible setting. *Applied Economics, 41*(12), 1513–1523.

Munda, G., & Saisana, M. (2011). Methodological considerations on regional sustainability assessment based on multicriteria and sensitivity analysis. *Regional Studies, 45*(2), 261–276.

Munksgaard, J., Wier, M., Lenzen, M., & Dey, C. (2005). Using input-output analysis to measure the environmental pressure of consumption at different spatial levels. *Journal of Industrial Ecology, 9*(1–2), 169–185.

Murty, S., Russell, R. R., & Levkoff, S. B. (2012). On modeling pollution-generating technologies. *Journal of Environmental Economics and Management, 64*(1), 117–135.

Nardo, M., Saisana, M., Saltelli, A., Tarantola, S., Hoffman, A., & Giovannini, E. (2005). *Handbook on constructing composite indicators*. Paris: OECD.

Ness, B., Urbel-Piirsalu, E., Anderberg, S., & Olsson, L. (2007). Categorising tools for sustainability assessment. *Ecological Economics, 60*(3), 498–508.

Nuti, S., Daraio, C., Speroni, C., & Vainieri, M. (2011). Relationships between technical efficiency and the quality and costs of health care in Italy. *International Journal for Quality in Health Care, 23*(3), 324–330.

Ødegaard, F., & Roos, P. (2014). Measuring the contribution of Workers' health and psychosocial work-environment on production efficiency. *Production and Operations Management, 23*(12), 2191–2208.

Ou, C. H., & Liu, W. H. (2010). Developing a sustainable indicator system based on the pressure–state–response framework for local fisheries: A case study of Gungliau, Taiwan. *Ocean & Coastal Management, 53*(5), 289–300.

Paoli, C., Vassallo, P., & Fabiano, M. (2008). An emergy approach for the assessment of sustainability of small marinas. *Ecological Engineering, 33*(2), 167–178.

Parris, T. M., & Kates, R. W. (2003). Characterizing and measuring sustainable development. *Annual Review of Environment and Resources, 28*(1), 559–586.

Peres-Neto, P. R., Legendre, P., Dray, S., & Borcard, D. (2006). Variation partitioning of species data matrices: Estimation and comparison of fractions. *Ecology, 87*(10), 2614–2625.

Pérez, V., Guerrero, F., González, M., Pérez, F., & Caballero, R. (2013). Composite indicator for the assessment of sustainability: The case of Cuban nature-based tourism destinations. *Ecological Indicators, 29*, 316–324.

Picazo-Tadeo, A. J., Gómez-Limón, J. A., & Reig-Martínez, E. (2011). Assessing farming eco-efficiency: A data envelopment analysis approach. *Journal of Environmental Management, 92*(4), 1154–1164.

Picazo-Tadeo, A. J., Beltrán-Esteve, M., & Gómez-Limón, J. A. (2012). Assessing eco-efficiency with directional distance functions. *European Journal of Operational Research, 220*(3), 798–809.

Piot-Lepetit, I., & Vermersch, D. (1998). Pricing organic nitrogen under the weak disposability assumption: An application to the French pig sector. *Journal of Agricultural Economics, 49*(1), 85–99.

Pope, J., Annandale, D., & Morrison-Saunders, A. (2004). Conceptualising sustainability assessment. *Environmental Impact Assessment Review, 24*(6), 595–616.

Rashidi, K., & Saen, R. F. (2015). Measuring eco-efficiency based on green indicators and potentials in energy saving and undesirable output abatement. *Energy Economics, 50*, 18–26.

Reinhard, S., Lovell, C. K., & Thijssen, G. J. (2000). Environmental efficiency with multiple environmentally detrimental variables; estimated with SFA and DEA. *European Journal of Operational Research, 121*(2), 287–303.

Sánchez, M. A. (2015). Integrating sustainability issues into project management. *Journal of Cleaner Production, 96*, 319–330.

Sarkis, J. (2006). The adoption of environmental and risk management practices: Relationships to environmental performance. *Annals of Operations Research, 145*(1), 367–381.

Sarkis, J., & Cordeiro, J. J. (2001). An empirical evaluation of environmental efficiencies and firm performance: Pollution prevention versus end-of-pipe practice. *European Journal of Operational Research, 135*(1), 102–113.

Scheel, H. (2001). Undesirable outputs in efficiency valuations. European Journal of Operational Research, 132(2), 400–410. https://doi.org/10.1016/S0377-2217(00)00160-0

Schildt, H. A., Zahra, S. A., & Sillanpää, A. (2006). Scholarly communities in entrepreneurship research: A co-citation analysis. *Entrepreneurship Theory and Practice, 30*(3), 399–415.

Seiford, L. M., & Zhu, J. (2002). Modeling undesirable factors in efficiency evaluation. *European Journal of Operational Research, 142*(1), 16–20.

Sharma, K. R., Leung, P., Chen, H., & Peterson, A. (1999). Economic efficiency and optimum stocking densities in fish polyculture: An application of data envelopment analysis (DEA) to Chinese fish farms. *Aquaculture, 180*(3), 207–221.

Shi, P., Yan, B., Shi, S., & Ke, C. (2015). A decision support system to select suppliers for a sustainable supply chain based on a systematic DEA approach. *Information Technology and Management, 16*(1), 39–49.

Shieh, H. S. (2012). The greener, the more cost efficient? An empirical study of international tourist hotels in Taiwan. *International Journal of Sustainable Development & World Ecology, 19*(6), 536–545.

Simar, L., & Wilson, P. W. (2007). Estimation and inference in two-stage, semi-parametric models of production processes. *Journal of Econometrics, 136*(1), 31–64.

Simar, L., & Zelenyuk, V. (2006). On Testing Equality of Distributions of Technical Efficiency Scores. Econometric Reviews, 25(4), 497–522. https://doi.org/10.1080/07474930600972582

Soleimani-damaneh, M., & Zarepisheh, M. (2009). Shannon's entropy for combining the efficiency results of different DEA models: Method and application. *Expert Systems with Applications, 36*(3 PART 1), 5146–5150. https://doi.org/10.1016/j.eswa.2008.06.031

Song, M., Tao, J., & Wang, S. (2015). FDI, technology spillovers and green innovation in China: Analysis based on data envelopment analysis. *Annals of Operations Research, 228*(1), 47–64.

Speelman, S., Buysse, J., Farolfi, S., Frija, A., D'Haese, M., & D'Haese, L. (2009). Estimating the impacts of water pricing on smallholder irrigators in north West Province, South Africa. *Agricultural Water Management, 96*(11), 1560–1566.

Sueyoshi, T., & Goto, M. (2014a). Photovoltaic power stations in Germany and the United States: A comparative study by data envelopment analysis. *Energy Economics, 42*, 271–288.

Sueyoshi, T., & Goto, M. (2014b). Investment strategy for sustainable society by development of regional economies and prevention of industrial pollutions in Japanese manufacturing sectors. *Energy Economics, 42*, 299–312.

Sueyoshi, T., & Goto, M. (2015a). Environmental assessment on coal-fired power plants in US north-east region by DEA non-radial measurement. *Energy Economics, 50*, 125–139.

Sueyoshi, T., & Goto, M. (2015b). DEA environmental assessment in time horizon: Radial approach for Malmquist index measurement on petroleum companies. *Energy Economics, 51*, 329–345.

Sueyoshi, T., & Goto, M. (2015c). Japanese fuel mix strategy after disaster of Fukushima Daiichi nuclear power plant: Lessons from international comparison among industrial nations measured by DEA environmental assessment in time horizon. *Energy Economics, 52*, 87–103.

Sueyoshi, T., & Wang, D. (2014). Radial and non-radial approaches for environmental assessment by data envelopment analysis: Corporate sustainability and effective investment for technology innovation. *Energy Economics, 45*, 537–551.

Sueyoshi, T., & Yuan, Y. (2015a). China's regional sustainability and diversified resource allocation: DEA environmental assessment on economic development and air pollution. *Energy Economics, 49*, 239–256.

Sueyoshi, T., & Yuan, Y. (2015b). Comparison among US industrial sectors by DEA environmental assessment: Equipped with analytical capability to handle zero or negative in production factors. *Energy Economics, 52*, 69–86.

Sueyoshi, T., & Yuan, Y. (2016a). Marginal rate of transformation and rate of substitution measured by DEA environmental assessment: Comparison among European and north American nations. *Energy Economics, 56*, 270–287.

Tajbakhsh, A., & Hassini, E. (2018). Evaluating sustainability performance in fossil-fuel power plants using a two-stage data envelopment analysis. *Energy Economics, 74*, 154–178.

Tang, C. S., & Zhou, S. (2012). Research advances in environmentally and socially sustainable operations. *European Journal of Operational Research, 223*(3), 585–594.

Tone, K. (2001). A slacks-based measure of efficiency in data envelopment analysis. *European Journal of Operational Research, 130*(3), 498–509.

Tsolas, I. E., & Manoliadis, O. G. (2003). Sustainability indices of thermal electrical power production in Greece. *Journal of Environmental Engineering, 129*(2), 179–182.

Tyteca, D. (1997). Linear programming models for the measurement of environmental performance of firms—Concepts and empirical results. *Journal of Productivity Analysis, 8*(2), 183–197.

Tyteca, D. (1998). Sustainability indicators at the firm level. *Journal of Industrial Ecology, 2*(4), 61–77.

Ulanowicz, R. E. (2009). Quantifying sustainability: Resilience, efficiency and the return of information theory. *Ecological Complexity, 6*(1), 27–36.

Wang, H. (2015). A generalized MCDA–DEA (multi-criterion decision analysis–data envelopment analysis) approach to construct slacks-based composite indicator. *Energy, 80*, 114–122.

Wang, K., & Wei, Y. M. (2014). China's regional industrial energy efficiency and carbon emissions abatement costs. *Applied Energy, 130*, 617–631.

Wang, K., & Wei, Y. M. (2016). Sources of energy productivity change in China during 1997–2012: A decomposition analysis based on the Luenberger productivity indicator. *Energy Economics, 54*, 50–59.

Wang, K., Wei, Y. M., & Zhang, X. (2012). A comparative analysis of China's regional energy and emission performance: Which is the better way to deal with undesirable outputs? *Energy Policy, 46*, 574–584.

Wang, H., Zhou, P., & Zhou, D. Q. (2013a). Scenario-based energy efficiency and productivity in China: A non-radial directional distance function analysis. *Energy Economics, 40*, 795–803.

Wang, K., Lu, B., & Wei, Y. M. (2013b). China's regional energy and environmental efficiency: A range-adjusted measure based analysis. *Applied Energy, 112*, 1403–1415.

Wang, K., Wei, Y. M., & Zhang, X. (2013c). Energy and emissions efficiency patterns of Chinese regions: A multi-directional efficiency analysis. *Applied Energy, 104*, 105–116.

Wang, K., Yu, S., & Zhang, W. (2013d). China's regional energy and environmental efficiency: A DEA window analysis based dynamic evaluation. *Mathematical and Computer Modelling, 58*(5), 1117–1127.

Weber, C. A. (1996). A data envelopment analysis approach to measuring vendor performance. *Supply Chain Management: An International Journal, 1*(1), 28–39.

West, J. (2015). Capital valuation and sustainability: A data programming approach. *Review of Quantitative Finance and Accounting, 45*(3), 591–608.

Wey, W. M. (2015). Smart growth and transit-oriented development planning in site selection for a new metro transit station in Taipei, Taiwan. *Habitat International, 47*, 158–168.

Winfield, M., Gibson, R. B., Markvart, T., Gaudreau, K., & Taylor, J. (2010). Implications of sustainability assessment for electricity system design: The case of the Ontario Power Authority's integrated power system plan. *Energy Policy, 38*(8), 4115–4126.

Wu, H., Lv, K., Liang, L., & Hu, H. (2017). Measuring performance of sustainable manufacturing with recyclable wastes: A case from China's iron and steel industry. *Omega-International Journal of Management Science, 66*, 38–47.

Xie, X. M., Zang, Z. P., & Qi, G. Y. (2016). Assessing the environmental management efficiency of manufacturing sectors: Evidence from emerging economies. *Journal of Cleaner Production, 112*, 1422–1431.

Xie, H. L., Chen, Q. R., Wang, W., & He, Y. F. (2018). Analyzing the green efficiency of arable land use in China. *Technological Forecasting and Social Change, 133*, 15–28.

Yeh, C. C., Chi, D. J., & Hsu, M. F. (2010). A hybrid approach of DEA, rough set and support vector machines for business failure prediction. *Expert Systems with Applications, 37*(2), 1535–1541.

Yli-Viikari, A. (1999). Indicators for sustainable agriculture—A theoretical framework for classifying and assessing indicators. *Agricultural and Food Science in Finland, 8*(8), 265–283.

Ylvinger, S. (2003). Light-duty vehicles and external impacts: Product-and policy-performance assessment. *European Journal of Operational Research, 144*(1), 194–208.

Zeydan, M., Çolpan, C., & Çobanoğlu, C. (2011). A combined methodology for supplier selection and performance evaluation. *Expert Systems with Applications, 38*(3), 2741–2751.

Zhang, N., & Kim, J. D. (2014). Measuring sustainability by energy efficiency analysis for Korean power companies: A sequential slacks-based efficiency measure. *Sustainability, 6*(3), 1414–1426.

Zhang, B., Bi, J., Fan, Z., Yuan, Z., & Ge, J. (2008). Eco-efficiency analysis of industrial system in China: A data envelopment analysis approach. *Ecological Economics, 68*(1), 306–316.

Zhang, N., Kong, F., & Choi, Y. (2014). Measuring sustainability performance for China: A sequential generalized directional distance function approach. *Economic Modelling, 41*, 392–397.

Zheng, J., Liu, X., & Bigsten, A. (1998). Ownership structure and determinants of technical efficiency: An application of data envelopment analysis to Chinese enterprises (1986–1990). *Journal of Comparative Economics, 26*(3), 465–484.

Zhou, P., & Ang, B. W. (2008a). Decomposition of aggregate CO2 emissions: A production-theoretical approach. *Energy Economics, 30*(3), 1054–1067.

Zhou, P., & Ang, B. W. (2008b). Linear programming models for measuring economy-wide energy efficiency performance. *Energy Policy, 36*(8), 2911–2916.

Zhou, P., Ang, B. W., & Poh, K. L. (2007). A mathematical programming approach to constructing composite indicators. *Ecological Economics, 62*(2), 291–297.

Zhou, P., Ang, B. W., & Poh, K. L. (2008a). Measuring environmental performance under different environmental DEA technologies. *Energy Economics, 30*(1), 1–14.

Zhou, P., Ang, B. W., & Poh, K. L. (2008b). A survey of data envelopment analysis in energy and environmental studies. *European Journal of Operational Research, 189*(1), 1–18.

Zhou, P., Ang, B. W., & Han, J. Y. (2010). Total factor carbon emission performance: A Malmquist index analysis. *Energy Economics, 32*(1), 194–201.

Zhou, G., Chung, W., & Zhang, X. (2013). A study of carbon dioxide emissions performance of China's transport sector. *Energy, 50*, 302–314.

Zhu, Z., Wang, K., & Zhang, B. (2014). Applying a network data envelopment analysis model to quantify the eco-efficiency of products: A case study of pesticides. *Journal of Cleaner Production, 69*, 67–73.

Zhu, W. W., Yu, Y., & Sun, P. P. (2018). Data envelopment analysis cross-like efficiency model for non-homogeneous decision-making units: The case of United States companies' low-carbon investment to attain corporate sustainability. *European Journal of Operational Research, 269*(1), 99–110.

Zimmer, K., Fröhling, M., & Schultmann, F. (2016). Sustainable supplier management - a review of models supporting sustainable supplier selection, monitoring and development. *International Journal of Production Research, 54*(5), 1412–1442.

Zofio, J. L., & Prieto, A. M. (2001). Environmental efficiency and regulatory standards: The case of CO2 emissions from OECD industries. *Resource and Energy Economics, 23*(1), 63–83.

Chapter 5
Sustainable Manufacturing and Technology: The Development and Evaluation

Panitas Sureeyatanapas and Jian-Bo Yang

Abstract The industrial sector, which has been criticised as a major source of pollution and environmental degradation, is now moving towards facing the challenges to cope with pressures from society. Manufacturers and their internal stakeholders have realised that considering cost, quality and productivity only is no longer sufficient to maintain their competitiveness. Customers nowadays are more intensively concerned about green products and environmentally friendly production. There is shortage of natural resources and costs are increasing, while in the labour market workers need safe and secure workplaces for their health and well-being. Government regulations and the requirements of business practices are also stricter. Faced with these pressures, manufacturers have been driven to take environmental and social issues seriously in order to ensure their licences to operate in the long run. This leads to the topic of *sustainability*, which has been widely considered by modern industrialised societies and academics. As the world is moving towards meeting the demands of sustainability, this chapter provides an understanding of and guidelines for how a manufacturing company can develop and evaluate its sustainability in a practical manner.

5.1 Introduction

From the extensive focus on sustainable development throughout the industrial and business sectors, the development and assessment of corporate sustainability have become a topic widely discussed both in academic research and in practice. The

P. Sureeyatanapas (✉)
Supply Chain and Logistics System Research Unit, Department of Industrial Engineering, Faculty of Engineering, Khon Kaen University, Khon Kaen, Thailand
e-mail: panisu@kku.ac.th

J.-B. Yang
Alliance Manchester Business School, The University of Manchester, Manchester, UK
e-mail: jian-bo.yang@manchester.ac.uk

C. Chen et al. (eds.), *Pursuing Sustainability*, International Series in Operations Research & Management Science 301,
https://doi.org/10.1007/978-3-030-58023-0_5

benefits of initiating sustainable development for corporate performance has been studied and supported by many researchers. For instance, corporate image and attractiveness to prospective customers and investors are likely to be enhanced through public disclosure regarding efforts to manage environmental impacts (Azapagic 2003; Dixon et al. 2005; Henri and Journeault 2009; Turban and Greening 1997). Such efforts are also positively associated with companies' financial benefits and market advantage (Grolleau et al. 2013; Hart and Ahuja 1996; Henri and Journeault 2009; Pérez-Calderón et al. 2012). The public nowadays needs to be assured that ecological conditions do not deteriorate and that natural resources are adequate for the next generation, such that a good quality of life can be maintained (Brent and Visser 2005). The literature also suggests a positive relationship between companies' competitive advantages and the implementation of social responsibility programmes (Molamohamadi and Ismail 2013; Panayiotou et al. 2009; Turban and Greening 1997). Sustainability initiatives, furthermore, are the main driver in creating innovation in manufacturing (Dornfeld 2014) which finally leads to the improvement of both product and process quality (Molamohamadi and Ismail 2013; Rusinko 2007).

Performance evaluation is generally agreed as a prerequisite for continuous improvement and effective decision-making, and the complexity of sustainability evaluation has become an issue throughout the world. As the world is moving towards meeting the demands of sustainability, this chapter provides an understanding of and guidelines for how a manufacturing company can develop and assess its sustainability in a practical manner. This chapter includes six main sections. Following the introduction, Sect. 5.2 describes definitions and the main components of sustainability. Next, Sect. 5.3 provides guidelines for manufacturers to develop sustainability, based upon the three common and well-known strategies: green manufacturing, corporate social responsibility and green supply chain. Section 5.4 describes the concept of and the procedure for evaluating sustainability and ends with a discussion of various choices of evaluation methods. Finally, Sect. 5.5 demonstrates an industrial case study for the development of corporate sustainability.

5.2 Definitions and the Components of Sustainable Manufacturing

Since the introduction of *sustainable development* by the Brundtland Report in 1987 as 'the development that meets the needs of the present generation, without compromising the ability of future generations to meet their own needs' (WCED 1987), the term *sustainability* has been defined and adopted in various contexts. For the business world, the notion of *corporate sustainability* has been introduced. Dyllick and Hockerts (2002), one of the widely cited articles concerning corporate sustainability, interpret this term as "meeting the needs of a firm's direct and indirect stakeholders without compromising its ability to meet the needs of future stakeholders". At the same time, corporate sustainability has also been defined from a more straightforward angle as the capacity of a company to maintain its operations over a long period of time (Moldan and Dahl 2007; Perrini and Tencati 2006). The underlying definitions of these two interpretations are relatively consistent. Overall, they communicate that a firm is able to indefinitely continue its business when it satisfies all stakeholders, not only in the present, but also in terms of their future concerns (Sureeyatanapas et al. 2015).

When focussing on the industrial sector, the term *sustainable manufacturing* has been defined by many academics and organisations. One of the most frequently cited definitions, from the US Department of Commerce, is "the creation of manufactured products that use processes that minimize negative environmental impacts, conserve energy and natural resources, are safe for employees, communities, and consumers and are economically sound" (The U.S. Department of Commerce 2012). Based on the literature, although the various definitions differ according to context, their practical implications are identical, pointing out that it is the manufacturing system which minimises negative impacts on the environment and society (both within and outside of the organisation) while also making profits for its stakeholders.

To develop sustainability in general, John Elkington suggested a framework called the *triple bottom line (TBL)*, which encompasses three interrelated dimensions: economic, environmental and social (Elkington 1997). Since then the principle of TBL has encouraged organisations in all sectors to move beyond the consideration of merely financial benefits (i.e. profits, return on investment, etc.) to also take into account responsibility and concerns for people and the planet (Slaper and Hall 2011; Sureeyatanapas et al. 2015). For the business sector, the development and evaluation of corporate sustainability based upon the TBL has become a widely discussed topic during the past few decades. In addition to the TBL, Sureeyatanapas et al. (2015) has proposed *quality* as a separate element, which does not come under any of the dimensions of the TBL, and which complementarily reflects sustainability for manufacturers. Taking these elements into account in policy and strategic planning, as well as in performance assessment, could allow manufacturers to analyse and monitor their sustainability with a holistic view.

A number of factors have been proposed in scholarly articles as drivers for corporate sustainability. *Technology* and *innovation* are among the factors claimed as significant in enabling the performance of the TBL (Garetti and Taisch 2012; Joung et al. 2013; Molamohamadi and Ismail 2013; OECD 2009). According to the sustainability concept, a company should be proactive in planning its moves to fulfil prospective regulations and societal concerns. Proactive companies always integrate innovations into their products and processes as well as managerial strategies in response to environmental challenges, rather than simply react to these (Aragon-Correa 1998; Ramanathan et al. 2010). Innovation is viewed as a mechanism of sustainable development through modification, re-design, and replacement of existing products, processes, and organisational structures. It also enables the creation of entirely new elements and activities (OECD 2009). Technology advancement in information and communication, moreover, drives supply chain collaboration and the development of education and knowledge, which are the fundamentals of research and development initiatives. Apart from these, additional drivers have also been proposed in the literature, such as *leadership* or *management commitment* (Schneider and Meins 2012; Székely and Knirsch 2005) and *learning and education* (Garetti and Taisch 2012; Hubbard 2009; Molamohamadi and Ismail 2013). A number of articles, furthermore, praise the influence of *quality management practices* on corporate sustainability (Curkovic et al. 2000; Isaksson 2006; Jonker 2000; Kuei and Lu 2013; Rusinko 2005; Wiengarten and Pagell 2012).

5.3 The Development of Sustainable Manufacturing

Based on recent literature, the development of sustainable manufacturing has been encouraged to expand beyond internal processes to include the entire product life cycle (Jayal et al. 2010). Companies, in other words, need to consider all aspects of the supply chain, from the design of products to the disposal process or after-use management. This is due to the fact that policies for sustainable manufacturing cannot be successful only in the design of environmentally friendly products and processes. The assessment of and collaboration with suppliers are also critical to obtain green materials and support facilities. A new business model with innovative technology becomes necessary to reach the goal of sustainability (OECD 2009). More importantly, sustainability is not merely to be viewed internally, but should also be seen from the point of view of external acceptability and support. This means that all relevant activities which affect people, both within and outside of an organisation, need to be evaluated and reported. Based on the holistic view of sustainability, a framework which encompasses three interrelated principles is introduced, including *green manufacturing*, *corporate social responsibility* and *green supply chain*, as shown in Fig. 5.1. The details regarding the main objectives and values of each principle are briefly described in Sects. 5.3.1–5.3.3.

5.3.1 Green Manufacturing

Green manufacturing (hereafter GM) has been widely considered by industrial business corporations as a strategy to alleviate tremendous pressure to protect the environment and retain the availability of ecological resources. In general, GM means the process of transforming materials into products by minimising pollution and waste as well as minimising the consumption of non-renewable resources and toxic materials.

A change from a corrective to a preventive approach becomes necessary for the success of GM. Pollution controls, for example by trapping, treating, or disposing of pollution after it is generated, are only regarded as corrective actions at the end-of-pipe. They are viewed as non-value added and costly activities. Pollution

Fig. 5.1 Framework for the development of sustainable manufacturing

prevention is therefore encouraged instead (Rusinko 2007). The 3Rs concept, including reduce, reuse and recycle, is recommended for initiating preventive solutions. *Reduce*, the first element, encourages manufacturers to move in two directions. First, it implies minimisation of resource consumption (energy, water, material, and land). Regarding this, substitution and reduction in the use of non-renewable and toxic materials are particularly suggested. The second direction focusses on the reduction of waste generated from processes. *Reuse*, at the same time, drives the utilisation of what remains after manufacture, while *recycle* promotes the creation of processes which convert waste into usable materials or other products. Reuse and recycle encourage manufacturers to reduce the usage of original materials or natural resources.

Examples of operational strategies which follow the 3Rs concept are given below (Despeisse et al. 2013; Gunasekaran and Spalanzani 2012; Narayanan and Das 2014):

- Avoid usage of unnecessary resources.
- Optimise production plans and the input of resources to improve efficiency.
- Switch off or use standby mode when machines and processing facilities are not in use.
- Substitute non-renewable energy and toxic materials by renewable and non-toxic inputs.

- Employ cleaner technology.
- Enhance the capacity and effectiveness of waste collection systems.
- Manage hazardous waste in an appropriate way.
- Maximise the utilisation of water, waste, and by-products.
- Monitor material usage and set a target for reduction.
- Reduce specific emissions of greenhouse gases, dust, smoke, and effluents.
- Conduct an environmental impact assessment and create preventive actions.

To initiate GM strategies and action plans, typical work improvement techniques, such as Lean, Kaizen, or Quality Control Circle (QCC) activities, are still useful and to be recommended. When environmental concerns are integrated into such techniques, practitioners are encouraged to investigate where, when, and why waste is generated, and whether it can be reduced, reused or recycled. Root causes and action plans need to be determined, based upon brainstorming.

An obstacle for the implementation of GM might be undesirable investment. Since paying to be *green* is usually unavoidable, some companies may decide to delay or reject such projects. Evidence from the literature, however, shows that the costs can be recovered quickly, mostly within a few years (Dheeraj and Vishal 2012; Molamohamadi and Ismail 2013; O'Brien 1999). This indicates that, once a cost-benefit analysis and a GM activity are conducted in proper ways, the benefits obtained should outweigh initial expenditures in the near future. The literature also supports the idea that GM can decrease production costs and enhance efficiency in reducing the consumption of resources as well as in maximising outputs from processes through the minimisation of waste (Hart 1995; OECD 2011; Porter and Van Der Linde 1995; Srivastava 2007).

The term *green manufacturing* is sometimes used interchangeably with *sustainable manufacturing*. It is important to clarify here that a typical definition of GM focusses merely on environmental protection. This is regardless of any direct statement made to enhance a company's financial prosperity and the quality of life of society, although some scholarly articles have already extended its definition to be consistent with the context of sustainability. It is considered that positive impacts on the economic and social dimensions are only potential *by-products* from the implementation of GM. When not every aspect that firms need to satisfy in becoming truly sustainable (the TBL) is directly encouraged, rigidly focussing on GM might still be insufficient to guarantee sustainability for manufacturers.

5.3.2 Corporate Social Responsibility

Corporate social responsibility (CSR) is suggested here as another driver for fulfilling sustainable manufacturing. It has been defined by Kotler and Lee (2005) as 'a commitment to improve community well-being through discretionary business practices and contributions of corporate resources'. Similarly, the European Commission officially defines CSR as 'the responsibility of enterprises for their impacts on society'. Here, responsibility means that a company should integrate not only legal regulations, but also collective agreements between stakeholders, ethics, human rights, and consumer concerns, into their business manners as well as their core policies and strategies (The European Commission 2011).

By definition, the concept of CSR puts emphasis on the development of sustainability through the social dimension. The development of social sustainability at the corporate level should include two perspectives. The first is an effort to minimise and protect society from the negative impacts of a company's operations, and the second is to build a positive relationship with the community and other social institutions. Based on Dyllick and Hockerts (2002), social sustainability refers to the value added to communities in terms of both human capital and societal capital. Human capital is mainly associated with quality of life, knowledge, and the motivation of internal stakeholders, while societal capital relates to the quality of the external community, such as their education, infrastructure, and culture. That means CSR initiatives should not focus merely on customers and the community, but internal parties such as employees, suppliers, and business partners also need to be cared for.

Voluntary commitment has become a key phrase for CSR initiatives. However, in the business world, it might be argued that a company conducts and publicly reports its voluntary activities due to expectations for its economic outcomes and competitiveness. Based on the authors' experience, for example, six people in the top management positions of different sugar manufacturers in Thailand were interviewed regarding the reasons why they have conducted social responsibility and environmental protection activities. Although different answers were given by the participants, the underlying reasons were all linked to the companies' prosperity (i.e. sales, market share, brand image, and company reputation). Most managers explicitly stated that one of the main aims was to strengthen the relationship with the local community. When environmental issues were unintentionally generated by companies, a positive relationship could relieve the tension between them and the public. Only one manager mentioned that conducting social activities was an opportunity for employees to practice voluntary work, to encourage their social commitment, and to promote team building, while employees were proud to be a part of a company which contributed to the development of their own community. The literature also supports the notion that pursuing a social responsibility programme is strongly associated with an increase in companies' financial benefits (Crowther and Aras 2008; Kotler and Lee 2005; Panayiotou et al. 2009; Turban and Greening 1997). The positive image of an organisation making a significant contribution to social well-being and the environment leads to market advantages. A positive relationship with the local community also allows for a company to operate over the long term. Superior practices in human resources management, furthermore, build devoted employees and also make a company attractive to prospective employees (Azapagic 2003; Davis 1973; Fombrun and Shanley 1990; Kotler and Lee 2005; Turban and Greening 1997). In addition, the trust of customers and citizens, which is fundamental for a sustainable business model, can be obtained through an effective CSR programme (The European Commission 2011). This is not to argue that a company's genuine desire for social responsibility does not exist, though it is not really important whether it does or not. CSR is a kind of win-win strategy in which a company that satisfies stakeholders regarding investments from its own resources or efforts should receive such benefits in return. This provides support for why CSR has now become a necessary element in the sustainability of a business organisation.

Various options for a manufacturer to initiate a CSR programme, following Kotler and Lee (2005), include:

- Corporate cause promotion.
- Cause-related marketing.
- Corporate social marketing.
- Corporate philanthropy.
- Community volunteering.
- Socially responsible business practices.

Corporate cause promotion refers to a company's communications to increase awareness about a particular social issue. Fundraising is commonly seen, as is the recruiting of volunteers to participate in a social event organised or led by the company with the aim of alleviating the effects of such an issue. For example, A.P. Honda Company Limited, based in Thailand, persuaded people to donate blood during 1st–15th August 2014 in order to resolve the issue of an inadequate blood supply. *Cause-related marketing* is a company's commitment to donate all or part of its sales revenue during a particular period or event to a charitable organisation or for a specific cause. An example of this is the case of Thai Wacoal Public Company Limited, producer of the ladies' lingerie brand 'Wacoal', who donated a percentage of revenue from the sale of the lingerie for the purchase of mammograms (for breast cancer screening) by the National Cancer Institute of Thailand. *Corporate social marketing* uses a company's resources to campaign for a behaviour change in order to improve public health and safety as well as community well-being. These three options for CSR initiatives mainly employ external resources to develop the society, while a company's own resources become the core mechanism for the next three options. *Corporate philanthropy* is an option whereby a company donates from its own budget or supplies something to fulfil the needs and requirements of a particular social sector. CSR activities, according to this, can include supporting educational facilities and community infrastructure, sponsoring religious activities, or providing scholarships for poverty-stricken children. This is slightly different from the next option, *community volunteering*, in that a company encourages its employees to act voluntarily in community development, instead of providing monetary grants. Many times, both options (monetary donation and volunteering) are practised simultaneously. The last option, *socially responsible business practices*, includes programmes that a company puts in place to ensure the safety and quality of life of its customers, employees, the community, or other groups of stakeholders, as well as to protect the environment. For instance, Thai Airways has launched the 'Travel Green' programme through various initiatives such as reducing energy consumed by making all containers from low-weight materials, washing their aircraft regularly in order to reduce friction, maximising the use of organic materials in the food served on board, controlling the aircraft through eco-mode, putting effort into measuring and reducing CO_2 emission, etc.

Apart from GM and CSR programmes, in which manufacturers mainly add environmental and social concerns into their operations and strategies, in the modern business world the concept of the *green supply chain*, which encourages collaboration with external parties, has been discussed as an additional driver to pursue *actual* corporate sustainability (Garbie 2014; Jayal et al. 2010). The detail of this is provided in the next section.

5.3.3 Green Supply Chain

As mentioned, achieving sustainability in manufacturing firms needs consideration of a holistic view across the product life cycle, from downstream to upstream processes. Traditionally, in supply chain management, people generally start from the extraction of raw materials, continue with purchasing operations and end with the distribution of products to customers. For the green supply chain (hereafter GSC), the downstream and upstream processes must be extended. Srivastava (2007) has defined it as "integrating environmental thinking into supply-chain management, including product design, material sourcing and selection, manufacturing processes, delivery of the final product to the consumers as well as end-of-life management of the product after its useful life". According to this, Fig. 5.2 demonstrates the integration of all of these related activities for an effective GSC management. From this, downstream processes are extended to consider the design of green products and processes (green design) while upstream processes cover the use and consumption of a product in an environmentally friendly way (green consumption) as well as the management of the used product and waste (green recycling).

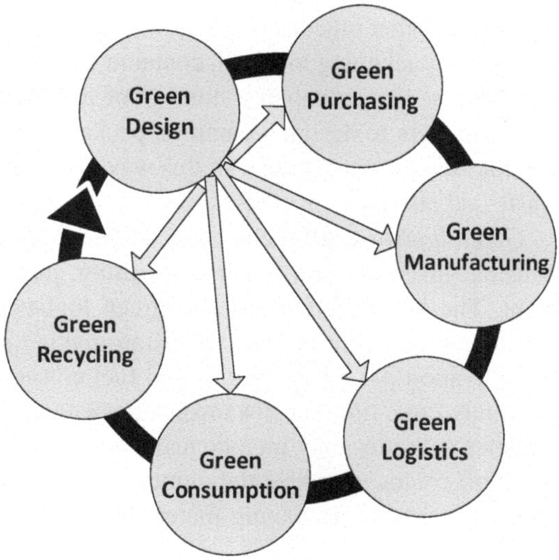

Fig. 5.2 Mechanisms of green supply chain management

Green design, as illustrated by the arrows in Fig. 5.2, is not only the first step of the GSC cycle but is also considered to be the control process for the other activities. In other words, the effectiveness of other activities is heavily dependent upon the design process (Jayal et al. 2010). It comprises the consideration of material usage, methods of material extraction, manufacturing processes, product storage, product delivery, consumption or usage processes for customers, and the pathway for managing product waste and end-of-life (collecting, disassembling, reusing, recycling, or remanufacturing). To strengthen the methodology of green design for sustainable manufacturing, the 6Rs concept, an extension of the 3Rs, has been introduced (Joshi et al. 2006). Additional elements include (a) recover, (b) redesign, and (c) remanufacture. *Recover* addresses the collection of used products and the processes that follow this (i.e. disassembly, inspection, and cleaning) which are needed for subsequent utilisation. *Redesign* includes the process of redesigning the product and the logistics networks that make the after-use processes easier and cleaner, mostly through the Design for Environment (DfE) technique. The last element is *remanufacture*, which refers to the process of restoring a used product to its original form without any loss of functionality. This could be done through a combination of reusing and repairing some of the original parts as well as replacing obsolete and/or dead parts with new ones (Jayal et al. 2010). By including these *Rs*, the concept of sustainable manufacturing has been extended to cover the whole supply chain, rather than focussing on the original 3Rs concept which mainly addresses only the production system. According to this, the evaluation of sustainability performance is encouraged to consider the entire product life cycle based upon the principle of life cycle assessment (LCA), and the procedure to perform this should also be considered during the design process.

Green purchasing, the next activity, addresses two objectives. The first is to support and encourage suppliers, through buyers' power, to align their operations with the approach of sustainable manufacturing. Auditing suppliers as to whether or not they have implemented any green and CSR programmes, as well as their effectiveness, is also recommended (Chiarini 2014). Setting up a green procurement process becomes another objective. Practitioners may consider reduction in the use of paper, electricity, packaging material, containers for parts, and other resources used during the process. For example, a number of electronic manufacturers collaborate with their suppliers to design a common package which fits both the raw material itself and the subsequent product. In this way, they can eliminate a substantial amount of waste and also save on costs.

Green logistics still focusses on the management of transporting of goods, warehousing, inventory and material handling, just as in traditional logistics management. The key difference is that green logistics activities include environmental considerations with regard to decision-making and strategic planning, such as the consideration of CO_2 emissions and fuel consumption as additional criteria, rather than only cost, time, or distance (Sureeyatanapas et al. 2018). Examples of green logistics activities may include using alternative fuels in transportation, optimising transport routes to reduce the number of miles and save energy, adopting environmentally friendly packaging materials, improving vehicle efficiency, encouraging eco-driving, changing mode of transportation from road to rail, and collaborating with partners to maximise vehicle utilisation (Lai and Wong 2012; Lau 2011; Murphy and Poist 2003; Pazirandeh and Jafari 2013; Sureeyatanapas et al. 2018).

The next activity is *green consumption* which highlights communication with customers in order to encourage them to use, consume or dispose of products in an environmentally friendly manner. Product packaging, labelling, instruction manuals, or advertising media can be utilised for such communication.

Green recycling is considered to be the final activity of the cycle. Its main practice is *reverse logistics* or *reverse distribution* which has been typically defined as "the movement of goods from a consumer back towards a producer in a channel of distribution" (Murphy 1986). Generally speaking, it is the process of planning, implementing, and controlling the collection and transportation of used products from the point of consumption back to a producer in order to recapture product value or to manage proper disposal (Ravi et al. 2005; Rogers and Tibben-Lembke 1998). The principle of reverse logistics actually covers the return of products due to other reasons, such as product damage, recalls, restock, or excess inventory (Rogers and Tibben-Lembke 1998). The context of green supply chain, however, usually mentions reverse logistics as an enabler in rehabilitating used products that are no longer required into usable ones, rather than in fulfilling other business strategies. Note that rehabilitation is only possible with a supportive design, mostly based upon the 6Rs concept mentioned earlier. Green recycling directly promotes sustainability in both economic and environment terms since it leads not only to internal cost saving, but also to the utilisation of used products as well as to the reduction of material usage and solid waste. Furthermore, this acts against scarcity of the natural resources used in producing new products (Gunasekaran and Spalanzani 2012).

The GSC concept encourages all parties within the supply chain to together integrate environmental concerns into their practices. Collaboration and innovation in supply chain management for the goal of environmental improvement have both become key success factors. Effective GSC management should bring significant benefits to both manufacturing firms and the environment, without sacrificing quality, cost, or other aspects of corporate performance (Srivastava 2007). A number of research studies have investigated the driving factors for sustainable and green supply chain practices based upon case studies and/or questionnaire-based surveys (Bhanot et al. 2015; Ghazilla et al. 2015; Kasim and Ismail 2012; Kulatunga et al. 2013; Mittal and Sangwan 2014; Nordin et al. 2014; Walker et al. 2008). Several drivers were reported, including future regulatory compliance; pressure from customers and society; financial benefits; company image; and quality improvement. GSC management, as presented in Fig. 5.2, should be performed as a cycle of continuous improvement. This means that performance and effectiveness need to be assessed and monitored once an activity is conducted. The design of the product and/or process should then be reconsidered in order to improve the most polluting or inefficient point within the chain.

To sum up, as illustrated in Fig. 5.1, the development of sustainable manufacturing should start from strategies which affect internal operations, including GM and CSR. While GM mainly focusses on the improvement of environmental sustainability, covering the minimisation of harmful effects to the environment and the uses of natural resources, CSR extensively addresses the social dimension through the development of quality of life for society and stakeholders, as well as through public relations and reporting in order to enhance the corporate image. Once the internal processes are *green* and social responsibility is part of corporate policy and strategy, it is suggested that a manufacturer should move forward to consider other sectors within the supply chain in order to reach *actual* sustainability. As mentioned previously, however, performance assessment is a key principle to maintain the success of sustainable development and to drive the improvement. The next section provides a guideline for how sustainability performance can be assessed in a practical manner.

5.4 The Evaluation of Sustainable Manufacturing

Measurement is fundamental for improvement. A challenge for businesses seeking to develop corporate sustainability is deciding how they will assess their progress and investigate weaknesses that they should improve. Sustainability assessment is complex because it is related to several criteria measured in different units. Furthermore, sustainability involves many qualitative aspects for which standard means of assessment remain unavailable. Mostly, subjective judgement is employed and, as such, the results are likely to be questionable or unacceptable to other parties. This also inhibits comparability between different assessors and over time. From these concerns, a challenge arises as to how to aggregate diverse information and rigorously monitor sustainability performance. Section 5.4.1 provides guidelines for identifying an appropriate set of sustainability indicators. Section 5.4.2 suggests a procedure for the assessment, and Sect. 5.4.3 offers discussions of various choices of aggregation and analysis methods. Section 5.5 illustrates the case of sugar manufacturers pursuing sustainable development initiatives.

5.4.1 Sustainability Criteria and Indicators

During the past decade, a number of frameworks for assessing corporate sustainability have been published, such as Global Reporting Initiative (GRI) guidelines, General Motors metrics for sustainable manufacturing, sustainability metrics from Britain's Institution of Chemical Engineers (IChemE), Ford product sustainability index, Dow Jones sustainability indices, OECD sustainable manufacturing toolkit, etc. In addition to these, many frameworks with sets of criteria and/or indicators, have also been proposed by academic researchers (Azapagic and Perdan 2000; Figge et al. 2002; Garbie 2014; Hubbard 2009; Joung et al. 2013; Keeble et al. 2003; Krajnc and Glavic 2005; Labuschagne et al. 2005; Rahdari and Rostamy 2015; Rosen and Kishawy 2012; Sangwan et al. 2018; Schneider and Meins 2012;

Table 5.1 Relevant aspects of corporate sustainability, reproduced from Sureeyatanapas et al. (2015)

Relevant aspects of corporate sustainability	
(1) Air emission	(12) Compliance with laws and regulations
(2) Liquid effluent	(13) Supplier support and collaboration
(3) Solid waste disposal	(14) Society and local community concerns
(4) Energy consumption	(15) Employee wages and benefits
(5) Water consumption	(16) Stakeholder involvement
(6) Land used	(17) Employee health and safety
(7) Material consumption	(18) Employee training and education
(8) Environmental management	(19) Employee satisfaction
(9) Financial benefits	(20) Quality performance
(10) Market advantage	(21) Conformance to international standards of business conduct
(11) Investment and expenditure on sustainable development	

Slaper and Hall 2011; Székely and Knirsch 2005; Tseng et al. 2009; Veleva and Ellenbecker 2001; Ziout et al. 2013). From these, several aspects of corporate sustainability, according to the TBL, have been synthesised as shown in Table 5.1. For companies starting an assessment, these aspects could be a starting point for them to develop their own set of criteria and indicators.

Indicators belonging to each criterion (or each aspect) need to be identified in order to break down the multidimensional nature of a criterion into a number of unidimensional and tangible terms. Note that it is difficult to define a standard set of indicators which is implementable in every industry and country since each sector always has its own concerns and characteristics (Keeble et al. 2003; Kolk and Mauser 2002). For a particular company or industrial sector, key suggestions for identifying appropriate indicators are listed below (Sureeyatanapas et al. 2015):

- Indicators should be associated with industrial or corporate characteristics, strategies, and culture.
- Indicators should be dynamic and adjustable for changes in any conditions and stakeholders' concerns.
- A set of indicators should cover the concerns of all relevant stakeholders.
- Words to name each indicator must be clear and readily understandable without risk of different interpretations among practitioners.
- Quantitative indicators must be implementable; this covers the requirements that the standard measurement procedure is available and the measurement system is reliable.
- For qualitative indicators, a logical way to assess them should be clearly defined and standardised; intuitive judgment should be minimised.
- A set of indicators should include not only historical performance measures, but also indicators which reflect the ability of a company to maintain or achieve favourable performances.

As suggested above, governance-related features, which are viewed as mechanisms put in place to manage and monitor corporate performance, should be added to a sustainability assessment framework (Schneider and Meins 2012). Examples of this could be the establishment of specific organisational units and goals relating to sustainability, or the determination of formalised codes of conduct, mission/vision statements, sustainability measurement indices, etc. Note that it is not necessary to start the assessment from a complete set of indicators covering all criteria posed by the literature. Manufacturers may start from their existing data in order to understand the current performance, and an initial improvement can be conducted based upon only a few indicators. Once they have gained more experience in collecting and analysing data, additional indicators and accounting systems can be developed. For each indicator, a specification, benchmark value, or ideal case should be definable in order to drive improvement.

5.4.2 Steps in the Evaluation

Figure 5.3 demonstrates a guideline for a company to evaluate its sustainability step-by-step. The first step recommends practitioners to clarify the objective of the evaluation which should be in accordance with the focus of the development policy. For example, some companies might conduct the evaluation for performance analysis and improvement internally, while benchmarking with competitors might be the main aim for others. Putting effort into alleviating a particular issue such as global warming, CO_2 emission, community complaints, or legal penalties, as well as

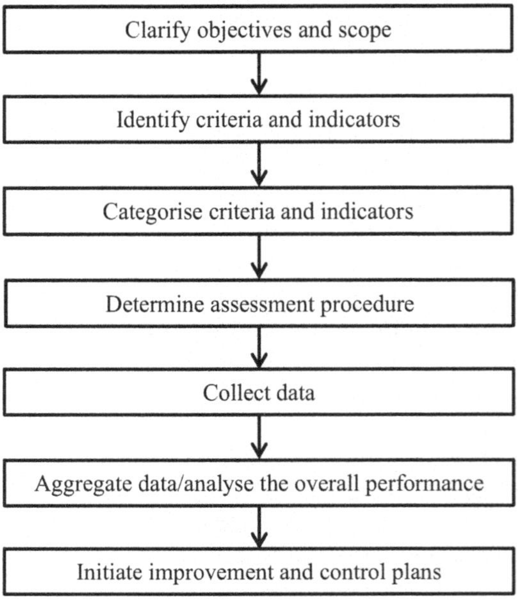

Fig. 5.3 Procedure for the assessment of corporate sustainability

promoting corporate image and reputation, might also be the case. At this stage, positive and negative environmental and social impacts from each of the company's operations need to be reviewed. The boundary of the evaluation then needs to be identified. The evaluation of sustainability, with regard to the manufacturing sector, can be conducted within various scopes (i.e. a particular operation, organisation, product life cycle or the whole supply chain). Conditions regarding available time, budget, and other resources need to be taken into account.

A set of criteria and indicators should then be identified according to the objective and scope defined earlier, following the guidelines provided in Sect. 5.4.1. Indicators should be grouped under a particular criterion, and a number of criteria should be put into each major dimension. These should be constructed as a hierarchy in order to simplify the analysis process.

The next step is to specify a clear measurement procedure for each indicator. Policy or internal regulation may be established to ensure that effective data collection and monitoring are conducted on a regular basis. An appropriate normalisation unit needs to be specified in order to allow standard assessment and comparison. The common normalisation units are: production volume, period of time, product mass, or units of value added, such as net sales or revenue. Normalisation units might be chosen based on the peer group against which a company plans to benchmark its own performance; or they may be based on standard factors applied in that sector (OECD 2011). For quantitative indicators, the measurement procedure should be practical and cost-effective. The measurement instrument and/or data accounting system should also be proved to be reliable. The challenge, as mentioned, comes with the assessment of qualitative performance. According to Sureeyatanapas et al. (2015), a company's performance regarding each qualitative criterion, such as '*Society and local community concerns*' or '*Management commitment*', is broken down into a number of practice items. The rating scales or the evaluation grades for each item are then developed, so that the assessor is able to select the option which best reflects the actual situation in the company being assessed. This is a way to transform qualitative judgement into a quantitative scale in order to enhance comparability. Standardisation is due to the evidence-based scale since tangible evidence is attached to each evaluation grade. It is recommended that, for each criterion, the best practice or the ideal case should be given the best grade. The procedure should be standardised throughout relevant stakeholders and then documented.

After the data-collection process, the literature suggests several techniques to facilitate the aggregation of performances against various indicators. In order to understand sustainability, combining various performance measures into one index (called a composite index) is helpful, and this better facilitates the strategic decision-making process (Garbie 2014). Computing a composite index for each major dimension and for overall sustainability is recommended, because it allows companies to determine focus points and to prioritise improvement plans. Recently, Thies et al. (2019) conducted a review of literature to explore mathematical methods commonly adopted for sustainability assessment. The results show that more than half of the reviewed articles employ multiple criteria decision analysis (MCDA) techniques (the term is used interchangeably with multi-attribute decision-making, MADM), followed by data envelopment analysis (DEA). Advantages and limitations of such commonly used techniques are briefly discussed in the next section.

5.4.3 Performance Analysis Techniques

This section provides an overview of advantages and conditions of using some popular techniques for sustainability assessment. First, for DEA, the main objective of the analysis is generally to compare sustainability of different decision-making units (DMUs) on the basis of relative efficiency. DMUs could concern products, processes, or organisations under study. Implementing DEA requires classifying various sustainability indicators as inputs or outputs for the efficiency evaluation. DEA ultimately provides a useful guideline for how to improve performance based on the best (i.e. the most efficient) DMU in the peer group. Some basic formulations of the DEA methodology, called the basic CCR model, appear below to briefly explain the determination of 'relative efficiency' for each DMU (Cooper et al. 2007).

Suppose there are n DMUs under study; let each DMU_j be evaluated with m inputs ($x_{1j}, x_{2j}, \ldots, x_{mj}$) and s outputs ($y_{1j}, y_{2j}, \ldots, y_{sj}$). Then, for the DMU_o being considered, the virtual output is maximised, and the virtual input is normalised to 1, as shown in the following linear program (LP_o), where v_i and u_r are the weights of each input i and output r ($i = 1, 2, \ldots m$ and $r = 1, 2, \ldots s$), respectively.

$$(\text{LP}_o) \quad \max \theta = u_1 y_{1o} + u_2 y_{2o} + \cdots + u_s y_{so} \tag{5.1}$$

$$\text{Subject to} \quad v_1 x_{1o} + v_2 x_{2o} + \cdots + v_m x_{mo} = 1 \tag{5.2}$$

$$u_1 y_{1j} + \cdots + u_s y_{sj} \le v_1 x_{1j} + \cdots + v_m x_{mj}, \quad (j = 1,2,\ldots n) \tag{5.3}$$

$$v_1, v_2, \ldots, v_m \ge 0 \tag{5.4}$$

$$u_1, u_2, \ldots, u_s \ge 0 \tag{5.5}$$

Haibo et al. (2018) have recently published a review article about the papers that have applied DEA in sustainability research from 1996 to 2016. Their findings enable classifying DEA applications in sustainability according to four objectives: to assess corporate sustainability; to assess regional sustainability; to form a sustainability composite index; and to analyse sustainability performance. Many of those studies applied DEA to analysing or comparing DMUs' eco-efficiency, focussed mainly on the combination of economic and environmental indicators. Although widely adopted among academics, users must recognise several limitations in implementing DEA (Sureeyatanapas et al. 2014). First, the significance of guidelines for improvement, which emerge after the process, depends on whether the peer group covers the real best performers. Second, each DMU is assigned a best set of weights with values that optimise its efficiency, and the weights may vary among DMUs. This concept tends to be unrealistic for sustainability. Third, one widely adopted rule of thumb suggests that the number of DMUs should be at least twice as large as the total number of criteria, in order to guarantee the discriminating power of DEA (Ramanathan 2006; Zhou et al. 2008). For sustainability involving a large number of criteria, this would be a tough requirement, which also means that DEA is not a practical method for company self-assessment in the absence of alternatives for comparison. Moreover, the assumption in DEA that outputs definitely increase with adding more inputs into the process may not always be true (Shimshak et al. 2009). This is still questionable, because effort and resources put into environmental and social development may not be substituted completely, and the associated results may not be apparent within a short period (Dyllick and Hockerts 2002).

Meanwhile, MCDA techniques focus on performance assessment and decision-making under multiple criteria and/or alternatives. In sustainability assessment problems involving various criteria measured in different units, MCDA techniques can be employed to aggregate those criteria and determine overall performance. Nevertheless, each MCDA technique has particular advantages and limitations. As such, the selection of MCDA techniques should depend on the objectives and the characteristics of the problem under consideration. For instance, Simple Additive Weighting (SAW), or the weighted-sum technique, is one of the most popular techniques for sustainability assessment. The basic formulation of SAW is shown in Eq. (5.6), where CI_j is a composite index of a company j, X_{ij} denotes normalised value of j for indicator i, and w_i is the relative weight of indicator i. Krajnc and Glavic (2005) and Ugwu et al. (2006) provide examples of applying SAW in sustainability assessment. Although its computational process is claimed to be simple and transparent, assessors should be aware that this technique is open to complete compensability and the loss of the original information. This is because a low value for one criterion might be compensated by a very high score for another criterion. Moreover, this technique relies on the additive value function approach, which assumes preferential independence or no overlap among the assessments of all criteria (Keeney and Raiffa 1976; Winston 2004). This condition may be considered unrealistic for the case of sustainability criteria.

$$\mathrm{CI}_j = \sum_{i=1}^{n} w_i X_{ij} \tag{5.6}$$

Another aggregation technique, the Weighted Geometric Mean (WGM) technique, is suitable when the assessors need to decrease the degree of compensability and agree that each criterion has an absolute power to decide the existence of any unit. Generally speaking, its combined value will be zero once the normalised value of at least one criterion is zero, no matter how great the importance of that criterion. Its formulation is shown in Eq. (5.7).

$$\mathrm{CI}_j = \prod_{i=1}^{n} X_{ij}^{w_i} \tag{5.7}$$

Table 5.2 An example of the pairwise comparison matrix using the scale 1–9 (a case of four alternatives A–D)

A particular indicator	A	B	C	D
A	1	5	2	4
B	1/5	1	1/2	1/2
C	1/2	2	1	2
D	1/4	2	1/2	1

The analytic hierarchy process (AHP) is one of the most popular MCDA techniques generally used in sustainability assessment, either for determining criteria weights or prioritising alternatives, or both. The main advantage of AHP is that for each criterion or indicator, the assessor does not have to fill in an absolute value of each alternative on the assessment sheet. Instead, pairwise comparisons between alternatives, using a ratio scale, are required. It is therefore suitable for relative assessments between criteria, or indicators that are difficult to assess or compare directly in absolute values. Prioritising alternatives consists of the following steps: (1) Define the problem in hierarchical structure; (2) For each specific indicator, construct a pairwise comparison matrix among all alternatives (as Table 5.2 shows) situated on an importance rating scale (the scale 1–9, ranging from 1 = equally important to 9 = extremely more important, is generally employed); (3) Use eigenvector of each matrix to yield alternatives' relative weights; (4) Check reliability of the result through the consistency index (CI) and consistency ratio (CR); (5) For each alternative, combine all relative weights from every indicator (using a particular technique such as SAW or WGM) in order to form a composite index; and (6) Make a final decision according to the composite scores. Saaty (1980) provides further details about how to compute weights, CI, CR, and all conditions that must be considered.

Ziout et al. (2013) present an example of AHP applied to assess and compare the sustainability of used and new manufacturing systems. For their study, 15 indicators were placed under the three major criteria: economic, environment, and social. Finally, all indicators were combined to form a composite sustainability index.

Nevertheless, it must be noted that AHP is not practical for company self-assessment (without another alternative with which to compare), due to the requirement for paired comparisons between two units. It might be useful only for ranking several companies or determining the most sustainable option.

A critical issue in green and sustainability performance assessment concerns the fact that most MCDA techniques are not typically designed to deal with uncertainty and/or incompleteness of the assessment information. Literature relating to environmental management implies that practitioners are always reluctant to report information concerning environmental issues, such as pollution and harmful impacts upon their processes (Nawrocka 2008; Walker et al. 2008). For some indicators, in addition, firms may not be able to report a precise score or level of their performances since the assessment of those indicators depends on a specific measurement technique or skilled personnel. Also, for qualitative aspects which are always found in the social dimension of sustainability performance, it might be difficult for assessors to give precise judgements because of the unavailability or incompleteness of evidence. Although analysers may try to transform subjective judgements into numerical data using various forms of rating scales or evaluation grades, criticisms can be made since people tend to interpret this kind of linguistic scale differently. Moreover, it is likely that a particular situation being assessed will not exactly match a particular grade, but two or more grades might better explain the current practice (Sureeyatanapas and Tawwan 2018). These issues affirm that an aggregation method for analysing overall sustainability performance needs to be able to manage the incompleteness and uncertainty of information in a sensible way.

The evidential reasoning (ER) approach, another MCDA technique, has been suggested as an effective tool to reasonably aggregate results from qualitative judgments with quantitative measurements in complex and uncertain situations without the need to satisfy the preferential independence condition among the criteria. The aggregation of scores is based upon the evidence theory through the degrees of belief assigned by assessors. The method is effective not only for performance comparison across several companies, but also for the self-assessment of a single company. Thorough details of the ER algorithm can be found in Yang and Singh (1994), Yang (2001), Wang et al. (2006), and Xu et al. (2006). Note that MCDA techniques always need users to assign weights, or degrees of importance, to each criterion and indicator. Various weighting techniques are available (i.e. Direct rating, Point allocation, Trade-off, AHP, SMART, SWING, Rank-based weighting, Entropy, etc.), and the selection of such techniques is mainly dependent upon the characteristics of data and agreement among decision makers. A number of research studies have exemplified the assessment of green and sustainability performances through the ER approach, using different weighting techniques. For instance, Sureeyatanapas et al. (2014) presented an assessment and comparison of corporate sustainability performances, based on the ER model, between three sugar manufacturing companies in Thailand. Sureeyatanapas and Tawwan (2018) then applied the ER algorithm to evaluate the environmental performances of two logistics service companies in Thailand based on the list of criteria from ISO14031. These two studies employed the direct rating technique to determine the weights of criteria. In Akhoundi and Nazif (2018), another example, the ER approach was used to prioritise several types of wastewater reuse applications and wastewater treatment technologies in Iran. The relative weights of the criteria were determined using AHP. These studies show that the ER approach provides a flexible and logical method for dealing with qualitative judgements and uncertainty in the scoring processes which are likely to be seen in green and sustainability assessment.

For the analysis of sustainability performance, as mentioned, not only does an aggregated score represent the overall sustainability, but a score for each major dimension and criterion should also be computed in order to allow a company to identify its weak points, develop further improvement plans, and define improvement targets. A control plan should also be determined for its strong points in order to maintain this superior performance.

5.5 An Industrial Case Study

This section offers discussions, based on a case study, regarding how sugar manufacturers could enhance their sustainability. Note that this case study focusses merely on the organisational level, rather than the whole product life cycle.

Based on the case study, a practical guideline for sustainable development within the cane sugar manufacturing sector is reported. Overall, the use of fossil fuels in manufacturing processes needs to be replaced by biofuels. The utilisation of bagasse, which is the major by-product from sugar production, should be maximised. A simple way of doing this is to use it as an energy source for boilers in the process of heat and steam generation. It can also be applied to the manufacturing of paper, panel board, or chipboard. Other by-products, such as molasses and filter cake, can be completely transformed into value-added products as well. This could help a company to reduce the amount of waste disposed of and to enhance economic profits. Other kinds of solid waste must be carefully managed in an effective and environmentally friendly manner, such as recycling or utilising as alternative sources of energy. In terms of water consumption and effluent, a close-loop system can be conducted through the reuse and recycle principles. For example, water from the water treatment system, being rich in organic matter, can be supplied to cane farming. Water recovered from the evaporation and crystallisation operations might be returned to the manufacturing process after cooling. Cleaner technology and materials should be promoted not only to lessen negative impacts on the environment, but also to prevent illnesses caused by the presence of chemicals and toxic substances. Workplaces and working procedures should be safe and ergonomically designed. The risks posed by physical agents need to be monitored and managed, particularly noise, dust, and heat, which are generally found in the working environment of sugar factories.

In terms of social responsibilities, community complaints should always be taken into account, especially about dust, smoke and odour problems. Sustainable sugar manufacturers should not only put efforts into avoiding a negative impact, but should also build a positive relationship with the community by performing external social activities. This enhances corporate image and reputation, as well as promotes unity and a voluntary mindset among their employees. Sustainable companies, in addition, should pay serious attention to making their practices conform to the international standards of business conduct. The development of employee knowledge and skills as well as the consideration of their welfare and involvement should always be considered in the company's policies. A strong relationship with the sugarcane farmers needs to be emphasised not only to express corporate social responsibility, but also to secure a satisfactory quality and quantity of the cane supply.

Quality is also viewed as another driver of sustainable manufacturing for the sugar business. Regarding this, sugar quality and customer satisfaction with the products and services offered should be given high priorities. Sugar yield and process efficiency should be closely monitored and continuously improved in order to maximise profit and to lower the requirement to consume original resources. Finally, manufacturers need to ensure that they still satisfy internal stakeholders in terms of financial profits and shared values. A win-win situation should always be sought in all aspects of business operations.

For the evaluation of sustainability performance based on Sureeyatanapas et al. (2015), the assessment framework was developed based on a survey using a questionnaire and multiple case studies. At the end of the process, the framework encompasses various criteria and indicators in a hierarchical form organised into five levels, shown in Fig. 5.4. Sustainability performance is at the top level and has four major dimensions (i.e. environment, economic, social, and quality) at the second level. Under each dimension, the third level embraces 12 criteria, and their sub-criteria are then placed at the fourth level. The criteria reflect key information about each dimension, and the suggested execution of the performance analysis occurs from the criteria level to the overall sustainability-performance level. Implementable indicators are positioned at the fifth level. Following the empirical study, 40 indicators can be identified to suit current situations of the Thai sugar industry. The list comprises both quantitative and qualitative indicators, as summarised in Table 5.3. Note that the list of indicators should be periodically adjusted to reflect changed circumstances. For instance, indicators for the sub-criterion *'Air emission'* can be CO_2 emission per year in units of kilograms of CO_2 equivalents, if and only if all companies in the peer group can adequately estimate the emission using the standard form of evaluation. Otherwise, other indirect indicators might be employed. The measure of concentration of total suspended particles in the unit mg/m^3 can also be used to reflect air pollution generated by a company. For some sub-criteria, both quantitative and qualitative indicators might be employed. For example, the sub-criterion '*Employee health and safety*' might be assessed through the rate of

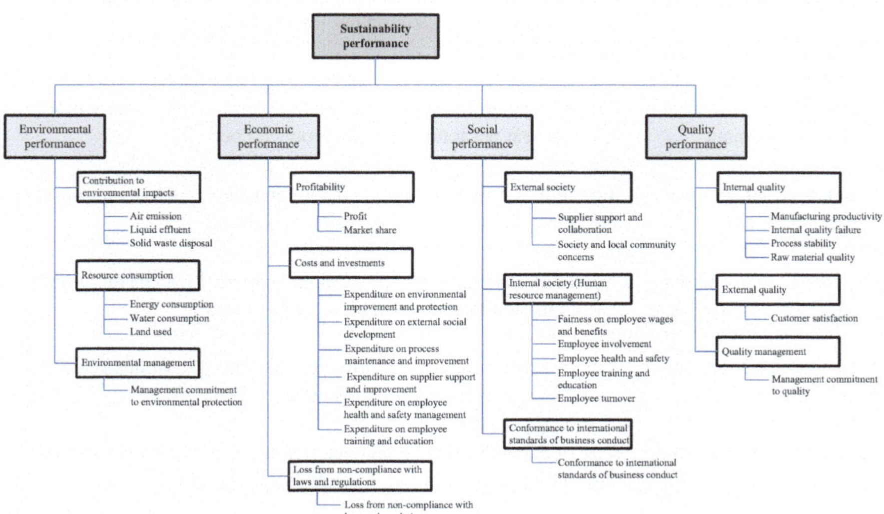

Fig. 5.4 A hierarchical framework of corporate sustainability assessment for sugar manufacturing (Sureeyatanapas et al. 2015)

Table 5.3 List of indicators belonging to each sub-criterion (Sureeyatanapas et al. 2015)

Sub-criteria	Indicators	Measurement units
Air emission	Rate of fossil fuels used by steam boilers relative to total amount of electricity produced per year	kg/kWh or L/kWh
	Concentration of total suspended particles (TSP)	mg/m^3
Liquid effluent	Rate of water discharged into the environment relative to a tonne of cane processed per year	m^3/t
Solid waste disposal	Rate of hazardous waste disposed of relative to a tonne of cane processed per year	t/t
	Rate of non-hazardous waste disposed of relative to a tonne of cane processed per year	t/t
Energy consumption	Rate of steam consumption relative to a tonne of cane processed per year	t/t
	Rate of electricity consumption relative to a tonne of cane processed per year	kWh/t
Water consumption	Rate of external water consumption relative to a tonne of cane processed per year	m^3/t
Land used	Rate of areas of sugar manufacturing sites relative to a tonne of cane processed per year	m^2/t
Management commitment to environmental protection	Management commitment to environmental protection	Qualitative evaluation (5 items)
Profit	Gross profit margin per year	%
Market share	Percentage of market share based on the quantity of sugar produced per year	%
Expenditure on environmental improvement and protection	Rate of expenditure on environmental improvement and protection per tonne of sugar produced per year	Monetary unit/t
Expenditure on external social development	Rate of expenditure on external social development per tonne of sugar produced per year	Monetary unit/t
Expenditure on process maintenance and improvement	Rate of expenditure on process maintenance and improvement per tonne of sugar produced per year	Monetary unit/t
Expenditure on supplier support and improvement	Rate of expenditure on cane farming support and improvement per tonne of sugar produced per year	Monetary unit/t
Expenditure on employee health and safety management	Rate of expenditure on employee health and safety management per tonne of sugar produced per year	Monetary unit/t
Expenditure on employee training and education	Rate of expenditure on employee training and education per tonne of sugar produced per year	Monetary unit/t

(continued)

Table 5.3 (continued)

Sub-criteria	Indicators	Measurement units
Loss from non-compliance with laws and regulations	Total amount of fines paid per year	Monetary unit
	Total number of non-monetary sanctions and warnings per year	Number
Supplier support and collaboration	Cane farmers support and collaboration	Qualitative evaluation (4 items)
Society and local community concerns	The number of complaints from the local community per year	Number
	Social responsibility	Qualitative evaluation (1 items)
	Social development and participation	Qualitative evaluation (3 items)
Fairness on employee wages and benefits	Internal fairness on employee wages and benefits	Qualitative evaluation (3 items)
	External fairness on employee wages and benefits	Qualitative evaluation (3 items)
Employee involvement	Employee involvement and empowerment	Qualitative evaluation (4 items)
	Employee communication	Qualitative evaluation (2 items)
Employee health and safety	Rate of work-related accidents relative to the total working hours in the working schedule per year	Number/h
	Percentage of working hours lost relative to the total working hours in the working schedule per year	%
	Employee health and safety provision	Qualitative evaluation (10 items)
Employee training and education	Employee training and education provision	Qualitative evaluation (7 items)
Employee turnover	Annual employee turnover rate	%
Conformance to international standards of business conduct	Conformance to international standards of business conduct	Qualitative evaluation (11 items)
Manufacturing productivity	The sugar yield at 96 POL—10 CCS equivalent (adjusted kilograms of sugar produced per tonne of cane processed)	kg/t

(continued)

Table 5.3 (continued)

Sub-criteria	Indicators	Measurement units
Internal quality failure	Percentage of reprocessing, derived from the weight of remelted sugar relative to total weight of the sugar produced per year	%
Process stability	Percentage of production shutdowns, derived from the total hours of unplanned shutdowns relative to the total operating hours per year	%
Raw material quality	Commercial cane sugar (CCS)	CCS
Customer satisfaction	The number of customer complaints and product returns per year	Number
Management commitment to quality	Management commitment to quality	Qualitative evaluation (5 items)

work-related accidents and the percentage of working hours lost, relative to the total working hours in the annual work schedule (quantitative assessment), in combination with the qualitative assessment of how well and effectively a company provides facilities and its policies for taking care of employee health and safety. Note that in terms of quantitative indicators for the sugar industry, the normalisation unit could be a tonne of sugarcane processed or a tonne of sugar produced within a year.

For the qualitative indicators, lists of practice items have been identified to assist the evaluation. For example, regarding the indicator '*Social development and participation*', overall performance can be reflected by combining three items altogether, as shown below.

(a) The company significantly contributes to a better quality of life for the local community through supporting education, health/medical, recreation, and public infrastructure and facilities.
(b) The company has been recognised as one of the major contributors to local employment.
(c) The company employs indicators or methods to assess the image of the company in terms of social contributions and external perceptions.

As stated previously, assessment grades (evidence-based scale) should be then specifically established for each item, allowing assessors to select the option which best reflects actual situations in their company. Each grade is defined clearly by referring to objective evidence or feasible situations. The number of grades for each item is not necessarily equal and depends on how many distinct practices exist which are related to that item. The evidence-based scale helps to enhance standardisation and to avoid bias from intuitive grading. An example set of assessment grades is given below. The grades are used to assess the item (a) '*The company significantly contributes to a better quality of life for the local community through supporting education, health/medical, recreation, and public infrastructure and facilities*' which partly represents the indicator '*Social development and participation*'.

(a) There is no evidence of a budget and activities for supporting education, health/medical, recreation, and public infrastructure and facilities in the local community; and there is no evidence of plans to do this in the near future.
(b) The company is planning to do something that contributes to a better quality of life for the local community in the near future.
(c) Evidence shows that employees of the company have carried out some activities to support the education, health/medical, recreation, and public infrastructure and facilities of the local community by using their own resources or raising funds by themselves. No budget has been officially allocated by the company.
(d) Evidence shows that the company has officially made a contribution to a better quality of life for the local community. However, there is no evidence to show that the results and feedback have been followed up and reported.
(e) Evidence shows that the company has officially made a contribution to a better quality of life for the local community. Also, the operating results and feedback have been reported in management reviews.

From the defined set of grades, an improvement plan can be initiated according to the evidence required to achieve a better grade. The validity, reliability, and generalisability of the assessment items have been confirmed and discussed within Sureeyatanapas et al. (2015). Its applicability has been demonstrated using actual data from three sugar companies through the application of the ER approach. The details of the data collection process and the role of the ER algorithm in aggregating scores from various indicators are thoroughly described in Sureeyatanapas et al. (2014). The ER algorithm finally gives the aggregated interval degrees of belief for each criterion, each dimension (economic, environment, social, and quality), and the overall sustainability performance. They are then converted into the corresponding expected utilities to serve the purpose of comparison.

5.6 Conclusion

This paper presents an overview of sustainable manufacturing, which explains how a manufacturer can maintain its business and operations in long-term competition. Financial profits should still be the main focus, but a company should simultaneously consider external concerns regarding environment and social responsibility. This is a powerful way for a company to differentiate itself from its competitors while staying competitive. The combination of three principles: green manufacturing, corporate social responsibility, and green supply chain, is recommended for a manufacturer to initiate sustainable development. These principles need to be considered as an integrated business strategy with mutual collaboration by the entire organisation rather than having one team or division to assume responsibility. This also needs to be continuously driven by top management and agreed by relevant stakeholders. Since effective improvement needs to be made on the basis of reliable performance evaluation, the second half of the paper provides discussions about how criteria and indicators for the assessment of progress towards sustainable development can be established. The procedure for the evaluation is also suggested in a step-by-step fashion. The development and the assessment of sustainability performance within the Thai sugar industry are briefly discussed towards the end. Not only does this information provide a great contribution to sugar manufacturers, but practitioners in other industries can also apply these principles to draw up their own sustainable development policies and strategies, eventually enhancing overall sustainability performance.

It may be true that in many countries nowadays legislative pressure and the requirements of society might not be sufficient to tackle pressing issues. However, it is believed that once a manufacturer moves proactively to take environmental and social concerns into account and then initiates associated policies and conducts assessments, this will enable them to be in a much better position to encounter potential changes, as well as to maintain their competitive ability in the future as the market becomes increasingly competitive.

References

Akhoundi, A., & Nazif, S. (2018). Sustainability assessment of wastewater reuse alternatives using the evidential reasoning approach. *Journal of Cleaner Production, 195*, 1350–1376.

Aragon-Correa, J. A. (1998). Strategic proactivity and firm approach to the natural environment. *Academy of Management Journal, 41*, 556–567.

Azapagic, A. (2003). Systems approach to corporate sustainability: A general management framework. *Process Safety and Environmental Protection, 81*, 303–316.

Azapagic, A., & Perdan, S. (2000). Indicators of sustainable development for industry: A general framework. *Process Safety and Environmental Protection, 78*, 243–261.

Bhanot, N., Rao, P. V., & Deshmukh, S. G. (2015). Enablers and barriers of sustainable manufacturing: Results from a survey of researchers and industry professionals. *Procedia Cirp, 29*, 562–567.

Brent, A. C., & Visser, J. K. (2005). An environmental performance resource impact indicator for life cycle management in the manufacturing industry. *Journal of Cleaner Production, 13*, 557–565.

Chiarini, A. (2014). Strategies for developing an environmentally sustainable supply chain: Differences between manufacturing and service sectors. *Business Strategy and the Environment, 23*, 493–504.

Cooper, W. W., Seiford, L. M., & Tone, K. (2007). *Data envelopment analysis: A comprehensive text with models applications, reference and dea-solver software*. New York: Springer.

Crowther, D., & Aras, G. (2008). *Corporate social responsibility*. London: Ventus Publishing Aps.

Curkovic, S., Melnyk, S. A., Handfield, R. B., & Calantone, R. (2000). Investigating the linkage between total quality management and environmentally responsible manufacturing. *IEEE Transactions on Engineering Management, 47*, 444–464.

Davis, K. (1973). The case for and against business assumption of social responsibilities. *Academy of Management Journal, 16*, 312–322.

Despeisse, M., Oates, M. R., & Ball, P. D. (2013). Sustainable manufacturing tactics and cross-functional factory modelling. *Journal of Cleaner Production, 42*, 31–41.

Dheeraj, N., & Vishal, N. (2012). An overview of green supply chain management in India. *Research Journal of Recent Sciences, 1*, 77–82.

Dixon, R., Mousa, G. A., & Woodhead, A. (2005). The role of environmental initiatives in encouraging companies to engage in environmental reporting. *European Management Journal, 23*, 702–716.

Dornfeld, D. A. (2014). Moving towards green and sustainable manufacturing. *International Journal Of Precision Engineering and Manufacturing-Green Technology, 1*, 63–66.

Dyllick, T., & Hockerts, K. (2002). Beyond the business case for corporate sustainability. *Business Strategy and the Environment, 11*, 130–141.

Elkington, J. (1997). *Cannibals with forks: The triple bottom line of 21st century business*. Oxford: Capstone.

Figge, F., Hahn, T., Schaltegger, S., & Wagner, M. (2002). The sustainability balanced scorecard – linking sustainability management to business strategy. *Business Strategy and the Environment, 11*, 269–284.

Fombrun, C., & Shanley, M. (1990). What's in a name? Reputation building and corporate strategy. *Academy of Management Journal, 33*, 233–258.

Garbie, I. H. (2014). An analytical technique to model and assess sustainable development index in manufacturing enterprises. *International Journal of Production Research, 52*, 4876–4915.

Garetti, M., & Taisch, M. (2012). Sustainable manufacturing: Trends and research challenges. *Production Planning & Control, 23*, 83–104.

Ghazilla, R. A. R., Sakundarini, N., Abdul-Rashid, S. H., Ayub, N. S., Olugu, E. U., & Musa, S. N. (2015). Drivers and barriers analysis for green manufacturing practices in Malaysian Smes: A preliminary findings. *Procedia Cirp, 26*, 658–663.

Grolleau, G., Mzoughi, N., & Pekovic, S. (2013). Is business performance related to the adoption of quality and environmental-related standards? *Environmental and Resource Economics, 54*, 525–548.

Gunasekaran, A., & Spalanzani, A. (2012). Sustainability of manufacturing and services: Investigations for research and applications. *International Journal of Production Economics, 140*, 35–47.

Haibo, Z., Yi, Y., Yao, C., & Joe, Z. (2018). Data envelopment analysis application in sustainability: The origins, development and future directions. *European Journal of Operational Research, 264*, 1–16.

Hart, S. L. (1995). A natural-resource-based view of The firm. *The Academy of Management Review, 20*, 986–1014.

Hart, S. L., & Ahuja, G. (1996). Does it pay to be green? An empirical examination of The relationship between emission reduction and firm performance. *Business Strategy and the Environment, 5*, 30–37.

Henri, J. F., & Journeault, M. (2009). Eco-efficiency and organizational practices: An exploratory study of manufacturing firms. *Environment and Planning C: Government and Policy, 27*, 894–921.

Hubbard, G. (2009). Measuring organizational performance: Beyond the triple bottom line. *Business Strategy and the Environment, 18*, 177–191.

Isaksson, R. (2006). Total quality management for sustainable development: Process based system models. *Business Process Management Journal, 12*, 632–645.

Jayal, A. D., Badurdeen, F., Dillon, O. W., Jr., & Jawahir, I. S. (2010). Sustainable manufacturing: Modeling and optimization challenges at the product, process and system levels. *CIRP Journal of Manufacturing Science and Technology, 2*, 144–152.

Jonker, J. (2000). Organizations as responsible contributors to society: Linking quality, sustainability and accountability. *Total Quality Management, 11*, 741–746.

Joshi, K., Venkatachalam, A., & Jawahir, I. S. (2006). A new methodology for transforming 3R concept into 6R concept for improved product sustainability. In *4th Global Conference on Sustainable Product Development and Life Cycle Engineering*, Sao Paolo, Brazil.

Joung, C. B., Carrell, J., Sarkar, P., & Feng, S. C. (2013). Categorization of indicators for sustainable manufacturing. *Ecological Indicators, 24*, 148–157.

Kasim, A., & Ismail, A. (2012). Environmentally friendly practices among restaurants: Drivers and barriers to change. *Journal of Sustainable Tourism, 20*, 1–20.

Keeble, J. J., Topiol, S., & Berkeley, S. (2003). Using indicators to measure sustainability performance at a corporate and project level. *Journal of Business Ethics, 44*, 149–158.
Keeney, R. L., & Raiffa, H. (1976). *Decision with multiple objectives: Preferences and value tradeoffs*. New York: John Wiley & Sons.
Kolk, A., & Mauser, A. (2002). The evolution of environmental management: From stage models to performance evaluation. *Business Strategy and the Environment, 11*, 14–31.
Kotler, P., & Lee, N. (2005). *Corporate social responsibility: Doing the most good for your company and your cause*. Wiley: Hoboken.
Krajnc, D., & Glavic, P. (2005). How to compare companies on relevant dimensions of sustainability. *Ecological Economics, 55*, 551–563.
Kuei, C., & Lu, M. H. (2013). Integrating quality management principles into sustainability management. *Total Quality Management & Business Excellence, 24*, 62–78.
Kulatunga, A. K., Jayatilaka, P. R., & Jayawickrama, M. (2013). Drivers and barriers to implement sustainable manufacturing concepts in Sri Lankan Manufacturing Sector. In *The 11th Global Conference on Sustainable Manufacturing, 23rd-25th September 2013*, Berlin, Germany.
Labuschagne, C., Brent, A. C., & Van Erck, R. P. G. (2005). Assessing the sustainability performances of industries. *Journal of Cleaner Production, 13*, 373–385.
Lai, K. H., & Wong, C. W. Y. (2012). Green logistics management and performance: Some empirical evidence from Chinese manufacturing exporters. *Omega, 40*, 267–282.
Lau, K. H. (2011). Benchmarking green logistics performance with a composite index. *Benchmarking: An International Journal, 18*, 873–896.
Mittal, V. K., & Sangwan, K. S. (2014). Prioritizing drivers for green manufacturing: Environmental, social and economic perspectives. *Procedia Cirp, 15*, 135–140.
Molamohamadi, Z., & Ismail, N. (2013). Developing a new scheme for sustainable manufacturing. *International Journal Of Materials, Mechanics And Manufacturing, 1*, 1–5.
Moldan, B., & Dahl, A. L. (2007). Challenges to sustainability indicators. In T. Hák, B. Moldan, & A. L. Dahl (Eds.), *Sustainability indicators: A scientific assessment*. Scope: Washington.
Murphy, P. (1986). A preliminary study of transportation and warehousing aspects of reverse distribution. *Transportation Journal, 25*, 12–21.
Murphy, P. R., & Poist, R. F. (2003). Green perspectives and practices: A "comparative logistics" study. *Supply Chain Management: An International Journal, 8*, 122–131.
Narayanan, R. G., & Das, S. (2014). Sustainable and green manufacturing and materials design through computations. *Journal of Mechanical Engineering Science, 228*, 1581–1605.
Nawrocka, D. (2008). Environmental supply chain management, Iso 14001 and Rohs. How are small companies in The electronics sector managing? *Corporate Social Responsibility and Environmental Management, 15*, 349–360.
Nordin, N., Ashari, H., & Hassan, M. G. (2014). Drivers and barriers in sustainable manufacturing implementation in Malaysian manufacturing firms. In *2014 IEEE International Conference on Industrial Engineering and Engineering Management*, 9-12 December 2014, Bandar Sunway, Malaysia (pp. 687–691).
O'Brien, C. (1999). Sustainable production—A new paradigm for a new millennium. *International Journal of Production Economics, 60–61*, 1–7.
OECD. (2009). *Sustainable manufacturing and eco-innovation: Framework, practices and measurement* [Online]. Paris: Oecd Publishing. Retrieved December 17, 2014, from http://www.oecd.org/innovation/inno/43423689.pdf
OECD. (2011). *OECD sustainable manufacturing toolkit: Seven steps to environmental excellence* [Online]. Retrieved December 17, 2014, from http://www.Oecd.Org/Innovation/Green/Toolkit/
Panayiotou, N. A., Aravossis, K. G., & Moschou, P. (2009). A new methodology approach for measuring corporate social responsibility performance. *Water, Air, & Soil Pollution: Focus, 9*, 129–138.
Pazirandeh, A., & Jafari, H. (2013). Making sense of green logistics. *International Journal of Productivity and Performance Management, 62*, 889–904.

Pérez-Calderón, E., Milanés-Montero, P., & Ortega-Rossell, F. J. (2012). Environmental performance and firm value: Evidence from Dow Jones sustainability index Europe. *International Journal Of Environmental Research, 6*, 1007–1014.

Perrini, F., & Tencati, A. (2006). Sustainability and stakeholder management: The need for new corporate performance evaluation and reporting systems. *Business Strategy and the Environment, 15*, 296–308.

Porter, M. E., & Van Der Linde, C. (1995). Green and competitive: Ending the stalemate. *Harvard Business Review, 73*, 120–134.

Rahdari, A. H., & Rostamy, A. A. A. (2015). Designing a general set of sustainability indicators at the corporate level. *Journal of Cleaner Production, 108*, 757–771.

Ramanathan, R. (2006). Data envelopment analysis for weight derivation and aggregation in the analytic hierarchy process. *Computers and Operations Research, 33*, 1289–1307.

Ramanathan, R., Black, A., Nath, P., & Muyldermans, L. (2010). Impact of environmental regulations on innovation and performance in the UK industrial sector. *Management Decision, 48*, 1493–1513.

Ravi, V., Shankar, R., & Tiwari, M. K. (2005). Analyzing alternatives in reverse logistics for end-of-life computers: ANP and balanced scorecard approach. *Computers and Industrial Engineering, 48*, 327–356.

Rogers, D. S., & Tibben-Lembke, R. S. (1998). *Going backwards: Reverse logistics trends and practices*. Pittsburgh: Reverse Logistics Executive Council.

Rosen, M. A., & Kishawy, H. A. (2012). Sustainable manufacturing and design: Concepts, practices and needs. *Sustainability, 4*, 154–174.

Rusinko, C. A. (2005). Using quality management as a bridge to environmental sustainability in organizations. *SAM Advanced Management Journal, 70*, 54–60.

Rusinko, C. A. (2007). Green manufacturing: An evaluation of environmentally sustainable manufacturing practices and their impact on competitive outcomes. *IEEE Transactions on Engineering Management, 54*, 445–454.

Saaty, T. L. (1980). *The analytical hierarchy process*. New York: McGraw Hill.

Sangwan, K. S., Bhakar, V., & Digalwar, A. K. (2018). Sustainability assessment in manufacturing organizations: Development of assessment models. *Benchmarking: An International Journal, 25*, 994–1027.

Schneider, A., & Meins, E. (2012). Two dimensions of corporate sustainability assessment: Towards a comprehensive framework. *Business Strategy and the Environment, 21*, 211–222.

Shimshak, D. G., Lenard, M. L., & Klimberg, R. K. (2009). Incorporating quality into data envelopment analysis of nursing home performance: A case study. *Omega, 37*, 672–685.

Slaper, T. F., & Hall, T. J. (2011). The triple bottom line: What is it and how does it work? *Indiana Business Review, 86*, 4–8.

Srivastava, S. K. (2007). Green supply-chain management: A state-of the-art literature review. *International Journal of Management Reviews, 9*, 53–80.

Sureeyatanapas, P., & Tawwan, K. (2018). Environmental performance assessment using the evidential reasoning approach: The case of logistics service providers. *Matec Web of Conferences, 192*, 1–9.

Sureeyatanapas, P., Yang, J. B., & Bamford, D. (2014). Evaluation of corporate sustainability. *Frontiers of Engineering Management, 1*, 176–194.

Sureeyatanapas, P., Yang, J. B., & Bamford, D. (2015). The sweet spot in sustainability: A framework for corporate assessment in sugar manufacturing. *Production Planning & Control, 26*, 1128–1144.

Sureeyatanapas, P., Poophiukhok, P., & Pathumnakul, S. (2018). Green initiatives for logistics service providers: An investigation of antecedent factors and The contributions to corporate goals. *Journal of Cleaner Production, 191*, 1–14.

Székely, F., & Knirsch, M. (2005). Responsible leadership and corporate social responsibility: Metrics for sustainable performance. *European Management Journal, 23*, 628–647.

The European Commission. (2011). *Communication from the commission to The European Parliament, The Council, The European Economic And Social Committee And The Committee Of The Regions: A renewed EU strategy 2011–14 for corporate social responsibility* [Online]. Brussels. Retrieved December 21, 2014, from http://Ec.Europa.Eu/Enterprise/Policies/Sustainable-Business/Files/Csr/New-Csr/Act_En.Pdf

The U.S. Department Of Commerce. (2012). *How does commerce define sustainable manufacturing?*. [Online]. Retrieved July 28, 2012, from http://www.Trade.Gov/Competitiveness/Sustainablemanufacturing/How_Doc_Defines_Sm.Asp

Thies, C., Kieckhäfer, K., Spengler, T. S., & Sodhi, M. S. (2019). Operations research for sustainability assessment of products: A review. *European Journal of Operational Research, 274*, 1–21.

Tseng, M.-L., Divinagracia, L., & Divinagracia, R. (2009). Evaluating firm's sustainable production indicators in uncertainty. *Computers & Industrial Engineering, 57*, 1393–1403.

Turban, D. B., & Greening, D. W. (1997). Corporate social performance and organizational attractiveness to prospective employees. *Academy of Management Journal, 40*, 658–672.

Ugwu, O. O., Kumaraswamy, M. M., Wong, A., & Ng, S. T. (2006). Sustainability appraisal in infrastructure projects (Susaip) part 1. Development of indicators and computational methods. *Automation in Construction, 15*, 239–251.

Veleva, V., & Ellenbecker, M. (2001). Indicators of sustainable production: Framework and methodology. *Journal of Cleaner Production, 9*, 519–549.

Walker, H., Di Sisto, L., & Mcbain, D. (2008). Drivers and barriers to environmental supply chain management practices: Lessons from The public and private sectors. *Journal of Purchasing and Supply Management, 14*, 69–85.

Wang, Y. M., Yang, J. B., Xu, D. L., & Chin, K. S. (2006). The evidential reasoning approach for multiple attribute decision analysis using interval belief degrees. *European Journal of Operational Research, 175*, 35–66.

WCED. (1987). *Our common future*. New York: Oxford University Press.

Wiengarten, F., & Pagell, M. (2012). The importance of quality management for the success of environmental management initiatives. *International Journal of Production Economics, 140*, 407–415.

Winston, W. L. (2004). *Operations research: Applications and algorithms*. Toronto: Thomson Learning.

Xu, D. L., Yang, J. B., & Wang, Y. M. (2006). The evidential reasoning approach for multi-attribute decision analysis under interval uncertainty. *European Journal of Operational Research, 174*, 1914–1943.

Yang, J. B. (2001). Rule and utility based evidential reasoning approach for multiple attribute decision analysis under uncertainty. *European Journal of Operational Research, 131*, 31–61.

Yang, J. B., & Singh, M. G. (1994). An evidential reasoning approach for multiple attribute decision making with uncertainty. *IEEE Transactions on Systems, Man, and Cybernetics, 24*, 1–18.

Zhou, P., Ang, B. W., & Poh, K. L. (2008). A survey of data envelopment analysis in energy and environmental studies. *European Journal of Operational Research, 189*, 1–18.

Ziout, A., Azab, A., Altarazi, S., & Elmaraghy, W. H. (2013). Multi-criteria decision support for sustainability assessment of manufacturing system reuse. *CIRP Journal of Manufacturing Science and Technology, 6*, 59–69.

Chapter 6
Circular Economy and New Research Directions in Sustainability

Hwong-Wen Ma, Hsiu-Ching Shih, and Meng-I Liao

6.1 Introduction

6.1.1 Circular Economy and Sustainability

Circular economy presents a new practical approach to sustainable development. Circular economy refers to redesigning, reusing, recycling, remanufacturing, or redistributing waste products, or prolonging product life through maintenance, to increase the added value of products by different circular setups. This new development pattern is beneficial to boosting economic development, facilitating resource efficiency, and reducing environmental pollution. Hence, many countries such as EU countries have taken circular economy as a pathway to sustainability, and have further formulated targets and measures for circular economy. In addition to environmental and economic benefits, circular economy can also support job creation, counteracting social problems arising from unemployment, and promoting social sustainability. Circular economy has a strong potential to transform the society toward environmental, economic, and social sustainability.

In 2000, the United Nations (UN) proposed 8 Millennium Development Goals (MDGs) in the Millennium Summit, and the 189 UN member countries undertook to pursue the 8 MDGs. When the MDGs were due in 2015, only preliminary progress had been made and the world was still confronted with the challenges from climate change and other environmental issues. In the Earth Summit held in Rio de Janeiro (Rio20+) in 2012, the UN placed more emphasis on the impacts of climate change, as well as planetary boundaries and social equality. Eventually, the member countries resolved to use 17 Sustainable Development Goals (SDGs) as the critical directions to address sustainable development from 2016 to 2030. The 17 SDGs

H.-W. Ma (✉) · H.-C. Shih · M.-I. Liao
Graduate Institute of Environmental Engineering, National Taiwan University, Taipei, Taiwan
e-mail: hwma@ccms.ntu.edu.tw

C. Chen et al. (eds.), *Pursuing Sustainability*, International Series in Operations Research & Management Science 301,
https://doi.org/10.1007/978-3-030-58023-0_6

comprise 169 sub-goals. Among the sub-goals, SDG 12 (Responsible Consumption and Production) is closely associated with circular economy, because circular economy serves to improve resource efficiency, slows down the exploitation and consumption of resources, and thus accomplishes sustainable production and consumption. In addition, the development of circular economy has directly resulted in the progress of several SDGs globally, including SDGs 6, 7, 8, 11, 12, 14, and 15 (Schroeder et al. 2018). SDG6 (Clean Water and Sanitation) mainly focuses on sustainable management of water and sanitation. SDG7 (Affordable and Clean Energy) mainly focuses on access to energy sources, microbial anaerobic digestion, and utilization of methane gas generated by anaerobic digestion; SDG7 is highly promising in terms of the utilization of agricultural wastes. SDG8 (Decent Work and Economic Growth) is to build a sound and productive employment environment through economic growth. SDG11 (Sustainable Cities and Communities) focuses on the tolerance, security, resilience, and sustainability of cities and communities. SDG12 (Responsible Consumption and Production) is to ensure sustainable consumption and production. SDG14 (Life below Water) focuses on conservation and sustainable utilization of oceans and marine resources. SDG15 (Life on Land) is to protect, maintain, and promote sustainable use of terrestrial ecosystems, manage forest resources, combat desertification, stop and reverse land degradation, and prohibit the loss of biodiversity. Table 6.1 lists the actions that could be taken by circular economy to attain the SDGs, and the influence of such actions.

6.1.2 Concepts and Principles of Circular Economy

For a long time, mainstream economic development has been based on the linear economy pattern. In an industrial production and consumption system, the exploitation, manufacture, use, and scrapping of resources follow a linear mode across the whole process from cradle to grave. Under the linear economy pattern, many resources are used only once, thus losing their utility and value. In contrast, "circular economy" highlights optimized use and consumption of natural resources, or to be specific, innovates the traditional production-supply pattern and creates a new consumption pattern. Rather than waste reduction in a traditional sense, circular economy places more emphasis on innovative design in new aspects (including technological, organizational, and social innovation), so as to affect the value chain of an economic system (EC 2014). The ultimate goal of circular economy is to decouple global resource consumption and environmental impact from economic development.

By assimilating related concepts and principles, such as industrial ecology, cradle to cradle, and performance economy, circular economy changes the economic logic, create alternative routes of materials, and becomes an important and practical mechanism to pursue sustainable development goals (Saavedra et al. 2018; Kirchherr et al. 2017). Industrial ecology is concerned with the material and energy flows through the industrial system and the associated impact, and provides a holistic

Table 6.1 Actions by circular economy and their association with the SDGs (Schroeder et al. 2018)

SDG 6 Clean Water and Sanitation
Development of water recycling technology, reduction in irrigation water by precision agriculture, pipeline leak detection by smart water meter, monitoring the change in water quality, and tracking pollution sources by Internet of Things (IoT)
SDG 7 Affordable and Clean Energy
Incinerators as energy recycling centers, agricultural wastes as bioenergy, transforming organic matter into methane gas by wastewater treatment plants, encouraging use of renewable energy
SDG 8 Decent Work and Economic Growth
New employment opportunities and economic growth impetus are sourced from the following factors: business models of product servitization, circular logistics systems, circle-oriented product and manufacturing process design, associated network service development, R&D of recycling technology, marketing, and quality inspection and certification.
SDG 11 Sustainable Cities and Communities
Circular cities, sharing cities, effective material sorting and collection facilities, reuse markets, urban mines, and roof and park-style farmlands
SDG 12 Responsible Consumption and Production
Extended producer responsibility (EPR) in recycling, recyclable design, naturally decomposable material design, integration of arterial and venous industries, reduction in the use of nonrecyclable materials, reduction in industrial wastes and pollution, reduction in carbon footprints and water footprints in manufacturing processes, and integration and reuse of energy and resources by industrial symbiosis
SDG 14 Life below Water
Reducing and removing marine plastics, and reducing seawater eutrophication caused by nitrogen and phosphorus nutrients
SDG 15 Life on Land
The improvement in resource efficiency brings about reduction in forest exploitation, mine exploitation and related pollution, wetland exploitation, and soil loss and soil degradation arising from intensive agriculture.

framework for guiding the transformation of the industrial system to a sustainable basis. The central goal of IE is to move from a linear to a closed-loop production and consumption system (Lowe and Evans 1995). The cradle-to-cradle looks at the life cycle of a product or a system and looks for optimal ways of closing the loop of materials. With emphasis on the design phase, it serves a measurable form of circular economy at the product level and provides a detailed list of practices to navigate through higher levels of circularity for sustainable development (Ünal and Shao 2019). The performance economy represents a full shift to servitization, with revenue obtained from providing services rather than selling goods. Key elements of the performance economy are reuse and remanufacturing as well as innovative business models, to maintain the quality of stock and extend its service life by reducing material intensity (Stahel and Clift 2016).

The Ellen MacArthur Foundation (EMF), a British organization committed to promoting circular economy, formulates three principles for circular economy: (1) preserve and enhance natural capital by controlling finite stocks and balancing natural resource flows; namely, if resources are needed, the circular system can choose technologies or processes involving recycled or better resources; (2) promote the recycling of products, parts, and materials through the technical or biological cycle, to maximize utility and improve resource benefits; and (3) boost system efficacy, and reveal and remove externalities with negative impacts; namely, reduce the loss of the circular system and manage the related external factors (EMF 2015a). The characteristics of circular economy include zero waste design, stability of the circular system by diverse circular paths, use of renewable energy, systems thinking, and prices and other mechanisms that reflect real costs. Circular economy advocates that resources should maintain their highest value; resources should be used continuously in a more efficient manner. Raw materials, products, and wastes are the forms that resources assume in different stages of their life cycle. EMF (2012) proposes four principles (as shown in Table 6.2), to specify the stages in which businesses can create value and expound the implications of such value to businesses.

As shown in Fig. 6.1, the whole system of material flows comprises two circular subsystems, including the technical material circular subsystem and biological material circular subsystem. The system is integrated with the original linear economy, namely, throughout the whole life cycle of materials from raw material exploitation and manufacturing, parts manufacturing, product manufacturing, service and product provision, product use and consumption, energy recycling, and waste and landfill. For resource recycling in the traditional sense (3R, including Reduce, Reuse, and Recycle), most materials will be treated in three ways at the end of their life cycle: (1) recycled as raw materials; (2) heat recovery through incinerators; and (3) final landfill. In contrast, circular economy highlights other circulating methods.

For the technical material circular subsystem, the circulating methods to be added include renovation or remanufacturing, reuse or redistribution, and repair. For the innermost circulating method, the value of products or materials can be kept (including the use for its original purposes) through the least processes, namely, with minimized costs and environmental impact. For the biological material circular subsystem, most of the materials can be naturally decomposed and recycled in the environment. For material circulation in the traditional sense, materials return to natural soils, but the potential utilization value for materials would be nearly lost. The lower part of Fig. 6.1 shows other valuable circulating methods for biological

Table 6.2 Four principles for value creation (EMF 2012)

	Principles for value creation	Description
1	The power of the inner circle	A circulating path as short and compact as possible is beneficial for retaining the maximum product (resource) value and creating service value.
2	The power of circling longer	It is advisable to extend the duration and times of circulating, thus saving the energy and resources required for manufacturing new products or parts and creating service value.
3	The power of cascaded use	Waste can be reused across different industries, thus accomplishing industrial symbiosis and creating new value for resources.
4	The power of pure input	In the manufacturing process, nontoxic materials should be used, and composite materials should be avoided. Clean and simple raw materials are beneficial for retaining resource value.

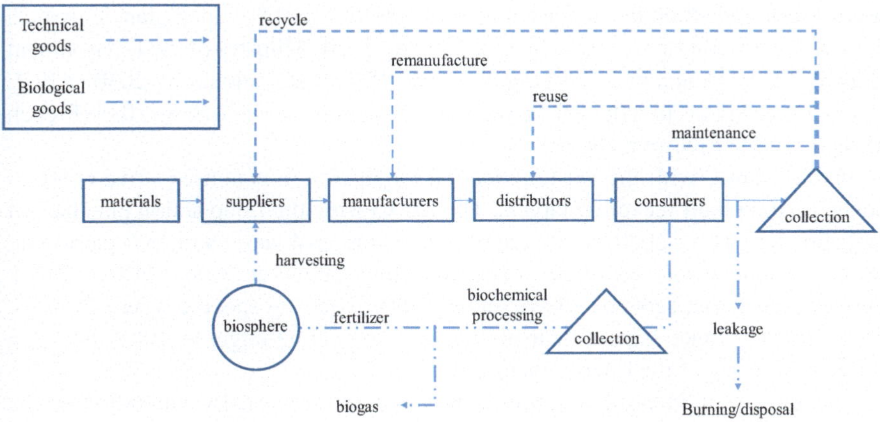

Fig. 6.1 A material flow model of circular economy. (Adapted from EMF 2012)

materials. Multilevel cascaded circulating methods can be developed. Some biological waste materials contain constituents with functional biochemical properties, so valuable biochemical materials could be extracted from them.

6.2 Literature Review

6.2.1 Goals and Strategies for Developing Circular Economy

According to the goals of circular economy set from top-down, many countries develop strategies that drive the transition of market operation to circular economy. The related goals and strategies on circular economy can be described in three aspects: target setting, governmental strategy, and market operation.

6.2.1.1 Target Setting

The management strategies related to circular economy have evolved through several stages, including waste management in the early stage, subsequent focus on sustainable material management, and the consummating circular economy in the recent stage. By extending the early goals of waste management to pursue maximization of the resource values of the overall system, many countries try to strengthen waste management with the purpose of increasing resource productivity. The goals have been expanded to consider their consistency with sustainability, and the strategies are designed to address the life cycle of products and materials.

Many countries have upgraded their waste management targets. Specifically, they not only reduce waste generation, but also restrict landfills of waste. The EU set its waste reduction target for landfills at 10% (EC 2015). Japan plans to reduce their waste generation to 25% of that in 2000 by 2020 (Ministry of the Environment (Japan) 2017). Denmark is working to reduce 15% waste amount by 2030 (ABCE 2017). France plans to reduce their wastes to 50% of those in 2010 by 2025 (French Ministry of the Environment 2018).

In accordance with the waste reduction targets, these countries strive continuously to attain their recycling targets. The EU Closing the Loop action plan has set recycling targets for 2030, which are 65% for municipal waste and 75% packaging waste. Denmark has set their waste recycling target at 80% (ABCE 2017). Furthermore, certain countries have set recycling targets for specific types of wastes. For example, France has set their plastic waste recycling target as 100% for 2025 (French Ministry of the Environment 2018).

The targets of increasing resource productivity are usually embodied in the improvement of national-level resource productivity or the reduction of raw materials consumption. For example, the short-term targets of China for circular economy include a 15% improvement in resource productivity by 2020 as compared to 2015 (NDRC of the People's Republic of China 2017). The transition targets of Denmark include a 40% improvement in resource productivity by 2030 (ABCE 2017). The use of raw materials is often included in material management targets for many countries. For example, the EU expects to reduce 20% of raw material consumption in the process of food production by 2020, and the Netherlands has set an intermediate target of reducing 50% consumption in mineral, fossil, and metal materials by 2030 (Ministry of Infrastructure and the Environment 2014).

As for circular economy targets, Denmark stood out to set direct targets. By 2035, (1) 75% of Danish industries and the service industry should take an active part in circular economy, to improve the added value by 15%; (2) 50% of the consumer demand should be met by sharing economy; and (3) the revenue from biolabeled products and services should quadruple to promote circular consumption (ABCE 2017). Although many other countries have not set direct targets, they have taken actions to improve product design in durability and recyclability, to foster the maintenance, reuse, and recycling markets, and to increase circular consumption behaviors.

6.2.1.2 Governmental Strategy

To accomplish the targets for circular economy, national-level strategies such as infrastructure reinforcement, industrial investment and taxation are carried out to impel industries to transform themselves toward circular economy. Infrastructure reinforcement involves treatment facilities for resource recycling. For example, UK's infrastructure reinforcement involves the collection, sorting, and repair or retreatment of products and materials, reducing the costs of secondary materials. Infrastructure reinforcement also involves urban or regional construction for behavioral change on the demand side. In China, the government coordinates the overall planning of urban production, citizen living, nature conservation, waste treatment, and green infrastructure reinforcement. To attain the circular economy targets for 2050, India advocates the development of smart cities and industrial corridors, speeding up the construction of urban infrastructure based on the principles for circular economy and preventing enduring inefficient resource systems (EMF 2016a).

Increasing the investment and funds for circular economy is also an important measure taken by many countries and related organizations. In conjunction with the European Commission, the European Investment Bank, and financial market participants and businesses, the EU is building and optimizing a financial support platform for circular economy, to increase the attraction of circular economy projects to investors (EC 2017). The Netherlands is building a turnover capital platform for circular economy (Ministry of Infrastructure and the Environment 2014). Japan has set up the Coordination Funds for Active Environmental Research, to finance research on circular economy and R&D of related technologies and equipment. South Korea attracts private investment by government intervention, by means of green finance products (Jin 2016).

Taxation includes tax levies and breaks. Tax levies are mostly used to facilitate government supervision, whereas tax breaks are used to invigorate market operation. The EU increases levies on landfill and incineration taxes to reduce waste disposal, levies on unrecyclable commodities and products containing toxic substances to reduce their usage, and set tax rates according to the recycling degree of products to promote the reuse of parts of second-hand goods. The Netherlands plans to levy value extracted tax (VET) instead of value added tax (VAT), which levies according to the type and quantity of brand-new raw materials used, and in turn encourages the use of renewable raw materials. In addition, the Netherlands sets differentiated tariffs (DIFTAR) for waste treatment according to the type of waste. The differentiated tariffs mechanism is intended to create incentives for businesses and individuals to reduce the generation of high-rate wastes for the purpose of tax saving. The UK encourages the use of durable and recyclable products and materials by levying tax on raw materials and resources, and landfill taxes as an economic incentive tool to control the quantity of waste disposal.

Government control measures include the following: (1) preventing the generation of waste; (2) reducing the quantity of waste entering landfills; (3) perfecting resource recycling systems; and (4) strengthening the EPR. These are waste management measures often taken by countries.

6.2.1.3 Operation Mechanism of Circular Economy

Among the circular economy policy viewpoints of various countries, it has been found that platform building and procurement strategies serve to strengthen the supply capacity and demand motives. Platform building is beneficial for (1) expanding sources of funding, (2) diversifying material supply, and (3) promoting the exchange and sharing of technical information; it is intended to bolster the supply-side capacity. For example, the EU plans to enhance financial support platforms for circular economy (EC 2017), and the Netherlands has designed a funding turnover platform

(Ministry of Infrastructure and the Environment 2014). Diverse material supply simultaneously includes diverse unused resources (including material sources, machinery, and services) and demands on the same platform. For example, the EU suggests that platforms be built for second-hand goods sale, sharing, and provision of maintenance and repair services. The Netherlands's packaging innovation system will build a packaging center and report platform for interactions between producers and consumers. In addition, the UK's circular economy strategy also suggests that a free information platform be applied to provide small and medium enterprises with the skills and tools for improving material use efficiency; this system will enable small and medium enterprises to know the quantity of waste they generate and the effect of such waste on their profits, as well as provide methods for waste reduction and cost saving (WRAP 2013). The UK also provides a platform to match demanders and suppliers, wherein consumers purchase the right to use rather than the ownership of commodities, namely, substituting purchasing with leasing; compared with the traditional purchase of commodities, this serves to maximize the utilization of commodities and reduce the possibility of disposal. Exchange and sharing of technical information can be seen on the EU platform built for circular economy stakeholder engagement, allowing stakeholders to brainstorm and conceive the objectives and execution contents of the platform; the platform currently presents related cases, dialogs between stakeholders, and related knowledge and strategies (EC 2017). Various communities in the UK also exchange information on resource efficiency through platforms to conduct technical exchange and cooperation. For example, the EMF has set up the Circular Economy 100 (CE100) platform committed to facilitate cooperation between leading businesses, local governments, and advanced technologies; it offers three types of support: (1) building a problem-solving mechanism; (2) establishing a database for optimal operation guides promptly accessible to businesses; and (3) providing a mechanism for businesses to ponder the development of circular economy, in order to speed up the transition.

In another aspect, procurement strategies can strengthen the motives of the supply side to adjust in response to demand. Procurement strategies can be roughly classified into mode and commodity approaches. Mode approaches include sharing economy and product-as-a-service, while commodity approaches imply setting restrictions on commodity properties, such as green-labeled or recyclable. The EU's main strategies include product-service system and green public procurement, and has issued a new green public procurement directive in order to oversee its member countries to fulfill the target of a 50% green public procurement rate, thus promoting the market development of circular economy through the enormous government procurement power. In response to the foresaid procurement choices, there are three schemes for executing product-design requirements in the EU's existing regulations: (1) reinforce the existing Eco-design Directive, and ultimately extend it to non–energy-related products; (2) guide commodity design through the Waste Framework Directive and Packaging and Packaging Waste Directive; and (3) integrate and simplify issues on circular economy through existing technologies and regulations, such as green labeling, green government procurement, eco-design, and energy labeling. The EU needs to establish mechanisms on eco-labeling and green public procurement to further encourage the use of recycled goods, and develop proper certification methods at the same time (EEB 2015). The UK Department for Environment, Food and Rural Affairs promulgates the decree of Government Buying Standards (GBS) to regulate the procurement of various commodities (including furniture, electric and electronic appliances, and transport vehicles) by central government agencies. The GBS considers the circulating concept by design and highlights the reparability, upgradeability, and recyclability of products; it intends to make government agencies play an exemplary role and promote the market development of circular economy through their procurement power.

6.2.2 Business Models for Transition Toward Circular Economy

From the perspective of supply chain management, circular economy promotes the sustainability of supply chains with the creation of self-sustaining production systems from revaluing materials (Genovese et al. 2017; Geissdoerfer et al. 2019). In addition, innovative business models, the key enablers for circular economy, are crucial to the success of business operation, and will contribute greatly to the transition toward green growth as well as strengthen the competitiveness of enterprises and the supply chain (OECD 2019). Existing business models cannot satisfy the requirements of circular economy because most enterprises create economic value based on a linear economy thinking.

EMF (2013) defines the circular business model as an important tool to promote the core concept of circular economy at a micro level, namely, establishing a renewable closed system; the greatest difference from those of linear economy is that the latter lacks the thought of reverse logistics returns products to the manufacturing processes or markets. Many international think tanks, research institutes, and consulting firms, such as Accenture, EMF, Forum for the Future, Circle Economy, and BSI, have proposed various types of business models for circular economy (Pieroni et al. 2019). The classification and presentation by Accenture are clear and lucid; explanations accompanied by the life cycle chart (Fig. 6.2) enable enterprises to contemplate more intuitively strategies and business models applicable to various stages in life cycles. Therefore, many international organizations have adopted Accenture's show-how to describe business models of circular economy.

The typical life cycle includes seven stages: product design and material procurement, manufacturing, logistics, sales and marketing, product use, end of life, and reverse logistics. By analyzing 120 successful commercial cases of successful improvement in resource productivity, Accenture sums up five categories of circular business models: (1) circular supplies: future circularity of the whole system should be considered at the very origin of the life cycle (product design and raw material procurement), to ensure healthy circulation in the whole supply chain; (2) resource recovery: the possibilities of economic value re-creation should be considered of

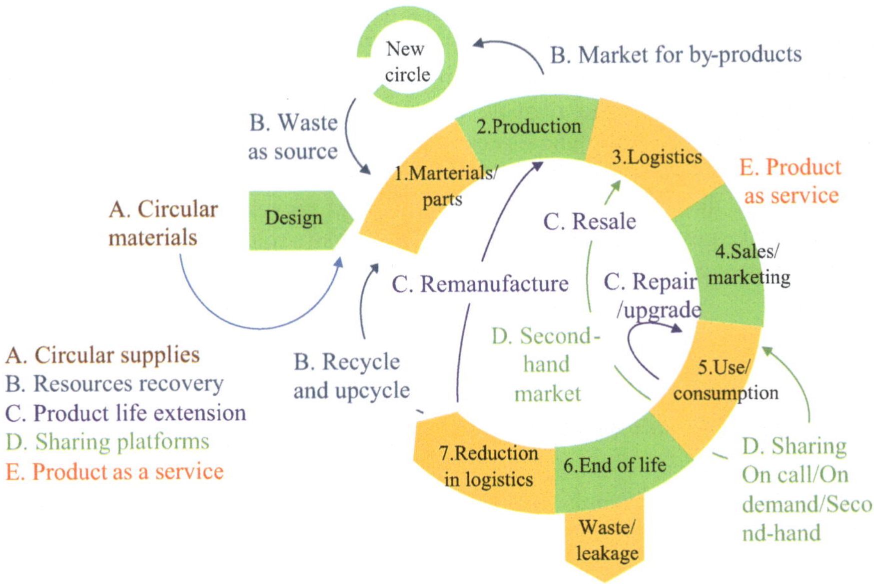

Fig. 6.2 Types of business models for circular economy (Accenture 2014)

residues from manufacturing processes and wastes from final consumption, to return discarded resources into circulation; (3) product life extension: to prolong service life of products through maintenance service, resale, and remanufacturing; (4) sharing platforms: information and transaction platforms for improving utilization rates of various idle assets at the use stage; and (5) product as a service: providing service of using a product rather than selling the product itself through leasing or pay-per-use mechanism. In addition, Accenture emphasizes that popularization of digital, physical, and biological technologies could promote circular economy if properly used.

EMF and McKinsey proposed the ReSOLVE as a strategic architecture for enterprises to develop circular business models (EMF 2013). ReSOLVE stands for Regenerate, Share, Optimize, Loop, Virtualize, and Exchange. The business models proposed by Forum for the Future mainly fall into two categories, one coined "circular" and the other "enabling." A circular business model aims at improving circularity of the overall commercial system through disruptive innovation as well as feasibility and popularity of product reuse through mechanisms and commercial design, mainly examining business opportunities of reverse logistics through product life cycle; an enabling business model refers to one that serves to promote or reinforce a circular business model. With the Value Hill framework, Circle Economy (2016) explains how to rethink and create value in a circular economy throughout three major stages in terms of product life cycle: (1) creating value, (2) maintaining value, and (3) retaining value.

The British Standard Institute (BSI) released BS 8001:2017 Framework for implementing the principles of the circular economy in organizations—Guide. This document proposed six circular business models, including on-demand production, dematerialization, product life cycle extension/reuse, recovery of secondary raw materials/by-products, product as a service/Product Service System, as well as sharing economy and collaborative consumption (BSI 2017).

6.3 Assessment Tools

6.3.1 Material Flow Analysis and Related Analytical Approaches

Material flow analysis (MFA) is a systematic assessment method used in environmental management to evaluate the flow and stock of materials within a certain spatiotemporal range (Brunner and Rechberger 2004; Wang and Ma 2018; Lavers et al. 2019). Materials can refer to substances or goods. In chemistry, substances are defined as types of matter that comprise uniform homogeneous units. If such units are atoms, the substances will be referred to as elements (e.g., C and Fe); if such units are molecules, the substances will be referred to as chemical compounds (e.g., CO_2 and $FeCl_2$). Goods refer to valued entities that comprise one or multiple substances. Some materials possess positive value (e.g., automobiles, fuels, and wood), while some possess negative value (e.g., wastes and sludge). A material flow system comprises flow, process, and stock; MFA is primarily based on the law of mass conservation, meaning that within a system, the mass remains unchanged and substances cannot be created or consumed with the lapse of time. Often used as a decision support tool in resource management, waste management, and environmental management, MFA can present complete and consistent information about the sources, pathways, intermediates, as well as final sink of the target substances.

The steps of MFA are shown in Fig. 6.3 (Baccini and Brunner 2012), mainly including problem definition, system definition (determination of system boundaries and selection of substances, goods and processes), determination of flow and stock (measurement of material flows, balancing of goods, inventory and calculation of concentrations, and balancing of substances), and result interpretation. Note that these steps need not be executed continuously in a strict manner, but should be optimized repeatedly so that they can meet the research objectives. Generally speaking, a desired practice is to first estimate data roughly and then improve the system and related data until the data quality meets a certain standard. MFA is an iterative procedure, and the selected components (substances, processes, goods, and boundaries) and exact data need to be repeatedly confirmed with the targets.

The goods and material concentration data for MFA can be empirically measured, and the estimates of this part depend on the available financial resources. For a large system or a long time period, this type of assessment may incur high costs. Therefore, flows, stocks and concentrations are preferably measured within a smaller system, such a waste treatment plant, a company, a farm, or a single

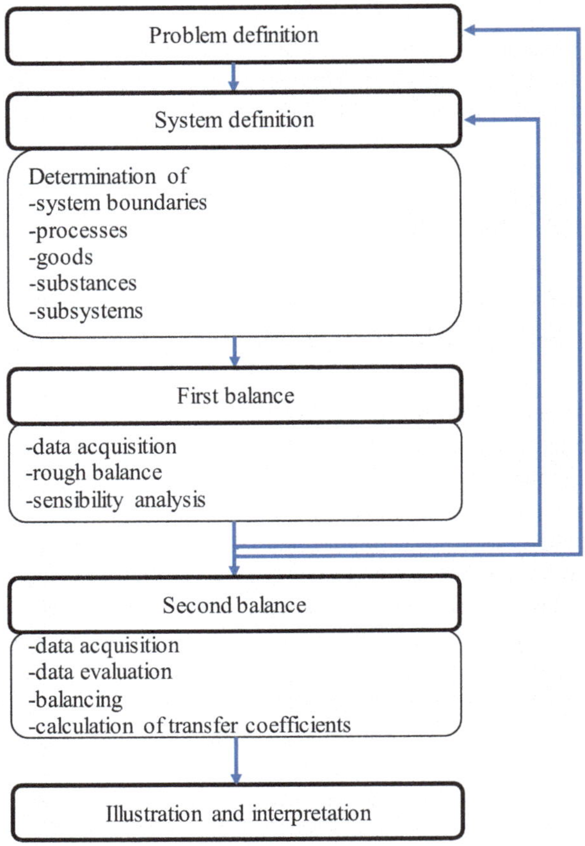

Fig. 6.3 Processes of material flow analysis (Baccini and Brunner 2012)

household. Such research entails an intensive and time-consuming inventory and calculation procedure. Notwithstanding recent studies aimed at improving inventory and measurement methods in MFA, there have been very few systems with balance error between inflow and outflow less than 10% of the total flow.

MFA has become an indispensable tool for research on transition paths toward circular economy at a national, urban, or corporate level. In such applications, MFA performs the following functions:

1. Understanding by inventory overall material flows of the observation objects (e.g., overall national resources, single industry, or company), as well as system status, in order to find out key factors and map out transition paths.
2. Pinpointing key locations of material losses within the target system. Locations with severe material losses are also hotspots that decision makers can aim at to plan and implement appropriate management measures (e.g., optimize manufacturing processes and formulate alternative schemes).
3. Recognizing distinctly sources of resource flows and the units to which the resources belong. This serves to obtain a bird's eye view of the system, and

connect related departments to maximize reuse rates (e.g., the wastes of one sector may become raw materials of another) and improve circularities of resources.
4. Identifying driving factors of material flow directions, in order to improve the overall utilization of resources starting from the design stage of circular economy.
5. Assessing the current extent of circular utilization of the target system through inventory results which can be used as data source for evaluating circular economy indicators.
6. Supporting evaluation of the economic value of an enterprise's material flows.

In addition to analysis of material flows in the anthroposphere and the environment, the evaluation of the impact derived from the flows is also needed to formulate management strategies. From the product life cycle perspective, input of energy and resources as well as generation of related pollutants at the raw material extraction, product manufacturing, consumption, circular utilization, and final disposal stages are all focal points of follow-up tracking and assessment. Life cycle assessment (LCA) is an objective process to assess the environmental load of products, manufacturing processes, activities or policies throughout the whole life cycle, from cradle to grave; it identifies and evaluates usage of energy and resource as well as emissions to the environment, and assesses in turn opportunities for further environmental improvement (Curran 1996; Haupt and Zschokke 2017; Ingrao et al. 2018). LCA comprises four steps: goal and scope definition, life cycle inventory, life cycle impact assessment, and interpretation; as a common method for evaluating environmental performance, LCA is especially suitable for assessing circular business models because circular economy calls for a systematic view (WBCSD 2016).

Cost–benefit analysis is also a common tool for research on circular economy (Lacovidou et al. 2017; Gigli et al. 2019; Huysveld et al. 2019). In a systematic manner, this tool can present the economic, social, and environmental impacts, within a specified time period and in terms of monetized costs and benefits, from the practices advocated by the circular economy concept. The intent is to select the best solution with maximized benefits and minimized costs. However, the disadvantage of this tool lies in the monetization process. It is not easy to convert impacts on various aspects into money as the sole unit of measurement; in addition, the element of time must be considered by incorporating the discounting factor into the monetization process. To promote the development of circular economy on a global scale, it is nevertheless necessary to motivate enterprises and even the whole society to transform from linear economy into circular economy by giving prominence to the higher cost efficiency and sustainability of the latter; it is thus necessary to rely on cost–benefit analysis to present the outlook for circular economy transition in a monetized manner. Like all public policies, governments or enterprises will invest only in projects with costs-beating benefits or minimal costs. In the process of transition toward circular economy, enterprises are concerned about not only environmental sustainability, but also operational costs and profits; Material Flow Cost Accounting (MFCA) is hence commonly used for analyses on business. By integrating data on material flows and costs, MFCA is capable of depicting clearly the physical and cost flows in manufacturing processes, and enables enterprises to attach equal importance to corporate development, resource efficiency and ecological environment; flow directions of raw materials and energy can be firmly grasped, revealing raw material losses and material treatment costs that are easily ignored.

Based on material flow analysis, relevant assessment indicators can be used to evaluate the effectiveness of circular transition, such as the Emergy indicators that summarize the flows in a common basis (Geng et al. 2013). The goal of developing a circular economy is to facilitate the decoupling of economic development from resource consumption. Resource productivity, defined as "GDP/natural resource consumption," is an important indicator for evaluating such decoupling and applied by governments to assessing the implementation of sustainable materials management. For businesses, effectiveness of the transition to a circular model can be evaluated using circularity indicators (EMF 2015b); the products of businesses can be assessed based on each stage of their life cycle, and scored in terms of their circularity. This assessment approach can be used with risk indicators to consider additional risks, such as material prices fluctuation, supply chain risk, and toxicity.

6.3.2 Evaluation and Decision-Making Processes

International organizations advocating circular economy have put forth business models with categorization and generalization approaches varied with their respective emphases (Accenture 2014; EMF 2015c; Circle Economy 2016; Forum for the Future 2016; BSI 2017). Actually, business models are merely one of the essential procedures within the transition course toward circular economy, be it the five steps proposed by 2Cbizz in 2013 to guide conversion of linear economy issues into business opportunities for circular economy, the five steps proposed by WBCSD in 2016 to facilitate an action plan on circular economy (WBCSD 2016), or the 8-step process of transition proposed by BSI in 2017, wherein the circular transition path is described with a complete flowchart in accordance with project management thinking (BSI 2017). For these processes it is recommended that diverse assessment tools should be used in different procedures to explain quantitatively the circular economy transition path.

Take BS 8001:2017 proposed by BSI in June 2017 as an example. It provides a guide to the flexible framework for transition toward circular economy, and is the first standard for circular economy in the world. To build a relatively complete transition framework, it attempts to incorporate the steps of project management thinking, to integrate the concepts and analysis tools regarding circular economy. The primary objective of BS 8001 is to clearly define and unify the concepts, terms, scopes, and business models regarding circular economy, and develop a common language through unified standards and definitions, in order to accelerate cooperation and development of circular economy. To help enterprises transition from business operation modes of linear economy into those of circular economy, the guide offers principles and execution strategies for enterprises irrespective of scale, geographical location, or product or service form, and proposes fairly complete circular business models. BS 8001:2017 Framework for implementing the principles of the circular economy in organizations—Guide proposed a flexible framework for circular transition, including eight steps guiding enterprises to transition toward circular business models. "Flexible" means that enterprises can decide by themselves which step to begin with according to their own maturity of circulation. The eight steps are sequentially enumerated only for the purpose of providing a clear descriptive framework; in practice, depending on their maturity of circulation, enterprises may iterate and/or amend certain steps. Specifically, the framework comprises the following eight steps: (1) framing, (2) scoping, (3) idea generation, (4) feasibility evaluation, (5) business case development, (6) piloting, (7) implementation, and (8) review and reporting. At each step, BS 8001 provides a check function, which serves as a reminder that the outputs of each step should comply with the six principles of circular economy and the core idea of circular business model, and that progression to the next step requires approval or authorization by superior executives. The objective and output for each of the eight steps are summarized as shown in Table 6.3 (Hung 2018).

Centering on ideas of the transition path toward circular economy, the foresaid eight steps can be generalized into three stages (as shown in Fig. 6.4):

1. Explore ideas: how to identify problems and find their solutions? (how to explore the business opportunities of circular economy).

   In face of the innovation as well as systems thinking and operational changes necessitated by circular economy, the primary challenges come from the three earliest steps of transition: framing, scoping, and idea generation. These steps deal with how an enterprise reexamines itself, evaluates the current status, finds problems, and defines opportunities, when confronted with innovative issues, knowledge systems, or domains. To satisfy the needs of this exploratory stage, qualitative methods can be applied such as checklist or rating table, stakeholder analysis, brainstorming, expert consultation, and case studies, while the current extent of circulation can be quantified by applying MFA, MFCA, and circularity indicators.
2. Formulate ideas: how to evaluate which idea should be adopted? (how to select a circular idea).

   After exploring various ideas, feasible ones can be formulated. At this stage, idea feasibility and business cases can be described qualitatively by applying circular business model or prototyping. In addition, the economic, environmental, and social impacts of feasible ideas or technologies can be quantified by applying LCA or its extension, cost–benefit analysis, and health risk assessment.
3. Implement ideas: how to evaluate the effectiveness of an idea? (how to evaluate sustainability performance).

For the stages of piloting, implementation and review, the circular economy performance indicators at the corporate level are yet to be developed and defined; however, the economic, environmental, and social impacts after transition can still be explored through LCA and its extension. Results thereof can serve as negotiation

Table 6.3 Steps of transition toward circular economy in BS 8001 (BS 8001, 2017; Hung 2018)

Step	Objective	Output
(1) Framing	Understand the relationship between the enterprise and circular economy, to decide where to begin with	1. Identify key resources
		2. Draw a stakeholder relationship diagram
		3. Create communication documents on circular economy
(2) Scoping	Understand the potential opportunities and requirements for circular economy, set the boundaries, establish visions, and formulate strategies	1. Create a system diagram
		2. Create a circulation vision, strategies, and a roadmap
		3. Form a project team
(3) Idea generation	According to the problems or opportunities identified at Step (2), make a list of ideas and sort them according to the circulation vision, targets, and strategies	1. Generate a list of prioritized ideas
(4) Feasibility evaluation	Evaluate the feasibility of the ideas generated at Step (3)	1. Prepare a feasibility evaluation report
		2. Confirm ideas
(5) Business case development	Develop business cases to ensure the availability of resources for piloting, and then implement, scale up, and officially launch the business model	1. Construct detailed and complete business cases
(6) Piloting and prototyping	Verify feasibility of the idea by piloting small-scale tests	1. Produce a Piloting plan and obtain piloting results
(7) Delivery and implementation	Scale up or officially launch the proven transition method	1. Indicate follow-up review indicators
(8) Review and reporting	Build a follow-up mechanism to ensure smooth implementation and continued circular transition	1. Control report
		2. Regular progress report

information regarding transition difficulties and market acceptance consultation, thus facilitating transition toward circular economy.

It is advisable that different evaluation tools be applied to reach varying goals of the above stages; quantitative evaluation tools mainly include MFA, LCA, cost–benefit analysis, and health risk assessment.

6.4 Applications

To demonstrate the applications of the framework presented in earlier sections, a case study of the transition paths of polypropylene in the automotive industry toward circular economy is described here. As a lightweight, multifunctional, and

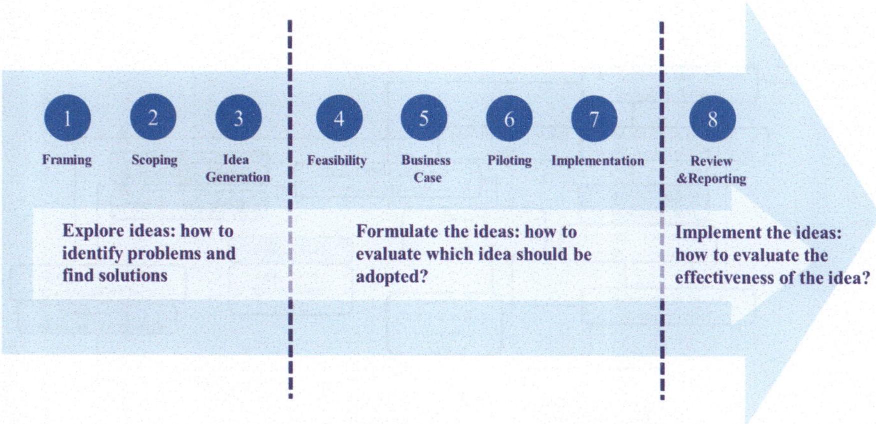

Fig. 6.4 Eight steps at three stages of transformation toward circular economy (Hung 2018)

durable material, plastic plays an important role in the development toward sustainability and resource efficiency. It acts as a critical material in the packaging, building materials, automotive, and renewable energy industries. Despite its various advantages, the drawbacks of today's plastics economy are becoming increasingly prominent. Take plastic packaging materials as an example. Economic losses due to one-time use amount to nearly 95% of the value of plastic packaging materials, equivalent to US$8–12 billion annually. Only 5% of the plastic is recycled and reused (EMF 2016b), most of which ending up in low-value products. To effectively improve the circularity of plastic, strategic solutions should be devised from the product life cycle perspective.

Plastic comes in many different types, among which polypropylene (PP) enjoys the largest demand. According to the survey of the Plastics Europe Market Research Group in 2015, the global demand for PP accounted for 23% of all plastics, surpassing that for PVC, the second most popular plastic, by 7% (PEMRG 2016). This section illustrates the transition process mentioned above by discussing the transition paths of PP in the automotive industry toward circular economy.

6.4.1 Step 1 Framing

The reuse of PP is crucial to the reuse of automotive plastics because automotive PP accounts for around 25–50% of all automotive plastics (Satoru et al. 2010; European Commission DG ENV 2011; Plastics market watch 2016). The PP used in each vehicle weighs approximately 36 kg (American Chemistry Council 2019). PP is mainly used in making car bumpers (including the side moldings on the car body) and lead acid battery cases, which respectively take up 23% and 1% of the total weight of automotive plastics (Nyamekie 2012). Due to its hydrophobicity, large

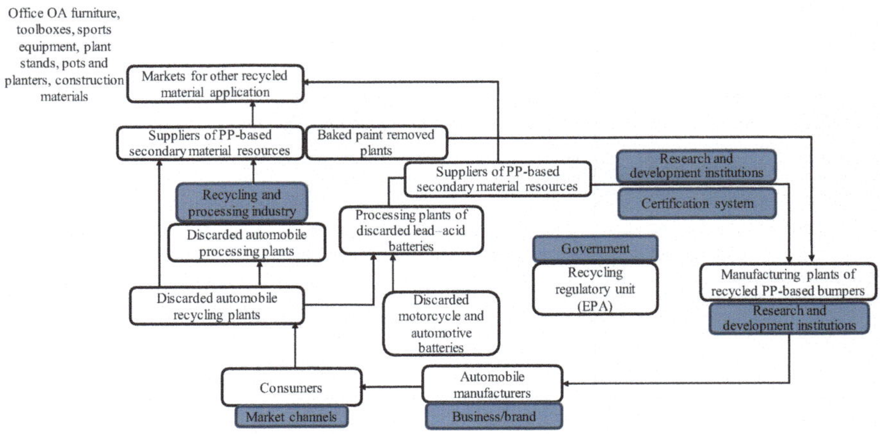

Fig. 6.5 The stakeholder ecosystem map

molecular mass, and lack of functional group, PP cannot be easily decomposed and is categorized as a nonbiodegradable plastic; hence, waste PP tends to accumulate in the natural environment, leading to negative environmental impacts (Tokiwa et al. 2009; Arkatkar et al. 2009).

Data on PP usage in the USA and Canada revealed that the amount of PP used in vehicles only increased slightly during the period of 2006–2016, but the proportion of PP in the automobile showed an upward trend; this indicates that the relative importance of PP has gradually increased compared with other types of automotive plastics.

Figure 6.5 summarizes the stakeholders related to PP-based car bumpers in the automotive industry. Besides the automotive industry and consumers that use PP-based bumpers, recycling of PP also involves discarded automobile processing plants, lead acid battery processing plants, and manufacturing plants of recycled PP-based bumpers. The government should play active roles in, to name a few, supervising the industry and promoting a certification system, to facilitate subsequent formulation of communication documents and bring synergy along the path toward circular economy.

6.4.2 Step 2 Scoping

The main approaches for processing waste PP include sanitary landfills, incineration (energy recovery), and resource recycling. With respect to sanitary landfills, although PP is not susceptible to biodegradation and hence does not emit greenhouse gases, landfills spoil the visual landscape, occupy land resources, and unnecessarily squander PP material resources; hence, they are not the optimal method for long-term waste management. Meanwhile, incineration can recycle the chemical energy in plastics, and incinerators that meet the waste emission and treatment standards, in most cases, exert minimal impact on the environment; however, this approach still emits a large amount of carbon dioxide and spends PP resources for disproportionate return (European Commission DG ENV 2011). Lastly, resource recycling also results in environmental impacts such as greenhouse gas emission and water pollution, as energy input is necessary for both transportation and processing; however, in comparison with landfilling and incineration, its environmental impact remains relatively mild. The overall environmental impact associated with recycled PP is 10–89% less severe than that caused by virgin PP. For instance, recycled PP incurs 50% less CO_2 emission, 65% less PO_4 emission, 29% less water consumption, and 78% less petroleum consumption than virgin PP (Yin et al. 2016). Based on the consideration of overall environmental impact, recycling waste PP does less harm; in addition, the economic value of recycled PP is retained by mixing it into car bumpers. Therefore, as an illustration, the system boundaries only consider the possible impact and benefits associated with the recycling of PP-based bumpers, as shown in Fig. 6.6.

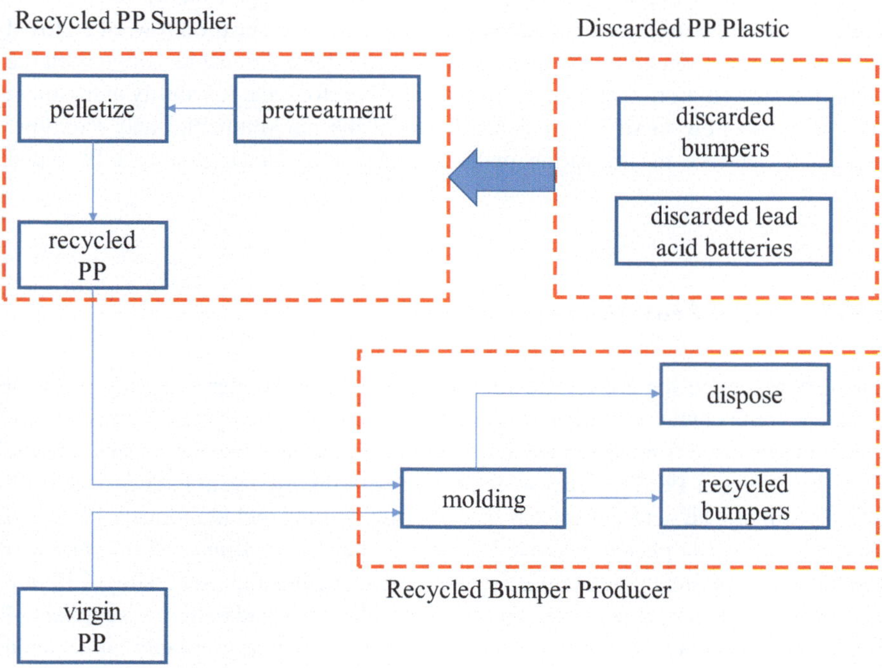

Fig. 6.6 The system boundary of PP-based car bumpers recycling

6.4.3 Step 3 Idea Generation

Of the five types of circular business models described previously, the value of recycled PP-based bumpers is created through practices of reverse logistics such as resource recovery, product life extension, and sharing platforms. At present, resource recovery is the most common business model. The circular economy of PP-based bumpers also adopts this prevalent business model, which contemplates the possibility of recreating economic value of recycling of discarded lead acid battery casings and scrapped bumpers to produce remade bumpers. Remade bumpers may be manufactured from discarded lead acid batteries, whose casings provide the waste PP material to be mixed with virgin PP for manufacturing recycled PP. As for the proportion of recycled materials in remade bumpers, interviews with manufacturers indicated that the maximum percentage is set as 30% to ensure quality of the latter.

In regard to remanufacturing scrapped bumpers into remade ones, the technical difficulties of the process lie in separating the baked paint and coatings. The separation technologies have been matured and applied to practical manufacturing process of remade bumpers. Due to poorer mechanical strength of waste PP, addition of virgin PP during the recycling process is also necessary.

Whereas the value chain of remade bumpers using discarded lead acid batteries as feedstock is essentially an open loop, that using scrapped bumpers would be more consistent with the recent appeal for enterprises to transition toward a closed loop. Such transformation may change the stakeholder relationships along with the economic and environmental impact involved. Therefore, the feasibility assessment will stress on how to strike a balance between the environmental and economic aspects to achieve the transition paths toward a circular economy of PP-based bumpers.

6.4.4 Step 4 Feasibility Assessment

The differences in the above transition paths toward a circular economy result in different value chains and impacts. Figure 6.7 shows the life cycle of remade bumpers manufactured from discarded lead acid battery casings, the arrows denoting the flow directions of PP. The main material inputs in the life cycle include virgin PP and discarded lead acid battery casings, and the outputs include remade PP-based bumpers and waste plastic generated during the machine milling and pressure and torsion testing in the manufacturing plants. The recycling and processing of PP are mainly undertaken by processing plants of discarded lead acid batteries, suppliers of PP-based secondary material resources, and manufacturing plants of remade PP-based bumpers. The PP casings of lead acid batteries are first separated and stored temporarily by processing plants. Afterward, suppliers of PP-based secondary material resources acquire the casings from processing plants for further processing, which includes pulverization, dehydration, drying, and granulation into

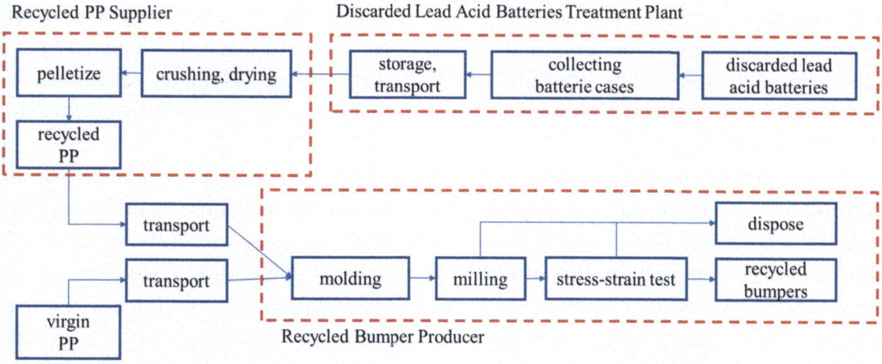

Fig. 6.7 Life cycle of the manufacture of recycled bumpers using discarded lead acid battery casings as raw materials

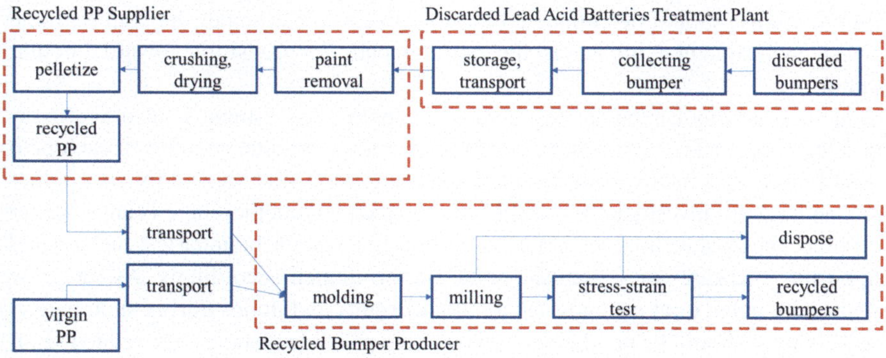

Fig. 6.8 Life cycle of the manufacture of recycled bumpers using scrapped bumpers as raw materials

recycled PP granules. The granules are then transported to the manufacturing plants as feedstock to be mixed with virgin PP granules, and some of them are discarded at the machine milling and pressure and torsion testing stages; the remaining PP is manufactured into remade bumpers (European Commission DG ENV 2011; Kozderka et al. 2017). There are five stakeholders involved in this value chain—the processing plants of discarded lead acid batteries, suppliers of PP-based secondary material resources, manufacturing plants of virgin PP, manufacturing plants of remade PP-based bumpers, and the transportation industry (European Commission DG ENV 2011; Kozderka et al. 2017).

Figure 6.8 illustrates the life cycle of the remanufacture of scrapped bumpers into remade bumpers. This transition path differs from the former in the supply of raw materials—its recycled materials come from scrapped bumpers dismantled by discarded automobile recycling plants; thus, the processing procedure also differs accordingly. Suppliers of PP-based secondary material resources must remove the baked paint and coatings on the surface of PP materials before subsequent processing, in order to prevent contamination of the recycled PP and resultant damage in mechanical properties (Zhao and Chen 2015). While the stakeholders involved in this process are almost identical to those in the recycling of discarded lead acid battery casings, the distinction lies in the suppliers of materials, as the recycled materials in this case come from waste from dismantling at discarded automobile recycling plants.

To conduct life-cycle assessment, we can define the demand for bumper PP as the functional unit of each transition path, to evaluate respective weights of the stakeholders' roles. Based on the life cycle of each path, an inventory of resource inputs, energy demands, pollutant emission and waste generation during the recycling process should be established. Both the immediate recycling process and the acquisition of energy, water, and other raw materials or additives may lead to air pollution, emission of water pollutants and waste generation (midpoint level), resulting in corresponding human health issues, ecosystem damages, climate change, and resource and energy depletion (endpoint level). A systematic analysis of these impact hot spots in each path, along with indicators like resource efficiency and circularity, should be conducted in order to explore whether the path with the lowest environmental impact is the optimal path for transition toward circular economy.

Apart from environmental assessment, economic assessment is also necessary for analyzing the potential sources and effectiveness of value creation in each path toward circular economy. For instance, regarding the value created by cost reduction, an in-depth investigation should be performed on whether there is any transfer or reduction of stakeholders' interests within the system boundaries, and related supporting measures or strategies should be formulated accordingly. As for value creation by behavioral changes, attribution of rights and duties during such behavioral changes should be pondered to prevent any possible disputes; the related regulations and supporting measures must also be considered in the meantime. If necessary, the supporting measures and strategies can be incorporated into the assessment system for a more comprehensive and systematic analysis of the pros and cons of each path.

6.4.5 Step 5 Business Case Development

According to the problems or opportunities identified in the previous steps and the feasibility analysis of each path, the circular business model selected for launch is proposed. As one of the tools for developing business models, business model canvas, a visualized and qualitative business development tool proposed by Osterwalder and Pigneur (2010), can be used. To further meet the need of the systemic design of a circular economy, augmented versions have been proposed by incorporating the concepts and principles of circular economy (Lewandowski 2016; Mentink 2014; Antikainen and Valkokari 2016). Elements of a business model canvas include the following: (1) business ecosystem and organizational culture, which are ramified into external factors such as market trends, domestic regulations and policies, technologies, and social acceptance and internal factors such as organizational culture and acceptance; (2) key partnerships, which include networks and types of collaboration; (3) key activities, which include optimization of product performance, product design, engagement of related partners and stakeholders, product remanufacturing and recycling, and related alternative technologies; (4) key resources, which include materials with better performance, regeneration and restoration of natural capital, virtualization of materials, and recovered resources (products, components, and materials); (5) value proposition, which includes product-service systems, circular products, virtualized services, and customer incentives for take-back; (6) customer relationships, which encompass production orders, customer advice, and social marketing strategies and community partnerships; (7) channels, such as virtualized channels; (8) take-back systems, which include take-back management and channels; (9) customer segments; (10) cost structure, which includes evaluation standards, preferential prices for customers, and material flow cost accounting criteria; (11) revenue streams such as input, availability, utility, performance, and value of recovered resources; (12) sustainability benefits, such as environmental benefits (resource productivity, environmental impacts, etc.) and social benefits (reinforcement of local communities, creation of employment opportunities, etc.). The project team should list as many required elements as possible and eventually construct a business model canvas to facilitate subsequent tryout of the transformation path.

Having the business case established, it calls for **Step 6. Piloting and prototyping**, **Step 7 Delivery and implementation**, and follow-up evaluation with **Step 8 Review and reporting** to repeatedly promote transition of the automotive industry toward a circular economy of PP bumpers. The tryout plan and verification of feasible idea during the process of piloting and prototyping can increase the chance of success in official implementation. Nevertheless, there might be unexpected consequences after official implementation; through control report and regular progress report to build subsequent control and follow-up are key to ensuring smooth implementation and sustained transition toward a circular economy.

6.5 Conclusion

We are facing the grand challenge of resource depletion and environmental impact. The key to reducing the pressure of resource and environmental issues while maintaining economic and societal growth is to change the way resources are used in our economic system. This calls forth change of the governance mechanism, integration of industries, redesign of infrastructure, and formation of partnerships. The Resource Efficiency Flagship Initiative proposed by the EU emphasized that following the existing linear economic model leads to a dead end; transitioning toward a circular economy is the only way to enhance the efficiency of resource utilization, open up opportunities for economic development, improve productivity, drive down production costs, and maintain competitiveness.

It has been estimated that the EU's continuous development toward a circular economy—through reuse, remanufacturing, and product recycling systems in industries, as well as technological revolution—can increase resource productivity by 3% annually. By 2030, the primary resource expenditure is estimated to be cut by €0.6 trillion per year. In addition, nonresource and environmental external benefits worth €1.2 trillion would be generated; overall, the annual benefits created by a circular economy could total up to €1.8 trillion (EMF 2015a). The EU's development of a circular economy can enhance environmental resilience; by 2030, CO_2 emission and resource consumption are estimated to be reduced by 48% and 32%, respectively. In the future, environmental benefits can be significantly increased through sharing, optimization, recycling, virtualization, and innovative technologies; by 2050, resource consumption is estimated to be cut by 53%.

Transitioning toward a circular economy is an important path toward sustainability from the viewpoint of resource efficiency. The 17 SDGs can be considered as the most pressing problems of human civilization in need of solutions. They are also the key subjects of enterprises in the integration of social and economic responsibilities and creation of shared values. At the core of such quest for sustainability, realization of a circular economy facilitates promotion of multiple SDGs; both the governments and the enterprises should reflect on relevant strategies sooner rather than later. The transition procedure, framework, and related assessment methods discussed previously can serve as references for practical implementation.

Circular economy can be regarded as an important path toward sustainable urban governance; the principles of circular economy can help achieve optimal energy and resource management in cities (EMF 2016a). Sustainable cities are also an important topic in sustainable development. The Reference Framework for Sustainable Cities (RFSC 2016) established in Europe is an online toolkit designed to help cities evaluate the progress of their sustainable urban development. It covers 5 main pillars including the environment, economy, society and culture, space, and governance as well as 30 objectives. Cities can communicate and cooperate with one another through the evaluation and formulate comprehensive strategies for urban development.

Sustainable governance of cities should take the overall urban system perspective to analyze the status of urban development. The inflow, outflow and stock of resources and solid materials (construction materials in particular) may differ significantly in growing and shrinking cities; hence, different governance methods are required. In expanding cities, accumulated construction materials may bring substantial environmental impact, but these solid materials may also become secondary material resources. In light of these secondary material resources that exist in the urban environment, urban planner Jane Jacobs proclaimed that "cities are the mines of the future" (Jacobs 1961). This prospect was based on the fact that considerable quantities of materials were flowing from mines into cities, where they were used in construction and infrastructure. These urban mines might provide an alternative to conventional mines in the earth's crust. Since around 2010, the vision for urban mining has been upgraded in the field of resource and waste management; urban mining has become a synonym for the recovery of secondary resources from products concentrated in urban areas (Graedel 2011; Johansson et al. 2013; Que et al. 2018). It focuses on different types of urban mineral deposits such as constructions and infrastructure (Ergun and Gorgolewski 2015; Wallsten et al. 2013) and landfills (Bockreis and Knapp 2011) or specific materials (Simoni et al. 2015; Wen et al. 2015). The mining potential depends on a wide array of driving factors including the future usability of urban mine waste, waste-related legislation, as well as the production costs of secondary resources and revenues obtained from the commodity market.

The pursuit of resource efficiency and resilience to decouple resource consumption and environmental impact from economic activities leads to important research directions. The endeavors on circular economy and sustainable cities have great potential to inspire innovations and practices on the journey toward sustainability.

References

ABCE (Advisory Board for Circular Economy). (2017). The Advisory Board for Circular Economy Recommendations for the Danish Government.

Accenture. (2014). *Circular advantage: Innovative business models and technologies to create value in a world without limits to growth*. Chicago: Accenture.

Antikainen, M., & Valkokari, K. (2016). Framework for sustainable circular business model innovation. *Technology Innovation Management Review, 6*(7), 5–12.

Arkatkar, A., Arutchelvi, J., Sudhakar, M., Bhaduri, S., Uppara, P. V., & Doble, M. (2009). Approaches to enhance the biodegradation of polyolefins. *The Open Environmental Engineering Journal, 2*, 68–80.

American Chemistry Council (2019) Plastics and Polymer Composites in Light Vehicles. https://www.automotiveplastics.com/blog/2019-plastics-and-polymer-composites-in-light-vehicles/

Baccini, P., & Brunner, P. H. (2012). *Metabolism of the anthroposphere*. Cambridge: The MIT Press.

Bockreis, A., & Knapp, J. (2011). Landfill mining – Deponien als Rohstoffquelle. *Ö sterreichische Wasser- und Abfallwirtschaft, 63*(2), 70–75.

Brunner, P. H., & Rechberger, H. (2004). *Practical handbook of material analysis*. Retrieved from https://iwr.tuwien.ac.at/fileadmin/mediapool-ressourcen/MFA_Handbook/solutions.pdf

BSI (British Standards Institution). (2017). *BS 8001:2017. Framework for implementing the principles of the circular economy in organizations – Guide*. London: The British Standards Institution.

Circle Economy. (2016). *Master circular business with the value hill*. Amsterdam: Circle Economy.

Curran, M. A. (1996). *Environmental life-cycle assessment* (Vol. 1, p. 179). New York: McGraw-Hill Press.

EC (European Commission). (2014). *Scoping study to identify potential circular economy actions, priority sectors, material flows and value chains*. Luxembourg: DG Environment.

EC (European Commission). (2015). *Circular economy strategy: Closing the loop—An EU action plan for the circular economy*. Retrieved from http://ec.europa.eu/environment/circular-economy/index_en.htm

EC (European Commission). (2017). *On the implementation of the Circular Economy Action Plan*. Retrieved from http://ec.europa.eu/environment/circular-economy/implementation_report.pdf

EEB (European Environmental Bureau). (2015). Circular Economy Package 2.0: Some ideas to complete the circle.

EMF (Ellen MacArthur Foundation). (2012). *Towards the circular economy Vol. 1: An economic and business rationale for an accelerated transition.* Retrieved from https://www.ellenmacarthurfoundation.org

EMF (Ellen MacArthur Foundation). (2013). *Towards the circular economy Vol. 1: Opportunities for the consumer goods sector.*

EMF (Ellen MacArthur Foundation). (2015a). *Towards a circular economy: Business rationale for an accelerated transition.* Retrieved from https://www.ellenmacarthurfoundation.org

EMF (Ellen MacArthur Foundation). (2015b). *Circularity indicators: An approach to measuring circularity.* Cowes: Ellen MacArthur Foundation.

EMF (Ellen MacArthur Foundation), SUN, & McKinsey. (2015c). *Growth within: A circular economy vision for a competitive Europe.* Ellen MacArthur Foundation: Cowes.

EMF (Ellen MacArthur Foundation). (2016a). *Circular economy in India: Rethinking growth for long-term prosperity.* Retrieved from https://www.ellenmacarthurfoundation.org

EMF (Ellen MacArthur Foundation). (2016b). *The new plastics economy: Rethinking the future of plastics.* Retrieved from https://www.ellenmacarthurfoundation.org

Ergun, D., & Gorgolewski, M. (2015). Inventorying Toronto's single detached housing stocks to examine the availability of clay brick for urban mining. *Waste Management, 45*, 180–185.

European Commission DG ENV. (2011). *Plastic waste in the environment.* Paris: BIO Intelligence Service.

Forum of the Future & Unilever. (2016). *Circular Business Models.* Retrieved from https://www.forumforthefuture.org/project/circular-economy-business-modeltoolkit/verview

French Ministry of the Environment (2018) Circular economy roadmap: The French experience, French Ministry of the Environment.

Geissdoerfer, M., Morioka, S., de Carvalho, M., & Evans, S. (2019). Business models and supply chains for the circular economy. *Journal of Cleaner Production, 190*, 712–721.

Geng, Y., Sarkis, J., Ulgiati, S., & Zhang, P. (2013). Measuring China's circular economy. *Science, 339*, 1526–1527.

Genovese, A., Acquaye, A., Figueroa, A., & Koh, S. (2017). Sustainable supply chain management and the transition towards a circular economy: Evidence and some applications. *Omega, 66*, 344–357.

Gigli, S., Landi, D., & Germani, M. (2019). Cost-benefit analysis of a circular economy project: A study on a recycling system for end-of-life tyres. *Journal of Cleaner Production, 229*, 680. https://doi.org/10.1016/j.jclepro.2019.03.223.

Graedel, T. (2011). The prospects for urban mining. *Bridge, 41*, 43–50.

Haupt, M., & Zschokke, M. (2017). How can LCA support the circular economy? *The International Journal of Life Cycle Assessment, 22*(5), 832–837.

Hung, P. Y. (2018). *The transition process and tools of moving towards a circular business model.* Graduate Institute of Environmental Engineering, National Taiwan University Master Thesis.

Huysveld, S., Hubo, S., Ragaert, K., & Dewulf, J. (2019). Advancing circular economy benefit indicators and application on open-loop recycling of mixed and contaminated plastic waste fractions. *Journal of Cleaner Production, 211*, 1–13.

Iacovidou, E., Millward-Hopkins, J., Busch, J., Purnell, P., Velis, C., Hahladakis, J., Zwirner, O., & Brown, A. (2017). A pathway to circular economy: Developing a conceptual framework for complex value assessment of resources recovered from waste. *Journal of Cleaner Production, 168*, 1279–1288.

Ingrao, C., Messineo, A., Beltramo, R., Yigitcanlar, T., & Ioppolo, G. (2018). How can life cycle thinking support sustainability of buildings? Investigating life cycle assessment applications for energy efficiency and environmental performance. *Journal of Cleaner Production, 201*, 556–569.

Jacobs, J. (1961). *The death and life of great American cities.* New York: Random House.

Jin, I. (2016). Circular economy policy in Korea. In V. Anbumozhi & J. Kim (Eds.), *Towards a circular economy: Corporate management and policy pathways. ERIA research project report 2014-44* (pp. 163–184). Jakarta: ERIA.

Johansson, N., Krook, J., Eklund, M., & Berglund, B. (2013). An integrated review of concepts and initiatives for mining the technosphere: Towards a new taxonomy. *Journal of Cleaner Production, 55*, 35–44.

Kirchherr, J., Reike, D., & Hekkert, M. (2017). Conceptualizing the circular economy: An analysis of 114 definitions. *Resources, Conservation and Recycling, 127*, 221–232.

Kozderka, M., Rose, B., Bahlouli, N., Kočí, V., & Caillaud, E. (2017). Recycled high impact polypropylene in the automotive industry-mechanical and environmental properties. *International Journal on Interactive Design and Manufacturing, 3*, 737–750.

Lacovidou, E., Millward-Hopkins, J., Busch, J., Purnell, P., Velis, C., et al. (2017). A pathway to circular economy: Developing a conceptual framework for complex value assessment of resources recovered from waste. *Journal of Cleaner Production, 168*, 1279–1288.

Lavers, A., Kalmykova, Y., Rosado, L., Oliveira, F., Laurenti, R., & Rydberg, T. (2019). Combining material flow analysis with life cycle assessment to identify environmental hotspots of urban consumption. *Journal of Cleaner Production, 226*, 526–539.

Lewandowski, M. (2016). Designing the business models for circular economy—Towards the conceptual framework. *Sustainability, 8*(1), 43.

Lowe, E., & Evans, L. (1995). Industrial ecology and industrial ecosystems. *Journal of Cleaner Production, 3*(1–2), 47–53.

Mentink, B. (2014). *Circular business model innovation: A process framework and a tool for business model innovation in a circular economy.*

Ministry of Infrastructure and the Environment. (2014). *Implementation of the waste to resource programme.*

Ministry of the Environment (Japan). (2017). *Progress of the fundamental plan for establishing a sound material-cycle society.*

NDRC (National Development and Reform Commission) of the People's Republic of China. (2017). *Circular development and leading action.*

Nyamekie, P. (2012). *Re-use and re-cycling of automobile plastics – A step to manage plastic waste in Ghana.* Helsinki: Arcada University of Applied Sciences.

OECD. (2019). *Business models for the circular economy: Opportunities and challenges for policy.* Paris: OECD Publishing.

Osterwalder, A., & Pigneur, Y. (2010). *Business model generation: A handbook for visionaries, game changers, and challengers.* New York: Wiley.

PEMRG. (2016). *Consultic marketing & Industrieberatung.* Retrieved from Plastics Europe Market Research Group.

Pieroni, M., McAloone, T., & Pigosso, D. (2019). Business model innovation for circular economy and sustainability: A review of approaches. *Journal of Cleaner Production, 215*, 198–216.

Plastics Market Watch. (2016). *Automotive recycling devalued is now revalued.* Washington, DC: Plastics Market Watch.

Que, S., Awuah-Offei, K., Wang, L., Samaranayake, V., Weidner, N., & Yuan, S. (2018). Individual preferences for mineral resource development: Perspectives from an urban population in the United States. *Journal of Cleaner Production, 189*, 30–39.

RFSC. (2016). Retrieved from http://rfsc.eu

Saavedra, M. B. Y., Iritani, D. R., Pavan, A. L. R., & Ometto, A. R. (2018). Theoretical contribution of industrial ecology to circular economy. *Journal of Cleaner Production, 170*, 1514–1522.

Satoru, M., Tsuyoshi, W., & Susumu, K. (2010). Polypropylene compounds for automotive applications. *SUMITOMO KAGAKU, pp., 2010*, 1–16.

Schroeder, P., Anggraeni, K., & Weber, U. (2018). The relevance of circular economy practices to the sustainable development goals. *Journal of Industrial Ecology, 23*, 77. https://doi.org/10.1111/jiec.12732.

Simoni, M., Kuhn, E. P., Morf, L. S., Kuendig, R., & Adam, F. (2015). Urban mining as a contribution to the resource strategy of the Canton of Zurich. *Waste Management, 45*, 10–21.

Stahel, W. R., & Clift, R. (2016). Stocks and flows in the performance economy. In R. Clift & A. Druckman (Eds.), *Taking stock of industrial ecology.* Cham: Springer.

Tokiwa, Y., Calabia, B. P., Ugwu, C. U., & Aiba, S. (2009). Biodegradability of plastics. *International Journal of Molecular Sciences, 10*(9), 3722–3742.

Ünal, E., & Shao, J. (2019). A taxonomy of circular economy implementation strategies for manufacturing firms: Analysis of 391 cradle-to-cradle products. *Journal of Cleaner Production, 212*, 754–765.

Wallsten, B., Carlsson, A., Frändegård, P., Krook, J., & Svanström, S. (2013). To prospect an urban mine – Assessing the metal recovery potential of infrastructure "cold spots" in Norrköping, Sweden. *Journal of Cleaner Production, 55*, 103–111.

Wang, Y. S., & Ma, H. W. (2018). Analysis of uncertainty in material flow analysis. *Journal of Cleaner Production, 170*, 1017–1028.

WBCSD. (2016). *Unlocking more value with fewer resources a practical guide to the circular economy.*

Wen, Z., Zhang, C., Ji, X., & Xue, Y. (2015). Urban mining's potential to relieve China's coming resource crisis. *Journal of Industrial Ecology, 19*(6), 1091–1102.

WRAP. (2013). *On course for zero waste.*

Yin, S., Tuladhar, R., Sheehan, M., Combe, M., & Collister, T. (2016). A life cycle assessment of recycled polypropylene fibre in concrete footpaths. *Journal of Cleaner Production, 112*, 2231–2242.

Zhao, Q., & Chen, M. (2015). Automotive plastic parts design, recycling, research, and development in China. *Journal of Thermoplastic Composite Materials, 28*, 142–157.

Part II
Manufacturing, Logistics, and Value Creation

Chapter 7
Sustainable Supply Chain Management: Research Pathways Based on Empirical Evidence from Chinese Automotive Companies

Lin Wu and Nachiappan Subramanian

Abstract China's current development model presents an over-emphasis on short-term economic gains over long-term sustainable development. As business activities play a crucial role in creating and solving such problems, firms are increasingly encouraged to take the initiative to contribute to the future. However, most companies in emerging economies are competing in the world market based on cost advantages. These firms, being predominantly small and medium-sized enterprises (SMEs) with limited resources, tend to favour short-term financial gains over long-term sustainability. It is clear that firms with a short-sighted focus on economic gains are not truly sustainable and those aiming for sustaining success need to balance various sustainability initiatives to meet current targets without compromising future prosperity. This chapter aims to provide insights on the status quo of sustainable development in the context of emerging economies' manufacturing sector. A systematic review of theories and empirical studies was conducted to clarify the research pathways on sustainability. Case studies with eight Chinese automotive companies were also carried out, and future research directions are proposed based on the findings.

Keywords Sustainable practices · Capabilities · Performance · Empirical evidence · China · CASE study

L. Wu (✉)
Nottingham University Business School, Nottingham, UK
e-mail: Lin.Wu@nottingham.ac.uk

N. Subramanian
University of Sussex Business School, Brighton, UK
e-mail: N.Subramanian@sussex.ac.uk

C. Chen et al. (eds.), *Pursuing Sustainability*, International Series in Operations Research & Management Science 301,
https://doi.org/10.1007/978-3-030-58023-0_7

7.1 Theoretical Perspective on Corporate Sustainability (CS)

CS is defined as "development that meets the needs of a firm's direct and indirect stakeholders expectations (shareholders, employees, clients, pressure groups, the local communities, etc.), without compromising its ability to meet the needs of future stakeholders" (Dyllick and Hockerts 2002), in accordance with the official definition of sustainability by (World Commission on Environment and Development (WCED) 1987). From the definition, a sustainable company should have the ability to create profit for its shareholders while protecting the environment and improving the lives of all stakeholders (Savitz and Weber 2006). Three pillars jointly support sustainability, namely, environmental integrity, economic prosperity, and social equity (Bansal 2005) with the intersection being true sustainability. It is the balanced development and healthy interaction of these three dimensions that ensures the sustainable development of business and the society.

Organizational theories mainly play two types of roles in existing sustainability research. The first stream of theories, including the stakeholder theory and the institutional theory, are mainly used to explain firms' adoption of sustainability initiatives from an external perspective. The second group of theories, including the resource-based view (RBV), the natural resource-based view (NRBV), the practice-based view (PBV), and the dynamic capabilities view (DCV), are employed to explain the internal motivation of firms' adoption of sustainability. The following section presents a detailed review of these theories and how they have been used in studies on sustainability.

7.1.1 External Drivers of CS

7.1.1.1 Stakeholder Theory

Stakeholder theory proposes that firms survive and make profits by satisfying different stakeholder groups (Clement 2005). Stakeholder pressures are widely proven to be a major driver of CS. Clement (2005) believes that firms can improve their bottom line by properly responding to stakeholder concerns. According to Garces-Ayerbe et al. (2012), there is a positive relationship between stakeholder pressure and firms' adoption of environmental management practices. Darnall et al. (2010) also provides empirical support for the positive impact of stakeholder pressure on firm environmental proactivity. Further, Ademambo et al. (2013) report that stakeholder pressure, legislation, and perceived benefits have a direct influence on the implementation of environmental manufacturing practices and firm performance. Stakeholder pressure therefore plays a crucial role in the formation of corporate change toward sustainability.

7.1.1.2 Institutional Theory

Different from stakeholder theory, institutional theory explains the way in which external institutional forces influence firms' adoption of an organizational practice (Sarkis et al. 2011). Dimaggio and Powell (1983) identify three forms of mechanisms through which isomorphic change takes place, namely, coercion, mimesis, and norms. Institutional theory has been widely applied to studies on CS. For example, Zhu et al. (2013) found that both international and domestic coercive, normative, and mimetic pressures have forced Chinese firms to achieve their diverse objectives of sustainability.

7.1.2 Internal Motivations of CS

7.1.2.1 Resource-Based View (RBV)

As a dominant paradigm in strategic management, the RBV has been increasingly extended to other research fields such as operations management, marketing, and management sub-disciplines, like human resource management and entrepreneurship (Hitt et al. 2016). The origin of the RBV dates back to (Penrose 1959) who argued that a firm's internal growth and external expansion through merger, acquisition and diversification are due to the manner in which its resources are employed. Barney (1991) first formalizes the RBV and explains how valuable, rare, inimitable, and non-substitutable (VRIN) resources can provide organizations with a competitive advantage. It is believed that when internal resources of an organization can be leveraged to help guard against external forces that may have a negative influence on performance, this organization is likely to enjoy a competitive advantage over its competitors (Campbell and Park 2017).

The RBV is one of the most widely adopted theoretical lenses in sustainability-related management research. Bansal (2005) justifies the applicability of the RBV in the context of CS with three points. Firstly, it has been well proven that CS can affect firm performance. Secondly, investment of financial and/or human resources are required for firms to shift towards sustainability. Thirdly, new opportunities such as international experience, capital management capabilities, and organizational slack can be created through changes in technology, legislation, and market forces from corporate sustainable development. Researchers use the RBV to explain the relationship between the environmental/social sustainability and business performance (e.g. Schoenherr 2012). The RBV serves as the optimal theoretical basis for studies on the potential competitive effects of sustainability strategies.

7.1.2.2 Natural Resource-Based View (NRBV)

Hart and Dowell (2011) point out that the constraints imposed by the natural environment have been ignored by the RBV and on this basis the natural resource-based view (NRBV) is proposed. The NRBV argues that business organizations are dependent on, and in the meantime constrained by, resources offered by the natural environment to prosper and flourish (Wong et al. 2012). Firms need to develop the capability to exploit and preserve natural resources to achieve superior performance (Hart and Dowell 2011). The NRBV mainly connects with sustainability in terms of the environmental dimension. From the NRBV perspective, the development of the green capabilities motivates firms to shift from the reactive to the proactive end of environmental spectrum, in which they develop preventive approaches instead of using end-of-pipe solutions to deal with environmental problems (Fraj et al. 2013). However, from a sustainability perspective, the NRBV is fragmented because it ignores the social (Klassen and Vereecke 2012) and the economic dimensions. More importantly, the synergistic effect of a simultaneous implementation of different kinds of CS practices on capability development is not captured in the NRBV.

7.1.2.3 The Practice-Based View (PBV)

Recently researchers have started to question the applicability of the RBV in studies looking at contemporary issues. Bromiley and Rau (2016) believe that the RBV is not appropriate when explaining performance differences across firms that are a result of the implementation of the same practices. A practice refers to "a defined activity or set of activities that a variety of firms might execute" (Bromiley and Rau 2016). Different from the central argument of the RBV, which suggests that a firm's competitive advantage comes from the possession of VRIN resources, the PBV believes that operational and management practices are in no way valuable, rare, inimitable, or non-substitutable (Bromiley and Rau 2016) and their performance implications are not the same as sustained competitive advantage.

Practices as standard procedures (Wu et al. 2012b) are publicly available without any "isolating mechanisms" (Bromiley and Rau 2016). Literature provides empirical evidence that the effective implementation of some publicly available practices leads to tremendous performance improvement (e.g Hajmohammad et al. 2013; Yang et al. 2011). This contrasts with the proposed independent variable specified in the RBV. On the other hand, firm or business unit performance improvement does not equal "sustained competitive advantage", the dependent variable in the RBV. "Advantage" implies comparisons with peers, and "sustained" needs further specification on the exact time length. In fact, very few empirical studies have measured competitive advantage or sustained competitive advantage (Bromiley and Rau 2016). From a review of 55 empirical articles using the RBV as the theoretical lens, Newbert (2007) found that the majority of them (93%) used performance as the dependent variable. Only 16% of the reviewed articles used competitive advantage as the dependent variable and even fewer used sustained competitive advantage

(2%). This summary indicates the misunderstanding in the capture of RBV outcomes in existing empirical studies. Based on Bromiley and Rau (2016), the PBV is the optimal lens through which the effect of CS practices on performance is examined.

7.1.2.4 Dynamic Capabilities View (DCV)

Capabilities are "high-level routines or bundles of routines" which refer to "organizational processes that utilize clusters of resources to achieve desired outcomes" (Peng et al. 2008). The DCV was developed from the RBV of the firm. The RBV is criticized as being "static" which neglects the influence of the dynamic market (Eisenhardt and Martin 2000). That is, it generally relies on protecting and leveraging existing resources (Russo 2009) and fails to address how future resources can be created, and how the currently possessed resources can be refreshed according to the new requirements of the changing environment (Ambrosini et al. 2009). According to Carmeli (2004), in the current business environment which is dynamic and highly competitive, the real source of competitive advantage lies in a firm's ability to consistently meet environmental changes, as well as to change the industry structure. It is not the VRIN resources per se that create and maintain a competitive advantage; instead, it is the firm's capability to constantly update its resources and operational capabilities to keep pace with the changing environment that is the key to success (Cepeda and Vera 2007).

Teece et al. (1997) define dynamic capabilities as "the firm's ability to integrate, build, and reconfigure internal and external competencies to address rapidly changing environments". Such capabilities are path-dependent, and are shaped by the decisions made by an organization throughout its history, the assets it holds and its leaning activities (Ambrosini et al. 2009).

The DCV has been increasingly applied to the context of CS. Russo (2009) explains the impact of ISO 14001 on firm emission performance based on the DCV. It isfound that early adoption of ISO 14001 is associated with lower emissions; the longer a facility operates ISO 14001, the lower its emissions. As Russo further point out (2009), early adoption of ISO 14001 and the continued use of path dependent learning to strengthen the application are considered as crucial ways to build dynamic capabilities. Similarly, Hofmann et al. (2012) identify a positive relationship between firms' dynamic capabilities (the adoption of advanced technology, experiences with inter-firm relations, and capacity for product innovation) and environmental management practices. Wu et al. (2012a), through an in-depth case study of the Chinese leading Telecom company, Huawei, find empirical evidence that dynamic capabilities (dynamic scanning, identification, and reconfiguration) facilitate firms' strategic change toward sustainability (which involves the implementation of various sustainable practices). According to Wu et al. (2012a), firms with higher levels of dynamic capabilities are more effective in detecting stakeholder requirements, seizing sustainability opportunities to meet rapidly changing needs and reconfiguring the existing functional capabilities for corporate sustainability.

7.2 Review of Empirical Studies on CS

An increasing body of empirical research has investigated sustainability from the Triple-Bottom-Line (3BL) perspective (Elkington 1998). Instead of regarding CS as simply environmental or social issues, these studies acknowledge the coexistence of all three dimensions of sustainability. Some attempt to clarify specific practices in each dimension (Amini and Bienstock 2014). Others focus on the impact of environmental and social sustainability on economic or overall sustainability performance. In general, CS practices have been found to be positively related to sustainability performance with different levels of strength and time required (Chang and Kuo 2008; Gimenez et al. 2012; Kurapatskie and Darnall 2013), and firms need to properly balance the economic, environmental, and social dimensions of CS to maintain the optimal sustainable operations in the context of stiff market competition (Tomsic et al. 2015). A summary of such studies is presented in Table 7.1.

Table 7.1 examples of empirical research on CS

Authors	Relationship addressed	Main findings
Chang and Kuo (2008)	CS & FP	CS positively relates to FP.
Pullman et al. (2009)	EP and SP & performance	CS practices improve quality performance, which in turn lead to better cost performance.
Gimenez et al. (2012)	Green & 3BL Social & 3BL	Internal EP positively relate to 3BL performance while internal SP positively relate only to social and environmental performance.
Kurapatskie and Darnall (2013)	CS & FP	Both lower- and higher-order sustainability activities positively relate to FP.
Govindan et al. (2014)	Lean, resilient and green supply chain management and supply chain sustainability	Green supply chain management practices positively associate with supply chain sustainability.
Luzzini et al. (2015)	Sustainability commitment & 3BL	Sustainability commitment positively relates to firm collaborative capabilities, and firm collaborative capabilities positively relate to 3BL performance.
Tomsic et al. (2015)	CS & economic performance	CS positively relates to the economic performance of SMEs.
Wu et al. (2015)	Integrated lean, EP and SP & sustainability performance	The implementation of the integrated sustainability practices has stronger impact on the 3BL performance than individual practice implementation.

FP financial performance, *EP* environmental performance, *SP* social performance

7.3 Methodology

This study employs a qualitative approach based on case studies. The case companies were drawn from a population including companies of varying sizes, ownership types, and supply chain positions. In order to better capture the long-term CS development, firms with an operating time of less than 5 years are not considered.

A total of eight (8) samples were drawn from the population. Both primary and secondary data were obtained from the case companies. Primary data were collected from semi-structured interviews with managers and experienced operations staff. In addition to primary data, secondary data such as company annual reports (when available), CSR reports, and other internal materials were also provided by the case companies. Relevant news releases (both negative and positive) on the case companies were also used to supplement the collected information. Triangulation of data sources enhanced the reliability of the research findings (Eisenhardt and Martin 2000).

7.3.1 Case Company Profiles

For ethical reasons, and in order to encourage respondents' openness, names of the case companies and the interviewees are not revealed at any stage of the research. Therefore, A, B, C, D, E, F, G, and H are used to refer to them respectively. Brief introductions of the case companies are given below and in Table 7.2.

Company A is a Chinese private enterprise established in 1987. The automotive business is only one of the sectors in which it operates, and others include travel and construction. Hereafter Company A refers to the automotive section of the organization. Its main products include seating systems, exterior and interior decoration systems, air conditioning systems, and lighting systems. Currently, A has seven major production bases in China and four sales and aftersales service centres across the world. Over 1500 workers are employed and the company is ISO 14001, TS16949 and ISO 9002 certified.

Company B was established in 1992 and has grown steadily ever since. It was started as a small and humble factory in Ningbo City, the Yangtze River Delta. In 1997, the company amalgamated its branches to form the Group, which was later (in 2005) listed on the Hong Kong Stock Exchange. Currently, Company B is a leading auto supplier in China in design, manufacturing and sales of structural body parts, roof racks, seat frame systems, trims and decorative parts of passenger vehicles. It supplies to world-renowned automotive companies such as Ford, GM, and Nissan, and its customers represent over 80% of the total global auto market share. To date, Company B has established an extensive production network in mainland China with production facilities in eastern, southern, northern and central China. It has also established overseas design and sales centres in the USA, Germany, and Japan, and production plants in developing countries such as Mexico and Thailand.

Table 7.2 Case profiles

Case	Size	Age	Position	Location	Product	Interviewee
A	1500	33	Tier-1	Ningbo	Seating systems, exterior and interior decoration systems, air conditioning systems and lighting systems	CEO and operations manager
B	7300	28	Tier-1	Ningbo	Body structural parts, trims and decorative parts	Operations manager
C	18,000	23	End assembler	Ningbo	Passenger vehicles	Production and logistics staff
D	5000	18	End assembler	Beijing	Passenger vehicles	Experienced operations staff
E	5000	27	End assembler	Zhengzhou	Passenger vehicles	Operations manager; design staff; R&D staff; head of CSR department
F	16,000	23	Tier-1	Shanghai	Seating systems for passenger vehicles	Engineer and experienced operations staff
G	195	29	Tier-1	Ningbo	Clutch booster pump	CEO
H	100	14	Tier-2	Ningbo	Dashboard skin	Operations manager

Company B currently employs over 7000 workers and it is TS 16949, ISO 14001, ISO 18001, and ISO 9001 certified.

Company C is a private Chinese auto company whose headquarter is located in the Yangtze River Delta. The company began business in 1997 and is now ranked among the top 10 automotive enterprises in China with employee numbers of over 18,000. Company C has manufacturing bases distributed widely in China and overseas design centres in Sweden, Spain, and the USA. The company has obtained the Environmental Management System Certification ISO 14001:2004 and Quality Management Systems Certifications of ISO 9001 and TS16949:2009.

Company D, a joint venture established in 2002, has experienced continued growth in terms of sales and revenue. It is located in the Bohai Sea Region where it currently has three assembly plants, three engine plants, and a technological centre. The company pays special attention to the control of environmental impact of its production process by engaging in various environmental programmes such as cleaner production and solid waste management. In terms of social responsibility, Company D emphasizes support for education and sport, desert management and philanthropic donations. The company employs over 5000 workers and is ISO 14001 certified.

Company E, established in 1993 jointly by a Chinese vehicle assembler and a leading foreign automotive company, is located in Zhengzhou City, in central mainland China. Currently, it has over 5000 full time employees. Its main products are pickup trucks and sport utility vehicles (SUVs). The company has strong capabilities in R&D, supply chain management, production and manufacturing, and sales and service. Company E is ISO 14001 and ISO 9001 certified.

Company F is a leading producer of vehicle seating systems in China with employees numbering over 16,000 all over China. It is a joint venture established in 1997 with headquarters located in Shanghai, the Yangtze River Delta. The company is QS 9000, ISO 9001 and ISO 14001 certified.

Company G is a small private enterprise located in Ningbo, in the Yangtze River Delta. Established in 1991, the company has been growing steadily. Its main product is clutch booster pumps. There are about 200 full time employees currently working for the company, 15 of which are technicians. The company has obtained ISO 14001 and ISO 9001 certificates.

Company H is a foreign company. Its China branches are located in Shanghai and Ningbo, in the Yangtze River Delta. The company has a history of more than 200 years, and its first China plant was established in 2006 in Ningbo. To date, the company has five plants across mainland China. In this study, Company H is used to refer to the Ningbo plant only, who has approximately 100 full time employees currently.

Semi-structured interviews were conducted following the guidelines shown in Table 7.3. General questions aim to capture the basic information of the case companies, which is useful for cross-case comparisons. After the general questions, firms were asked to provide information on motivations, benefits, and support achieved through the implementation of each sustainability practice. They were also

Table 7.3 List of interview questions

General questions:
1. What is the ownership type of your organization (private, joint venture, state-owned, or foreign)?
2. What is the number of full-time staff in your organization?
3. How long has your organization been in operation?
4. Is your organization implementing CS practices? If yes, what are they?
Sustainability practices, capabilities and performance (asked in the sequence of lean, green and social):
1. What do sustainability practices mean to you? What are the sustainability practices your organization is currently implementing?
2. Why did your firm begin to implement CS practices?
3. What are the specific CS practices your organization is currently implementing?
4. How does your organization implement the CS practice? (What support is given to the implementation, etc.)
5. What benefits have you realized from implementing each of these CS practices? (Both tangible and intangible).
6. (If the interviewee mentions benefits related to intangible aspects) How does it influence the performance of your organization?
7. How long did your firm realize the benefits of implementing CS practices? What has happened after your organization realized the benefits?.

asked to provide the time it took for them to realize the desired benefits of implementing these practices.

7.3.2 Data Analysis

Using content analysis, both within- and cross-case analyses were applied to the data collected from the case companies. Content analysis is defined as "a systematic, replicable technique for compressing many words of text into fewer content categories based on explicit rules of coding" (Montabon et al. 2007). The outcome of content analysis is simplified and organized information of each case company, which makes cross-case comparative analysis easier.

The procedures of data analysis are as follows. First of all, based on the research questions, common themes were developed.

- Motivations for sustainability practice implementation.
- Specific sustainability practices currently implemented.
- Supports for the better implementation of sustainability practices.
- Short-term benefits of sustainability practices.
- Long-term benefits of sustainability practices.
- Time length for the desired benefits to be realized.

The next step is the data coding process. The coding of qualitative data followed (Hennink et al. 2011). At the first stage, the interviews were transcribed and anonymized for code development. Then, both inductive and deductive approaches were used to develop codes. Deductive codes were first developed based on literature, including "motivations of CS practice implementation", "CS practices implemented", "operational improvement and innovation capabilities", and "short- and long-term benefits". After all deductive codes were identified, a codebook was developed where relevant information from the interview transcripts was summarized and entered.

The next stage of coding development is inductive. All the transcripts and notes were revisited for novel concepts and ideas. Inductive codes such as "how short- and long-term sustainability influence each other", "role of product in sustainability", "influence of companies' position in the supply chain on sustainable development" were further identified. Finally, only those closely relevant to the research objectives were added to the codebook.

The second round of coding is more structured. It is done in order to rate the companies in terms of sustainability practice implementation, capabilities and sustainability performance based on a comprehensive list of items identified from the literature (Table 7.4). Items in each category were divided into short- and long-term ones based on opinion of the interviewees. Codes of "high", "medium", and "low" are created for this purpose. For sustainability practices and capabilities, if a company reports more than 60% of the items in each category, it is coded as "high" in terms of this category. Similarly, a coverage of 40–60% of the items is coded as

Table 7.4 List of sustainability practices, capabilities, and performance

CS Practices	
Short-term CS practices (Gimenez et al. 2012; Shafiq et al. 2014)	• Lot size reductions. • JIT/continuous flow production. • Pull system. • Cellular manufacturing. • Cycle time reductions. • Focused factory production systems. • Agile manufacturing strategies. • Quick changeover techniques. • Bottleneck/constraint removal. • Reengineered production processes.
	• Competitive benchmarking. • Quality management programmes. • Total quality management. • Process capability measurements. • Formal continuous improvement programme.
	• Predictive or preventive maintenance. • Maintenance optimization. • Safety improvement programmes. • Planning and scheduling strategies. • New process equipment or technologies.
	• Self-directed work teams. • Flexible, cross-functional workforce.
	• ISO 14001 certification. • Reduce raw material quantity. • Reduce raw material variety. • Avoid hazardous/harmful/toxic materials. • Internal environmental policy.
	• Reduce emissions. • Reduce discharges.
Long-term CS practices (Gimenez et al. 2012; Penrose 1959; Schoenherr 2012)	• Solid wastes recycling. • Packages recycling. • Wastewater recycling.
	• Waste prevention. • Waste reduction. • Waste management.
	• Employee welfare. • Employee training and self-development. • Equality (female, the disabled, the minority). • Working and living conditions improvement. • Employee satisfaction.
	• Quality of the products. • Customer care. • Customer education on environmental and safety issues. • Customer satisfaction.
	• Fair and transparent transactions. • Supplier education on sustainability issues. • Effective communications.
	• Philanthropic activities.

(continued)

Table 7.4 (continued)

Capabilities	
Improvement (Ovchinnikov et al. 2014; Wu et al. 2012a)	• Continuously standardized production processes. • Continuously simplified production processes. • Continuously reduced waste and variance. • The use of the past successes and failures in improving processes continuously. • Continuous improvement. • Process management. • Leadership involvement.
Innovation (Ovchinnikov et al. 2014; Wu et al. 2012a)	• Innovations created that made the prevailing processes obsolete. • Innovations created that fundamentally changed the prevailing processes. • Innovations created that made the existing expertise in prevailing processes obsolete. • Search for new technologies. • Cross-functional product design. • Processes and equipment development.
Sustainability performance	
Short-term sustainability performance (Wu et al. 2012b)	• ROA. • ROS. • Cost reduction and sales increase from sustainable practices.
	• Reduction of air emissions. • Reduction of wastewater generation. • Reduction of solid waste disposal. • Reduction of consumption of hazardous/harmful/toxic materials. • Reduction of energy consumption.
Long-term sustainability performance (Lai et al. 2015)	• Improvement of long-term profitability. • Improvement of social reputation. • Improvement in number of social awards.

"medium" and that below 40% is "low". For sustainability performance, if a company reported an increase rate of more than 30%, it is coded as "high". Ten percent to 30% is "medium" and an increase of less than 10% is coded as "low". The coding results are summarized in Table 7.5.

7.4 Findings

7.4.1 Within-Case Analysis

Within-case analysis results reveal the status quo of the sustainable development of each case company. Questions on motivations for adopting sustainability practices reflect the firms' perceived stakeholder pressures and their desired benefits. Specific

Table 7.5 Summary of coding results

Company	Sustainability practices	Capabilities	Sustainability performance
A	Short-term: medium Long-term: high	Short-term: medium Long-term: low	Short-term: high Long-term: medium
B	Short-term: medium Long-term: high	Short-term: high Long-term: medium	Short-term: high Long-term: medium
C	Short-term: high Long-term: high	Short-term: high Long-term: high	Short-term: high Long-term: high
D	Short-term: high Long-term: high	Short-term: high Long-term: high	Short-term: high Long-term: high
E	Short-term: high Long-term: high	Short-term: high Long-term: medium	Short-term: high Long-term: high
F	Short-term: high Long-term: medium	Short-term: medium Long-term: medium	Short-term: high Long-term: medium
G	Short-term: low Long-term: medium	Short-term: medium Long-term: low	Short-term: high Long-term: medium
H	Short-term: high Long-term: medium	Short-term: high Long-term: medium	Short-term: high Long-term: medium

sustainability practices currently implemented show the depth and breadth of firms' sustainability involvement. Support given to the implementation of sustainability practices provides information on the capability-building efforts of the case firms. Benefits (both tangible and intangible) realized from the implementation of each sustainability practice reveal the causal relationship between sustainability practices and capabilities and/or performance. Lastly, the time it took for the firms to realize the desired benefits associated with the implementation of sustainability practices reflects the applicability of the time featured sustainability concept in the real business context. It also provides supporting evidence for the categorization of short- and long-term sustainability. Findings are summarized in Table 7.6.

Lean practices have been mentioned by all companies as an important part of CS. As shown in Table 7.6, almost all the case companies were motivated by the successful cases (mostly Japanese auto firms) of lean practices implementation. Firms implement various lean practices for the pursuit of waste reduction, efficiency enhancement, and profit improvement. The respondents reported that their organizations began to realize the desired benefits in 2 months to 1 year following the appropriate implementation. They also pointed out that the benefits of lean practices become more profound with the deepening of lean implementation. This is because improvement is a never-ending task and the more deeply a firm engages in lean practices, the more benefits it will realize. To better implement lean practices, companies usually need to learn by sending people to successful companies where they can observe best practice. Companies also engage in formal training programmes organized by specialized agencies. Lean experts are among the top talent that every firm seeks to employ. In addition to learning from successful companies and training agencies, information sharing sessions happen frequently among peer firms. Internal learning is also important for lean implementation. All case companies

Table 7.6 Summary of findings

		Motivation	Support	Benefit	Time
A	Short-term	Successful auto companies Environmental and safety regulations	Strong leadership support Training existing people Hiring external people who have relevant knowledge Visiting other auto companies regularly Frequent internal communications	Enhanced product quality More skilled and competent staff Gradually standardized process Cost savings	The firm has been experiencing the benefits continuously since implementation
	Long-term	Self-motivated, the firm's sense of responsibility	Effective information sharing with stakeholders A fixed budget for social purpose	Support from the local community Employee loyalty Stable cooperation with suppliers	Years
B	Short-term	Successful auto companies and the environmental and safety regulations	Leadership support Experts from outside Regular training programmes Frequent internal and external communications Formation of problem-solving team	Waste reductions Standardized processes Strengthened relationships with customers Cost savings Business increase	Within a year of implementation
	Long-term	Sense of responsibility of the firm	An increasing budget for social purpose The establishment of various facilities such as a gym for employees	Strong employee loyalty Better corporate reputation in the local community	No specific time length given
C	Short-term	Successful auto companies Environmental and safety regulations Competitors	Leadership support; External and internal learning and communications	Improved product quality Simplified processes Increased sales and productivity A better position in the world marketplace	Improvement can be seen every day since the implementation
	Long-term	Sense of responsibility and the belief in giving back to society The need to enhance the company's reputation	A fixed proportion of profit goes to social use	Better responsiveness to market changes The attraction of talent Strengthened stakeholder relations Increased sales and customer loyalty	Began to realize the desired benefits in 2–3 years

(continued)

Table 7.6 (continued)

		Motivation	Support	Benefit	Time
D	Short-term	The home company Environmental and safety regulations The competitors	Leadership support External and internal learning and communications	Standardized process and enhanced product quality Reduced costs and increased profit Strengthened customer and supplier relations	Immediately after the implementation and the benefits are on-going
	Long-term	The right thing to do The need to promote the brand	A fixed proportion of profit goes to social use	Strengthened stakeholder relations Increased sales High level of employee and customer loyalty	2–3 years
E	Short-term	Successful auto companies Environmental and safety regulations	Leadership support External and internal learning and communications	Standardized process and enhanced product quality Reduced costs and increased profit Strengthened customer and supplier relations	Improvement can be seen every day since implementation
	Long-term	Competitors Sense of responsibility The need to promote the brand	A fixed proportion of profit goes to social use	Strengthened stakeholder relations Increased sales High level of employee and customer loyalty Increased visibility of the brand in China	No specific time given
F	Short-term	Successful auto companies Environmental and safety regulations	Leadership support External and internal learning and communications	Standardized process and enhanced product quality Reduced costs and increased profit Strengthened customer and supplier relations	Benefits are perceived on a continuous basis
	Long-term	Stakeholder pressures Sense of responsibility	A fixed proportion of profit goes to social use	Strengthened stakeholder relations Increased sales High level of employee and customer loyalty	More than 5 years

(continued)

Table 7.6 (continued)

		Motivation	Support	Benefit	Time
G	Short-term	Successful auto companies	Training existing people Hiring professionals Frequent communications with other organizations	Constantly standardized processes Reduced costs	2–3 months
	Long-term	Stakeholder pressures (local government) Development stage of the firms	Leadership support	Satisfaction of government requirements	No benefits to the firm perceived so far
H	Short-term	Successful auto companies Environmental and safety regulations and internal environmental policy from the home company The highly polluting nature of the raw material (chemical)	Leadership support Frequently communications with other branches of the company	Standardized process and enhanced product quality Reduced costs and increased profit	Immediately after implementation
	Long-term	Self-motivated Requirements of the home company	A fixed budget for employee welfare	Strong employee loyalty	No definite length of time can be mentioned. Can be seen gradually

reported the establishment of a lean group composed of one staff member from each department. They meet regularly (once or twice a month) to discuss existing problems, achievements, and possible ways for further improvement. Past experience (both success and failure) plays an important role in internal learning. All these actions reflect the leadership support and resource allocations for lean implementation by the case firms. The benefits of lean implementation mentioned by the interviewees include: gradually standardized processes; enhanced efficiencies; reduced variances and waste; improved cost savings; the development of a responsible and saving organizational culture; improvement in product quality; quicker responsiveness to change; development of skilled personnel; improved problem-solving capability; simplified processes and procedures; increased driving forces to develop better products and processes; and enhanced competitiveness in the marketplace. While some of these benefits can be seen as tangible (such as performance improvement), others are more intangible (suck as capabilities improvement).

Another crucial aspect of CS reported by the interviewees is green/environmental management. The implementation of green practices by the case firms resulted from a combination of external pressures and an internal drive. Both national and regional environmental regulations have been pressuring auto companies to adopt reactive environmental activities such as the end-of-pipe solutions to emissions and sewage, noise control, and solid waste management. Environmental Management Systems (EMS) certifications such as ISO 14001 have been obtained mainly due to customer requirements, especially those from western countries. Some case companies such as C, D, and H are more proactive in the environmental aspect. They voluntarily design more environmental-friendly products, have replaced traditional fuel-consuming equipment with a natural gas equivalent, and engage in recycling wherever possible. For the reactive, regulation-driven environmental actions, firms reported that they can realize the benefits in 1–6 months. However, it is uncertain how proactive and voluntary environmental practices to be translated into actual benefits. Organizational support such as leadership attention and resource investment (funds, people, and time) are needed for the implementation of environmental practices. The realized and desired benefits of environmental actions reported by the case companies include the compliance of environmental regulations, stakeholder satisfaction and corporate image upgrading, the attraction of more environmentally conscious customers, more efficient processes, reduced waste, the development of new products and process which have environmental features, cost savings, and a better natural environment for humans, including for the next generation. Like lean, while some of the benefits are associated with enhanced tangible performance, others can be interpreted as operational improvement and innovation capabilities.

Social responsibility is also seen as an important element of CS, according to the interviewees. Similar to environmental practices, the implementation of socially responsible practices is mainly a result of stakeholder pressures, though some interviewees mentioned the social responsibility of relatively mature firms. The time needed for firms to realize the benefits of implementing social activities varies from half a year to "never", as reported by the interviewees. For employee-related practices such as training firms can experience enhanced worker productivity immediately after they are trained. The adoption of practices aimed at ensuring employee working safety has even faster benefits in terms of regulation compliance. However, for philanthropic donations made by the firms, though a better corporate image could be expected, only three case companies perceive a benefit (C, D, and E), while the remainder engage in this activity only because of coercive pressures. Benefits of social practice implementation reported by the interviewees include: increased worker productivity; the development of new (human- and environmental-friendly) products/processes; strengthened stakeholder relations; enhanced employee loyalty; improved customer satisfaction; improved corporate reputation; as well as increased sales and thus revenue.

Based on information provided by interviewees, lean practices and regulatory environmental practices are usually seen as short-term sustainability as they show quicker benefit realization. On the other hand, proactive environmental practices and social practices are long-term sustainability practices whose benefits are relatively longer-term and uncertain. Table 7.7 summarizes the short- and long-term CS practices currently implemented by the case companies.

7.4.2 Cross-case Comparison Analysis

The within case analysis results reveal differences among the case firms in terms of practice implementation, capabilities, and sustainability performance. Several company characteristics have been found to be associated with these differences, namely, position in the automotive supply chain, organization size, and age. Thus, cross-case comparative analysis was applied between end assemblers and parts suppliers, larger firms and smaller firms, as well as older companies and younger companies for more in-depth insights.

7.4.2.1 Auto End Assemblers vs. Parts Suppliers

In terms of position in the supply chain, C, D, and E are end assemblers, Companies A, B, F, and G are tier-1 suppliers and Company H is a tier-2 supplier. As presented in Table 7.8, differences were identified between these end assemblers and suppliers mainly in terms of motivations for long-term sustainability practice implementation, long-term sustainability practices commitment and the associated benefits.

Short-term CS practices are found to be important for both automobile manufacturers and parts suppliers, which is understandable because to reduce cost remains the fundamental goal of any for-profit organization. Both groups of companies acknowledge the importance of lean practices in continuous improvement and waste minimization, and they are all making efforts to implement the various practices in the lean management system.

It has been found that automobile manufacturers and parts suppliers differ in terms of motivation, focus, and benefit for implementing long-term sustainability practices. According to the interviewees, since automobile manufacturers face the end customers directly, and suppliers mainly deal with their downstream supply chain partners, they are confronted with different levels of sustainability challenges given the different level of coverage of stakeholder groups. As consumers are becoming more environmentally and socially conscious in recent years, the social image of the company is one of the most important factors to influence customers' purchase intentions (Newton et al. 2015). In response to this practical problem, automotive manufacturers were found to engage in various long-term sustainability practices, and through effective media exposure, they are also making efforts to promote and publicize such activities. However, parts suppliers face less intensive pressures in this aspect because they are assessed and selected by their customers (the upstream supply chain partners) based on factors including cost, quality, capabilities, efficiency, and strategic co-operation. Sustainability, especially long-term sustainability, is not a serious concern. As a result, facing less intensive stakeholder pressures, tier-1 and tier-2 parts suppliers tend to do the minimum in terms of long-term sustainability, that is, what is required by their customers. There is insufficient driving force for them to adopt long-term sustainability practices.

The different levels of emphasis on long-term sustainability result in different levels of operational capabilities, especially innovation capability developed between automobile manufacturers and parts suppliers. The ultimate performance outcomes also vary. In general, automobile manufacturers have reported a higher level of operational innovation capability than parts suppliers. Although there are different sources for organizational innovation capability, the implementation of practices which involve different stakeholders, such as long-term CS practices in this study, play a key role (Kemper et al. 2013). Another possible reason for the low level of innovation capability of parts suppliers identified in the case studies is the strict supplier management systems in which any changes made by the suppliers to any aspect of the production need to be reported to the focal companies for approval in advance. This, on the one hand, maintains the stability of the production and product quality; on the other hand, it hinders the development of the innovation capability of the suppliers. Parts suppliers usually do not perceive the positive effect of long-term sustainability practices on performance. As a result, they do not have sufficient motivation to implement long-term CS practices, and have not experienced any benefits associated with such implementation.

7.4.2.2 Large Companies vs. Small Companies

Among the eight case companies, A, B, C, D, E, and F are large companies with employee numbers over 1000, and Companies G and H are small companies with less than 300 employees. Differences in both short- and long-term sustainability commitment and the results of implementation were identified between these two groups of cases, and are summarized in Table 7.9.

In terms of short-term and long-term CS practices, in general larger firms were found to be more committed than smaller ones due to direct pressure from the end customers. Resource availability such as people and finance is found to account for the different levels of sustainability commitment and the subsequent capability and sustainability performance.

Larger firms are believed to have greater resource availability than their smaller counterparts (Shah and Ward 2007). Thus, in terms of investing in both short- and long-term CS practices, larger firms tend to be more proactive. On the other hand, smaller firms, with limited resources, have to be extra cautious when making such decisions. They tend to start small from the short-term CS practices as an attempt to minimize risks, which is supported by the findings. Consequently, the extent of capability development and performance improvement reported by larger firms is higher than that of the smaller firms.

Table 7.7 Summary of short- and long-term sustainability practices

Case company	Short-term CS practices implemented	Long-term CS practices implemented
A	• 6 sigma (failed) • TPM (regular checking and maintenance of equipment). • JIT. • Inventory management. • ISO 14000. • Cooperation with third-party organizations for the collection of usable parts.	• Regular systematic training for all employees. • Employee working condition improvement. • Accommodation, food and daily supplies for employees. • Constant introduction of talent. • Better compensation for multi-skilled workers. • Equal treatment to all employees. • Fair and transparent cooperation with suppliers; no delay in payment. • Provide monthly support for local residents over 60 years old.
B	• Problem and waste identification. • JIT. • 6 sigma • 5s • TQM. • An emphasis on continuous improvement. • Emissions and noise control. • The use of energy-efficient equipment. • ISO 14000, 18001.	• Regular training for employees. • Equal treatment of women and the disabled. • Employee caring Centre. • Multiple channels for employee voice. • Establishment of plants near customers. • Transparent cooperation with suppliers. • Customer first. • Special support for education, orphans and the disabled. • Philanthropic donations to disaster areas.
C	• Waste and problem identification. • TQM. • PULL. • SPC. • Regular maintenance of machines and other equipment. • Inventory management (3-day inventory). • Cell production. • ISO 14001. • The avoidance of harmful raw materials wherever possible. • Waste recycling (cooling water only so far). • Replacement of some fuel-using equipment by natural gas equivalent.	• The development of environmentally friendly interior parts. • Regular training for employees. • Constant improvement of the working conditions and accommodation. • Regular customer surveys. • Special customer caring programmes. • Strategic supplier management. • Special support for education. • Philanthropic donations to disaster areas. • Community services (free vehicle checking, public lectures, etc.) .

(continued)

Table 7.7 (continued)

Case company	Short-term CS practices implemented	Long-term CS practices implemented
D	• PULL. • TQM. • TPM. • SPC. • Equipment layout. • Inventory management (1–4-h inventory). • The reduction of emissions (CO_2 and VOC). • The reduction of harmful and toxic materials. • The reduction of solid waste. • The reduction of energy use. • Heavy investment on sewage management. • 100% recycling of industrial waste • ISO 14001.	• Multiple ways for employees to voice opinions/concerns. • Continuous improvement of employee welfare. • Various training for employees; 30 credits/year required. • A specialized department for customer to voice complaints and satisfaction. • Transparent purchasing procedures. • Philanthropic donations to disaster areas. • Special support for education, sport, and the arts.
E	• Strict quality management procedures. • JIT. • PULL. • TPM. • TQM. • SPC. • Inventory management. • ISO 14001.	• Regular employee training. • Employee satisfaction enhancement efforts. • Customer satisfaction enhancement efforts. • Transparent relationships with suppliers. • Tree planting. • Contributions to desert management. • Donation of products wherever needed. • Charity fund.
F	• Pull. • Layout. • Regular maintenance of equipment. • Quality management. • 5s • 0 inventory for finished products; a-weekly inventory for raw materials • JIT. • Multi-skilled workers. • Compliance with environmental laws and regulations. • Proper control of the use of harmful but necessary materials. • LCA. • Sewage and solid wastes handled by third-party organizations. • Filter applied to emissions. • Recycling of packages. • ISO 14000, 18000. • An emphasis on employee working safety.	• Employee welfare improvement based on performance. • Training for the self-development of employees. • Employee working condition improvement. • Multiple ways for employees to get their voice heard. • Specialized department for customer voices to be heard. • Transparent co-operation with suppliers; supplier selection based on capabilities and price. • Philanthropic donations to society and special support for education.

(continued)

Table 7.7 (continued)

Case company	Short-term CS practices implemented	Long-term CS practices implemented
G	• Just begun to implement lean practices. • JIT delivery (establish plants and transfer warehouses near customers). • Design the motions from a technical perspective. • The identification of wastes and problems. • Waste control. • A-monthly inventory (necessary). • ISO 14000. • Compliance with environmental laws and regulations. • Random environmental audit by local government. • Cooperation with local third-party organizations to deal with sewage.	• A whole set of salary management system. • Equal treatment of the disabled and female employees. • Regular training for employees. • Customer voice and demands heard. • Philanthropic donations and support for the local community.
H	• TPM. • Inventory management (safe inventory). • Compliance with internal environmental and government regulations. • Regular environmental audit from headquarters. • Systematic procedures to control harmful gas and sewage.	• A fixed budget for employee safety and welfare. • Specially designed training for all employees. • The company basically has no suppliers as a raw material producer itself. • No special activities for society or the local community because the company believes a responsible organization should first treat their own employees well.

7.4.2.3 Older Companies vs. Younger Companies

Companies A, B, E, and G are grouped as old companies with an age of over 25 years. Companies whose operating time is less than 25 years are grouped as young companies, and Companies C, F, D, and H belong to this group. As indicated in Table 7.10, old companies and young companies were found to differ in terms of sustainability commitment (both short- and long-term), and the benefits achieved from sustainability practice implementation.

It is not surprising that older companies exhibit higher levels of commitment to both short-term and long-term sustainability, nor that they have achieved more benefits from implementing such practices compared with their younger counterparts. As CS practices in this study are management systems rather than isolated practices, it is reasonable that they become more systematic as firms continue to invest in the systems. This is consistent with (Reverte et al. 2016) who hold that firms become more efficient as they grow; they gradually find out what they are good at and specialize accordingly.

A general belief held by the case companies is that companies should focus on capital accumulation and expansion in the early years of establishment. During this time, firms tend to be more reactive in terms of long-term sustainability due to resource constraints . They gradually move towards the proactive end of the spectrum as they grow. As expressed by the interviewee from Company G, society and the firm support each other. The firm initially takes from society, but when it reaches a steady phase, it will start to give back.

7.4.3 Propositions

According to the case study findings, implementing CS practices brings firms varying levels of benefits on different time scales. Specifically, practices with more certain and quicker benefits include various lean practices and reactive, regulation-driven environmental practices. On the other hand, practices with less certain and longer-term benefits include proactive environmental practices such as environmental design and most socially responsible practices. The benefits reported by the case companies as a result of the implementation of both types of CS practices can be interpreted as enhanced operational improvement and innovation capabilities and sustainability performance, though interviewees did not specifically use such terms. Thus, a causal relationship between the implementation of CS practices and the development of operational capabilities and sustainability performance of the firms can be proposed. In addition, several factors including firms' position in the supply chain, size, and age were identified as potential influences on the drivers, commitment levels as well as the results of practice implementation. Based on the findings, the following propositions are proposed.

P1: The implementation of short-term and long-term CS practices significantly contributes to the improvement of short-term and long-term sustainability performance of the firm.

P2: The implementation of short-term and long-term CS practices significantly contributes to the development of operational improvement and innovation capabilities.

P3: Operational improvement and innovation capabilities are significantly associated with the short-term and long-term sustainability performance of the firm.

P4: Firm characteristics including position in the supply chain, firm size, and age affect firms' sustainability commitment level and the benefits realized.

Table 7.8 Summary of major differences between end assemblers and tier-1 suppliers and tier-2 suppliers

Differences	End assemblers	Suppliers
Motivations for long-term sustainability practice implementation	The customer is a major driver to adopt long-term sustainability practices. They tend to be more proactive in order to satisfy the requirements of the environmentally and socially conscious consumers. At times their long-term sustainability activities even extend beyond the demand of customers, through which they try to communicate their sustainability values to those customers. Due to the higher levels of visibility of end assemblers' long-term sustainability efforts to their end customers, except for the intangible benefits (new product/ process development), they tend to have more tangible benefits, such as increased sales, than do suppliers.	Suppliers' implementation of long-term sustainability is mostly driven by coercive forces such as regulations and the local government because their customers are OEMs, for whom sustainability features are not a significant criterion in selecting suppliers. Compared with OEMs, suppliers experience less tangible and intangible benefits from the adoption of long-term sustainability practices.
Long-term sustainability practice implementation	It has been found that OEMs have a bigger coverage of long-term sustainability practices ranging from different stakeholders with a heavy emphasis put on customers. Instead of pure social responsibility, long-term sustainability is a way for OEMs to communicate with customers.	Suppliers are more focused on employees in terms of long-term sustainability. Besides, society-related practices such as philanthropic donations are the result of coercive forces.
Benefits of long-term sustainability practice implementation	Tangibles: increased productivity, sales and profitability. Intangibles: innovation capability development; enhanced customer and employee satisfaction and loyalty; strengthened stakeholder relations; better responsiveness to changes; and better corporate image.	Tangibles: not experienced. Intangibles: more skilled workers; higher employee loyalty; the satisfaction of governmental requirements; a more harmonious society.

7.5 Conclusion

Sustainability has come to one of the top priorities for all countries especially emerging economies in recent years. This study found empirical evidence that practices in lean, green, and social management systems are currently being implemented by Chinese automotive manufacturing firms as an attempt to be more sustainable. Firms realized varying levels of benefits from implementing these practices depending how they are implementing it, resource commitment, and other supporting structure and/or activities. It is obvious that manufacturing sector of emerging economies such as China is moving towards sustainability due to internal and/or external motivations, and both tangible and intangible benefits can be expected from implementing sustainability practices. However, firms do need to

Table 7.9 Summary of major differences between large firms and small firms

Differences	Large firms	Small firms
Sustainability commitment	Large firms exhibit higher levels of commitment to both short- and long-term sustainability.	Due to resource constraints, smaller firms tend to focus more on short-term sustainability and they usually start from the very basic level, then move forward gradually.
Results	Large firms were found to have more benefits from the implementation of sustainability practices, both tangible and intangible.	Small firms usually perceive no benefits from implementing long-term sustainability practices.

Table 7.10 Summary of major differences between old firms and young firms

Differences	Older firms	Younger firms
Sustainability commitment	Older firms were found to have higher levels of sustainability commitment, both in terms of breadth and depth.	Younger firms have a primary emphasis on short-term sustainability.
Benefits	Older firms have experienced more benefits from implementing long-term sustainability practices.	No benefits have been perceived from the implementation of long-term sustainability practices.

have a long-term mind-set when making decisions regarding sustainability so that short- and long-term benefits can both be secured.

References

Adebambo, H. O., Abdulkadir, R. I., Mat, N. K. N., Alkafaagi, A. A. J., & Kanaan, A. G. J. (2013). Drivers of sustainable environmental manufacturing practices and financial performance among food and beverages companies in Malaysia. *American Journal of Economics, 3*, 127–131.

Ambrosini, V., Bowman, C., & Collier, N. (2009). Dynamic capabilities: An exploration of how firms renew their resource base. *British Journal of Management, 20*, s9–s24.

Amini, A., & Bienstock, C. C. (2014). Corporate sustainability: An integrative definition and framework to evaluate corporate practice and guide academic research. *Journal of Cleaner Production, 76*, 12–19.

Bansal, P. (2005). Evolving sustainability: A longitudinal study of corporate sustainable development. *Strategic Management Journal, 26*, 197–218.

Barney, J. B. (1991). Firm resources and sustained competitive advantage. *Journal of Management, 17*, 99–120.

Bromiley, P., & Rau, D. (2016). Operations management and the resource based view: Another view. *Journal of Operations Management, 41*, 95–106.

Campbell, J. M., & Park, J. (2017). Extending the resource-based view: Effects of strategic orientation toward community on small business performance. *Journal of Retailing and Consumer Services, 34*, 302–308.

Carmeli, A. (2004). Assessing core intangible resources. *European Management Journal, 22*, 110–122.

Cepeda, G., & Vera, D. (2007). Dynamic capabilities and operational capabilities: A knowledge management perspective. *Journal of Business Research, 60*, 426–437.

Chang, D., & Kuo, L. R. (2008). The effects of sustainable development on firms' financial performance – An empirical approach. *Sustainable Development, 16*, 365–380.

Clement, R. W. (2005). The lessons from stakeholder theory for US business leaders. *Business Horizons, 48*, 255–264.

Darnall, N., Henriques, I., & Sadorsky, P. (2010). Adopting proactive environmental strategy: The influence of stakeholders and firm size. *Journal of Management Studies, 47*, 1072–1094.

Dimaggio, P. J., & Powell, W. W. (1983). The iron cage revisited: Institutional isomorphism and collective rationality in organizational fields. *American Sociological Review, 48*, 147–160.

Dyllick, T., & Hockerts, K. (2002). Beyond the business case for corporate sustainability. *Business Strategy and the Environment, 11*, 130–141.

Eisenhardt, K. M., & Martin, J. A. (2000). Dynamic capabilities: What are they? *Strategic Management Journal, 21*, 1105–1121.

Elkington, J. (1998). *Cannibals with forks: The triple-bottom-line of the 21st century*. Stoney Creek: New Society Publishers.

Fraj, E., Martinez, E., & Matute, F. (2013). Green marketing in B2B organizations: An empirical analysis from the natural-resource-based view of the firm. *Journal of Business & Industrial Marketing, 28*, 396–410.

Garces-Ayerbe, C., Rivera-Torres, P., & Murillo-Luna, J. L. (2012). Stakeholder pressure and environmental proactivity: Moderating effect of competitive advantage expectations. *Management Decision, 50*, 189–206.

Gimenez, C., Sierra, V., & Rodon, J. (2012). Sustainable operations: Their impact on the triple bottom line. *International Journal of Production Economics, 140*, 149–159.

Govindan, K., Azevedo, S. G., Carvalho, H., & Cruz-Machado, V. (2014). Impact of supply chain management practices on sustainability. *Journal of Cleaner Production, 85*, 212–225.

Hajmohammad, S., Vachon, S., Klassen, R. D., & Gavronski, I. (2013). Lean management and supply management: Their role in green practices and performance. *Journal of Cleaner Production, 39*, 312–320.

Hart, S. L., & Dowell, G. (2011). A natural resource based view of the firm: Fifteen years after. *Journal of Management, 37*, 1464–1479.

Hennink, M., Hutter, I., & Bailey, A. (2011). *Qualitative research methods*. London: Sage.

Hitt, M. A., Xu, K., & Carnes, C. M. (2016). Resource based theory in operations management research. *Journal of Operations Management, 41*, 77–94.

Hofmann, K. H., Theyel, G., & Wood, C. H. (2012). Identifying firm capabilities as drivers of environmental management and sustainability practices – Evidence from small and medium-sized manufacturers. *Business Strategy and the Environment, 21*, 530–545.

Kemper, J., Schilke, O., & Brettel, M. (2013). Social capital as a microlevel origin of organizational capabilities. *Journal of Product Innovation Management, 30*, 589–603.

Klassen, R. D., & Vereecke, A. (2012). Social issues in supply chains: Capabilities link responsibility, risk (opportunity), and performance. *International Journal of Production Economics, 140*, 103–115.

Kurapatskie, B., & Darnall, N. (2013). Which corporate sustainability activities are associated with greater financial payoffs? *Business Strategy and the Environment, 22*, 49–61.

Lai, W.-H., Lin, C.-C., & Wang, T.-C. (2015). Exploring the interoperability of innovation capability and corporate sustainability. *Journal of Business Research, 68*, 867–871.

Luzzini, D., Brandon-Jones, E., Brandon-Jones, A., & Spina, G. (2015). From sustainability commitment to performance: The role of intra- and inter-firm collaborative capabilities in the upstream supply chain. *International Journal of Production Economics, 165*, 51–63.

Montabon, F., Sroufe, R., & Narasimhan, R. (2007). An examination of corporate reporting, environmental management practices and firm performance. *Journal of Operations Management, 25*, 998–1014.

Newbert, S. L. (2007). Empirical research on the resource-based view of the firm: An assessment and suggestions for future research. *Strategic Management Journal, 28*, 121–146.

Newton, J. D., Tsarenko, Y., Ferraro, C., & Sands, S. (2015). Environmental concern and environmental purchase intentions: The mediating role of leaning strategy. *Journal of Business Research, 68*, 1974–1981.

Ovchinnikov, A., Blass, V., & Raz, G. (2014). Economic and environmental assessment of remanufacturing strategies for product + service firms. *Production and Operations Management, 23*, 744–761.

Peng, D. X., Schroeder, R. G., & Shah, R. (2008). Linking routines to operations capabilities: A new perspective. *Journal of Operations Management, 26*, 730–748.

Penrose, E. T. (1959). *The growth of the firm*. New York: Wiley.

Pullman, M. E., Maloni, M. J., & Carter, C. R. (2009). Food for thought: Social versus environmental sustainability practices and performance outcomes. *Journal of Supply Chain Management, 45*, 38–54.

Reverte, C., Gomez-Melero, E., & Cegarra-Navarro, J. G. (2016). The influence of corporate social responsibility practices on organizational performance: Evidence from eco-responsible Spanish firms. *Journal of Cleaner Production, 112*, 2870–2884.

Russo, M. V. (2009). Explaining the impact of ISO 14001 on emission performance: A dynamic capabilities perspective on process and learning. *Business Strategy and the Environment, 18*, 307–319.

Sarkis, J., Zhu, Q., & Lai, K. (2011). An organizational theoretic review of green supply chain management literature. *International Journal of Production Economics, 130*, 1–15.

Savitz, A. W., & Weber, K. (2006). *The triple bottom line*. San Francisco: Jossey-Bass.

Schoenherr, T. (2012). The role of environmental management in sustainable business development: A multi-country investigation. *International Journal of Production Economics, 140*, 116–128.

Shafiq, A., Klassen, R. D., Johnson, F., & Awaysheh, A. (2014). Socially responsible practices: An exploratory study on scale development using stakeholder theory. *Decision Sciences, 45*, 683–716.

Shah, R., & Ward, P. T. (2007). Defining and developing measures of lean production. *Journal of Operations Management, 25*, 785–805.

Teece, D. J., Pisano, G., & Shuen, A. (1997). Dynamic capabilities and strategic management. *Strategic Management Journal, 18*, 509–533.

Tomsic, N., Bojnec, S., & Simcic, B. (2015). Corporate sustainability and economic performance in small and medium sized enterprises. *Journal of Cleaner Production, 108*, 603–612.

Wong, C. W. Y., Lai, K., Shang, K., Lu, C., & Leung, T. K. P. (2012). Green operations and the moderating role of environmental management capability of suppliers on manufacturing firm performance. *International Journal of Production Economics, 140*, 283–294.

World Commission on Environment and Development (WCED). (1987). *Our common future*. Oxford: Oxford University Press.

Wu, Q., He, Q., Duan, Y., & O'Regan, N. (2012a). Implementing dynamic capabilities for corporate strategic change toward sustainability. *Strategic Change, 21*, 231–247.

Wu, S. J., Melnyk, S. A., & Swink, M. (2012b). An empirical investigation of the combinatorial nature of operational practices and operational capabilities – Compensatory or additive? *International Journal of Operations & Production Management, 32*, 121–155.

Wu, L., Subramanian, N., Abdulrahman, M. D., Liu, C., Lai, K., & Pawar, K. S. (2015). The impact of integrated practices of lean, green, and social management systems on firm sustainability performance – Evidence from Chinese fashion auto-parts suppliers. *Sustainability, 7*, 3838–3858.

Yang, M. G., Hong, P., & Modi, S. B. (2011). Impact of lean manufacturing and environmental management on business performance: An empirical study of manufacturing firms. *International Journal o Production Economics, 129*, 251–261.

Zhu, Q., Sarkis, J., & Lai, K. (2013). Institutional-based antecedents and performance outcomes of internal and external green supply chain management practices. *Journal of Purchasing & Supply Management, 19*, 106–117.

Chapter 8
Inventory Management in Multi-echelon After-Sales Service Networks

Sandeep Srivathsan

Abstract The rapid improvement in production technology over the last two decades has reduced the difference in quality of products produced by competitors. This has resulted in a paradigm shift in the outlook of firms from a product-centric view to a more customer-centric view. Consequently, after-sales service or warranty is high on the agenda of firms in a service supply chain. In general, firms offer service contracts that guarantee a quick turnaround time for repair of faulty components, while ensuring full functionality of their returned product. In this chapter, we focus on such contracts from an inventory management perspective in a multi-echelon service supply chain. The customer returns arrive at various service centers spread over a region. The repair activity essentially involves identifying the faulty component, and replacing it with a repaired component (if available) or with a new component. The faulty components are then collected and sent to a common repair center. The repaired components are then returned to the service centers, where they shall be used to satisfy demand generated by future returns. The problem is modeled using M/M/1/K queueing system, and we derive properties of the total cost function.

8.1 Introduction

For most part of the last century, manufacturing firms were focused on the quality of their finished products. However, with the rapid improvement in production technology over the past few decades, the difference in quality of products manufactured by different firms have drastically diminished. This has resulted in a shift in the firm's outlook from a product-centric view to a more customer-centric view (Srivathsan and Viswanathan 2017; Srivathsan and Kamath 2017). As a result, in a competitive setting, the after-sales service offered by manufacturing firms through service contracts and warranties have received significant attention. In summary, the perspective about after-sales service has seen a major transition from being seen as

S. Srivathsan (✉)
Great Lakes Institute of Management, Chennai, Tamil Nadu, India
e-mail: sandeep.sri@greatlakes.edu.in

C. Chen et al. (eds.), *Pursuing Sustainability*, International Series in Operations Research & Management Science 301,
https://doi.org/10.1007/978-3-030-58023-0_8

a cost center to being a driving force in a firm's profit. A study by Cohen et al. (2006) has reported that firms such as Caterpillar, GE, and Saturn have employed effective after-sales service to increase their customer loyalty. Additionally, an AMR survey conducted by Bijesse et al. (2002) reported that service contributed to about 24% of a manufacturing firm's revenue and about 45% of their profit (Jalil 2011).

In any after-sales service system, the service contracts promise a quick turnaround time (TAT) by ensuring readily available products. Usually with products that adopt a modular design, the after-sales service begins with a diagnostic step that focuses on identifying the faulty component in the product. This is usually followed by the repairing of faulty components. In reality, the customer returns arrive at different service centers that are secluded from the common repair facility. This gives rise to potential logistical issues as far as meeting TAT is concerned. For instance, a US-based firm, Xeptron, had a spare parts business unit in Singapore to cater to the needs of their Asia-Pacific customers. When a defective item arrived at this unit, it was sent to an engineer, who diagnosed the problem and replaced the defective item. The defective item was then sent to its US headquarters for repairs, and subsequently the repaired units returned to the Asia-Pacific region (Tan et al. 2003). This means that the lead time (comprising of time to ship the faulty component from the service center to the repair facility, time to repair a faulty component, and time to ship the repaired component back to the service center) is usually more than the promised TAT. As a consequence, the service centers resort to keeping stock of the repaired components as well as new ones to meet the TAT, and thereby improve customer loyalty.

With the recent global awareness about sustainable operations (Guide et al. 2000; Kleindorfer et al. 2009; Linton et al. 2007; Jayaraman et al. 1999), increased legislations on product take back (Esenduran et al. 2012; Klausner and Hendrickson 2000) and pricing on carbon emissions through carbon taxes (Fisher-Vanden et al. 1997; MacCracken et al. 1999) to counter the effects of global warming, the economics for using repaired component instead of new ones is getting stronger (Srivathsan and Viswanathan 2017). This demands ensuring sufficient inventory of repaired component at the service center, and highlights the paramount importance of inventory management in multi-echelon after-sales service systems.

The inventory management problem discussed in the chapter is a subset of the broad area of reverse logistics, and there is a huge body of literature in this area (Fleischmann et al. 1997; Govindan et al. 2015; Bazan et al. 2016). Furthermore, among the various streams of research in reverse logistics, the inventory management problem in an after-sales service system is closely related to the repairable item inventory systems (RIIS), inventory systems with product returns and inventory systems with remanufacturing. The focus of this chapter is to furnish a detailed review of literature in these areas and to demonstrate a modeling approach to tackle the after-sales service network problem in a multi-echelon setting.

8.1.1 Structure of the Chapter

The rest of the chapter is structured as follows. Section 8.2 presents a detailed review of literature on repairable item inventory systems, while Sects. 8.3 and 8.4 present the detailed review of literature on the inventory systems with product returns, and inventory systems with remanufacturing, respectively. Section 8.5 presents an analytical model developed for a multi-echelon after-sales service network with two service centers and one central repair facility. The section also discusses the scope for future research.

8.2 Literature on Repairable Items Inventory Systems (RIIS)

A seminal work in this area is the multi-echelon technique for recoverable item control (METRIC) model developed by Sherbrooke (1968). This is a closed-loop network model that was initially developed for the US Air Force with multiple bases and a central depot. In this model, the repair of faulty items (military equipment) can be conducted in the central depot. The model consists a fixed number of items that may be deployed at different bases or the central depot. An inventory of spare items was also maintained at the depots. In general, when an item being becomes faulty, it is sent for repair and replaced by a spare item. When there are no spare items, the repair orders are backordered. This gives rise to two cost components in the model, namely, holding cost and backorder costs.

The objective of the METRIC model was then to determine the optimal number of spare units to be held at different locations, in order to minimize the total cost (sum of the cost components) while ensuring that the expected number of backorders at a depot does not exceed their prespecified limit. The model assumed compound Poisson demand (failure rate) and arbitrary resupply times with ample repair capacity and large parts population. From the perspective of spare item inventory, the RIIS system is equivalent to a normal multi-echelon inventory system with one-for-one replenishment policy and a stochastic lead time. Note that the lead time is dependent on the inventory level of the spares and the time required for repair (Srivathsan and Viswanathan 2017).

The early model were focused on relaxing one or more assumptions made by Sherbrooke (1968). Sherbrooke (1971) reformulated the METRIC model to directly minimize the expected number of non-operationally ready (spare) aircraft. A major drawback of this approach was that the optimization model is not mathematically tractable, and only approximate solutions could be obtained.

The MOD-METRIC model, an extension of the METRIC model developed by Muckstadt (1973), considered multiple levels of indentures. A problem of obtaining the optimal number of spare aircraft engines and engine module stock that minimizes the total cost subject to system investment constraint was used to illustrate the developed model. For cases where the number of indentures increases, the solution of the problem became more difficult. Consequently, Muckstadt (1978) proposed a Lagrangian optimization technique to improve the computational speed. Muckstadt (1979) also extended the problem to address a three-echelon problem.

The restrictive assumptions on the demand and resupply times were relaxed, and closed-form solutions for optimal inventory levels were obtained by Simon (1971) and Kruse (1979). While both models assumed Poisson demand and constant resupply times, the former focused on a two-echelon problem and the latter focused on a multi-echelon problem (Sherbrooke 1986).

The VARI-METRIC model, developed by Slay (1984), provided an improvement on the METRIC model by fitting a negative binomial distribution for the first two moments of the number of items in repair. This approximation was used to more accurately represent variance in the part failure process. Sherbrooke (1986) refined the VARI-METRIC model to better estimate the number of backorders at the bases in case of a multi-indenture, multi-echelon RIIS, as well as to improve the computational speed of the model.

Graves (1985) developed a model for a two-echelon system with assumptions similar to those of the METRIC model. Further, they assumed that all repair takes place at the depot only. An exact model for the steady state distribution of inventory at each site was developed. An approximation was made for the steady state distribution for the case with ample servers at the repair depot. Extensive numerical experimentation was used to show the effectiveness of the model in predicting optimal spare inventory levels.

Lee (1987) considered a two-echelon system with lateral transshipment (transfer of spare parts from one base to another for satisfying demand during stock-out at that base) to satisfy the high service requirement. An approximate model was derived to obtain the optimal spare part inventory level that minimizes the total cost function that comprised of the inventory holding cost, backorder cost and transportation costs. A detailed procedure was furnished to obtain this the optimal solution.

Albright and Soni (1988) developed an exact solution methodology based on the continuous time Markov chain (CTMC) model for a two-echelon network. Given the complexity that arises due to large number of states even for small problems in the CTMC model, they resorted to the use of aggregation/disaggregation techniques to obtain the steady-state solution. Albright (1989) examined a two-echelon system using a queueing-based optimization model to minimize backorders. They furnished an approximate steady-state probability distribution of the system for a single-item under the assumption of no condemnation of faulty items, no lateral supply and deterministic travel time.

Gupta (1993) formulated a multi-indenture problem for a single base with finite population and finite servers using a CTMC model. The model considered a single product with repair only at the depot, and approximations were obtained for the steady-state characteristics of the system. These results were validated using an equivalent simulation model.

While most of the above models assumed ample repair capacity, that is, ample repair facilities and large parts population, there have also been models that have considered limited repair capacity. Shanker (1981) derived the exact solutions for the case where the repair times were assumed to be constant. The model also considered batch and unit inspection policies, where a faulty component could not be repaired. Gross et al. (1993) modeled a multi-echelon system as a finite-state space closed queueing network with finite population and limited servers using a CTMC model. They provided multiple methods to solve the steady-state balance equations of the non-Jackson Markov network. Given the complexity of the problem, they suggested simulation as a better option for solving large-scale problems.

Díaz and Fu (1997)furnished approximate solutions for a multi-echelon RIIS with limited repair capacity. They considered three scenarios, namely, single-class exponential repair time distributions, single-class general repair time distributions as well as multi-class general repair time distributions. Numerical experiments based on the models showed that the proposed models outperform the other traditional model that assumed ample repair capacity.

Perlman et al. (2001) investigated the congestion that results from the limited repair capacity in multi-echelon systems with two modes of repair. A novel expanding repair (ER) policy was prescribed for the bases to choose one of the two repair models and these were compared with different expediting policies that considered congestion. The ER policy that considered congestion were found to outperform the naive ER policy.

Sleptchenko et al. (2002) modified the VARI-METRIC model to allocate service part stocks in the network, and the repair shop was modeled as a multi-server queue. Sleptchenko et al. (2003) extended the work of Graves (1985) by modeling the repair shop as a multi-class multi-server queue. The model studied the trade-off between inventory and repair capacity. Avsar and Zijm (2003) modeled a two-echelon maintenance system with multiple bases and a central depot. The repair facilities were modeled as an open queueing network and the transit operations from the central depot to the bases were modeled as an infinite server queues (ample capacity). The model yielded near-perfect product-form solutions for the steady-state distributions, which were then used to estimate performance measures such as fill rate and stock-out probabilities.

Lau and Song (2008) considered a multi-echelon RIIS with limited repair capacity and non-stationary demands, and developed an optimization algorithm to solve the corrective maintenance problem in military logistics. Rappold and Roo (2009) developed an approach to solve the joint problem of facility location, inventory allocation and repair capacity investment in a two-echelon single-item, service supply chain with random demand. The model was then extended to the joint problem of facility location and multi-echelon inventory optimization. More recent work in the RIIS models with limited capacity include the work of Cohen et al. (2017), and Turan et al. (2018, 2020).

Mirzahosseinian and Piplani (2011, 2013) studied an inventory model for a repairable parts system operating under a performance-based logistics contract in which the service provider is compensated for the performance delivered by the system under contract. van Jaarsveld et al. (2015) studied spare parts inventory control for an aircraft component repair shop, where inventory of each component is controlled using independent (s, S) inventory policy. The cost minimization problem was formulated as an integer program with fill rate constraints, and column generation was applied to solve the optimization problem.

Although there are considerable similarities between the after-sales service network problem discussed in Sect. 8.1 and the RIIS model considered in literature, there are significant difference as well. In the RIIS model, the faulty products arrival basically comes from among the fixed number of items within the system. However, in our model, the faulty components arrive comes from a large customer/product base in the region. Consequently, the arrival rate of the failed components is independent of the number of components in repair. The high service requirement in our problem necessitates an immediate replacement of failed components (using either a repaired component or a brand new component). Another key difference is that the RIIS model is a closed system as all faulty components have to be repaired.

8.3 Literature on Inventory Systems with Product Returns

A common assumption in literature related to inventory system with product returns is that any returned item can be immediately used to satisfy demand for new products (i.e., they are considered to be as good as new). This means that the lead time and setup cost for repair (or remanufacturing) are negligible. Most papers in this stream consider stochastic product demand and return processes.

Heyman (1977) considered an inventory system with product return in which the arrival of a customer order into the system denotes arrival, while the arrival of product return signifies service completion. There is no condemnation of defective components, and an upper limit on the inventory level, K, is maintained. Assuming Poisson arrivals and exponential product returns, this system was modeled as an M/M/1/K queue.

Muckstadt and Isaac (1981) developed a solution methodology to estimate the parameter values of a continuous review procurement policy for a single-item, single location problem with return rate less than demand rate. They also provided an extension to a single-item two-echelon problem. Korugan and Gupta (1998) modeled a two-echelon inventory system with returns using an open queueing network with finite buffers and analyzed the model using an expansion methodology.

Fleischmann et al. (2002) focused on deriving the optimal control policy and computing their parameters for an inventory system with Poisson demand and returns. A comparison of the optimal control policy with the traditional EOQ formula indicated that the latter is a reasonable approximation in the return flow model when the reorder point is chosen appropriately. Fleischmann and Kuik (2003) considered a single inventory point with stochastic demand and returns, and analyzed the effect of exogenous inbound material flow. A Markov decision process was used to analyze the system, and it was established that the (s, S) order policy was optimal.

Ching et al. (2003) considered an inventory system where lateral transshipment of returns from one inventory system to another was allowed. A steady-state probability distribution was derived for large order quantity, Q, values for the (r, Q)

inventory policy, and it was shown that lateral transshipment helped in significantly bringing down rejection rate of returns. De Brito and Dekker (2003) explored the validity of the assumption on the demand and return following an independent Poisson process. The study concluded that there is a need to develop models for systems with seasonality in the product returns.

DeCroix et al. (2005) explored an inventory system with stationary costs and stochastic demand, where negative demand represented product returns. An exact procedure and a few approximations were provided to establish the operating characteristics for the stationary echelon base-stock policy. DeCroix and Zipkin (2005) analyzed a system where faulty returns could be condemned in an inventory system with an assembly structure. A major contribution of their effort is to identify conditions on the item recovery pattern and restrictions on the inventory policy under which an equivalent series system exists. De Brito and van der Laan (2009) studied the impact of unreliable return-related information on inventory management. The results indicated that models that consider information on returns are not guaranteed to work well in the presence of imperfect information.

Mitra (2009) relaxed the constraints on negligible setup and holding costs at various locations, and analyzed inventory control in a two-echelon system with returns. They provided the results for both a deterministic and stochastic version of the model under continuous review. Mitra (2012) extended the inventory management problem to a two-echelon system with correlated demand and returns under a generalized cost structure. The study demonstrated that a higher rate of return, and higher correlation between demand and return may not always lead to cost savings.

Wei et al. (2011) explored the integration of the product return process with a manufacturing process. In their methodology, they first proposed an inventory control model for the return and remanufacturing processes, and then employed a robust optimization approach to deal with the uncertainty in the system.

Srivathsan and Viswanathan (2017) considered an after-sales service network with one service center and one repair center with no delays in shipment within the centers. While their model has similarities to the model developed by Heyman (1977), their model considered two types of inventories, namely the inventory of components to be repaired at the repair center and the inventory of repaired components at the service center. They first assumed that all returned faulty components can be repaired, and modeled the repair center as a finite capacity multi-server queue ($M/M/c/K$ queue). They were able to prove interesting properties regarding the total cost function, including a convexity result when the utilization was less than 1, and exploited this property in designing algorithms to obtain the optimal values for the number of servers (c) and the queue length capacity (K). They later allowed for some of the faulty products to be condemned, and solved an underlying continuous time Markov chain (CTMC) model to obtain the expected inventory levels required for the total cost function. An algorithm was developed to obtain the optimal values of c and K.

The after-sales problem described in Sect. 8.1 falls under this category, and is similar in spirit to the model developed by Srivathsan and Viswanathan (2017). However, the current problem would represent a multi-echelon situation, where multiple service centers send faulty components to a centralized repair facility. In such a situation, an item waiting at the repair center may be from any one of the service center, and the repaired component will have to be sent back to the service center. This can be done by allocating capacity at the repair center to the service center based on the ratio of the arrival rates or by keeping track of information regarding the service center that placed the order. Also, there may be a need to model the transit operations from the service center to the repair facility and back.

8.4 Inventory Systems with Remanufacturing

The models for inventory system with remanufacturing generally consider that both manufacturing and remanufacturing processes are performed at the facility (henceforth this facility will be referred to as a hybrid system). The common assumption in these models is that the output (products) from both processes are as good as new, and can be used to service demand. Some models assumed deterministic demand (Schrady 1967; Nahmias and Rivera 1979; Teunter 2001, 2004; Rubio and Corominas 2008; Feng and Viswanathan 2014; Helmrich et al. 2014; Polotski et al. 2017; Sun et al. 2018), while others assumed stochastic demand (van der Laan et al. 1999; van der Laan and Teunter 2006; Alinovi et al. 2012; Pan et al. 2015; Ahiska et al. 2017; Turki et al. 2018; Fu et al. 2019). The models based on stochastic demand assume that the production rate for both the processes are infinite. Consequently, these problems reduce to a multi-item/multi-location inventory problem with setup costs and lead time for replenishment (Srivathsan and Viswanathan 2017).

The early models in this domain focused on deterministic demand. Schrady (1967) developed an Economic Order Quantity (EOQ) model in the context of remanufacturing. The manufacturing and remanufacturing processes were assumed to have infinite rate, and a $(1, R)$ policy was proposed. According to the policy, every manufacturing setup was followed by an integer number, R, of identical remanufacturing setups. The order quantity for manufacturing and remanufacturing was kept stationary. Nahmias and Rivera (1979) extended Schrady's (1967) model to finite remanufacturing rate, and suggested a class of heuristic policies.

Teunter (2001) extended the work of Schrady (1967) by considering two classes of control policies, a $(P, 1)$ policy in addition to the $(1, R)$ policy. In a $(P, 1)$ policy, every remanufacturing setup is followed by P identical manufacturing setups. Note that both P and R are integers. Teunter (2004) extended the work of Teunter (2001) to study finite rates for the manufacturing and remanufacturing processes. Heuristic solutions were presented for the two classes of control policies.

Rubio and Corominas (2008) moved away from the EOQ models, and considered a lean or just-in-time production environment. The study focused on analyzing the capacity, return rate and remanufacturing rate, and determining the optimal policies under which a company should implement reverse logistics. The major contribution of the study was the control on inventory generation by changing the manufacturing and remanufacturing capacity. Feng and Viswanathan (2014) developed a model that minimized the long run average cost per unit time by obtaining the optimal lot sizes and production schedule for manufacturing and remanufacturing. They developed heuristics which yielded guaranteed performance bounds for the hybrid system.

Helmrich et al. (2014) studied the economic lot sizing problem with a remanufacturing option. In each period, a manufacturing or a remanufacturing option could be chosen. Two Mixed Integer Programming (MIP) formulations were developed, one for the case with separate setup costs for both processes and the second for joint setup costs. Both formulations were shown to be NP-hard. Consequently, these models were reformulated to obtain tighter formulations that were shown to perform better than the original solution (in terms of both computation time and the linear program relaxation). Sun et al. (2018) extended the economic lot sizing problem to consider returns of different grades. The objective of the study was to minimize the average total cost by obtaining the optimal lot size and the optimal scheduling of the remanufacturing sequence.

Polotski et al. (2017) developed an analytical solution for optimal production and setup schedule along the production cycles in a hybrid system with no failures. The results indicated that the cycle consists of time periods when production happens at maximum rate, as well as periods of on-demand manufacturing and on-return remanufacturing. The study was extended to hybrid system with failure, and optimality conditions were developed. Polotski et al. (2019) enhanced their previous study by addressing time-varying demand and return.

Assid et al. (2019a) proposed an efficient structure of joint control policies bringing together the production and disposal activities, and the procurement of return and raw materials. A simulation-based optimization model was employed to estimate the optimal raw material supply, optimal storage space of finished products, raw materials and returns, while minimizing the total cost. Assid et al. (2019b) enhanced their previous model (Assid et al. 2019a) by considering setup operations to switch between the manufacturing and remanufacturing modes.

Shifting the focus towards stochastic demand, van der Laan et al. (1999) explored a hybrid system with a single component by comparing its performance under two strategies with a traditional manufacturing system. The two strategies that were considered are the push control (where all returned components are remanufactured as soon as possible), and the pull control (where the returned components are remanufactured as late as possible). The comparison study concluded that the hybrid system always had lower total costs. Additionally, it was shown that the pull strategy is more economical than the push strategy.

van der Laan and Teunter (2006) employed extensions of the (s, Q) policy based on the push and pull control strategies, and obtained approximate closed-form formulae for the optimal policy. Alinovi et al. (2012) developed a stochastic EOQ-based inventory control model for hybrid systems, where they first derived an optimal return policy to minimize the total cost. The optimal return policy proposes to incentivize customers to return products at its end of life. A simulation study was conducted on the model to assess the effect of stochastic demand on the optimal return policy.

Pan et al. (2015) explored a hybrid system with two products, which results in two return flows. First, an optimal solution for a single-period problem was characterized by a multilevel threshold policy. Based on this policy, an approximate dynamic programming approach was employed to solve the multi-period problem. The results showed that the threshold limit for each period depended only on the cost function.

Ahiska et al. (2017) demarcated the quality of the manufactured and remanufactured products in a hybrid system. Since the availability of the remanufactured product depends on the product returns, they considered substituting remanufactured products with manufactured ones in times of stock-out. The problem was formulated as a Markov Decision Process (MDP), and the focus was on determining the optimal manufacturing and remanufacturing decision under product substitution.

Turki et al. (2018) analyzed a hybrid system under carbon cap and trade policy. The model focused on optimizing storage and the manufacturing/remanufacturing plan, while explicitly considering the difference in quality between the manufactured and remanufactured products, machine failures and carbon constraints. Fu et al. (2019) extended the idea of hybrid systems to perishable products.

The after-sales service network discussed in Sect. 8.1 has some similarities with the inventory systems with remanufacturing. However, there are significant differences as well. While the former considers only a repair process, the latter considers the manufacturing and remanufacturing processes together. Additionally, there is an underlying queueing problem in the after-sales service network problem, while there is no such underlying problem in the inventory systems with remanufacturing.

8.5 Analytical Models for the After-Sales Service Network

In this section, we first present the after-sales service network under consideration, and then discuss the queueing-based optimization model developed to solve the inventory management problem. We conclude with the scope for extension and future directions.

8.5.1 After-Sales Service Network

The multi-echelon repaired component inventory system comprising of two service centers and a common repair center is depicted in Fig. 8.1. We consider the inventory system for a single component that can be a constituent of multiple products. The customer arrivals (i.e., arrival of faulty component) follows a Poisson process with rate λ_i at service center i, $i = 1, 2$. The failed product is first disassembled and

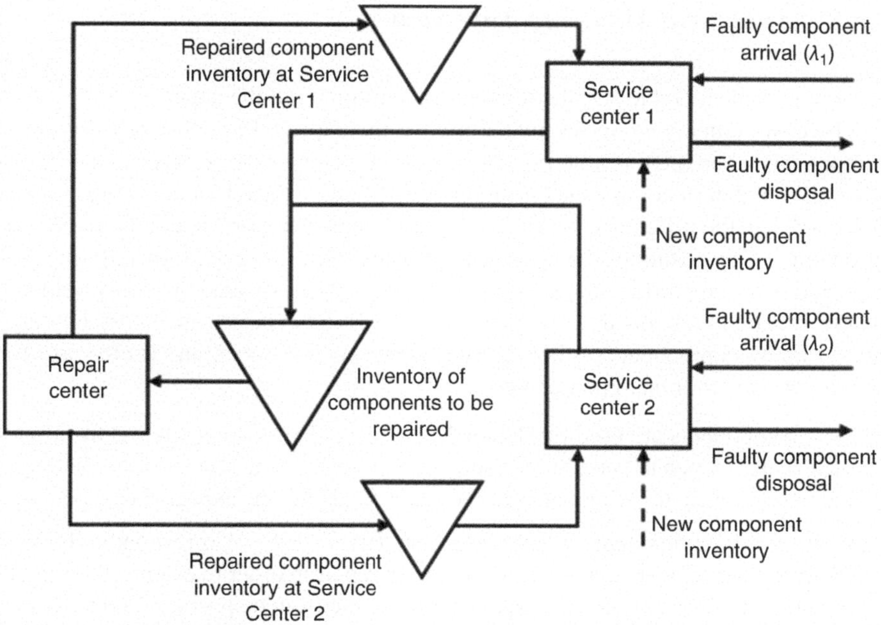

Fig. 8.1 The multi-echelon repaired component inventory system

diagnosed to identify the faulty component. The faulty component generates a demand for the repaired component at service center *i*. This demand is satisfied from the repaired component inventory at that service center if it is available or with a new component. We assume that sufficient inventory of new components is always available in the system or can be procured immediately from the supplier of the component. Note that it is assumed that all faulty components can be repaired.

When the demand at service center *i* is satisfied using a repaired component, the faulty component from the service center joins the queue for repair at the repair center. The repair times at the repair center follow an exponential distribution with mean μ. However, if the demand is satisfied using a new component, the faulty component is disposed of to ensure that the repair queue is stable. Consequently, the sum of the repaired component inventory at service center *i* and the inventory of faulty components from service center *i* at the repair center will be a constant, K_i. Note that as the inventory of faulty components and repaired components at service center *i* is restricted to K_i, the maximum repair queue length at the repair center will be $K = K_1 + K_2$. The variables, K_1, and K_2, are our decision variables as they are used to determine the total cost incurred by the system. The problem involves determining the optimal level of inventory of the repaired components at the two service centers and the production capacity of the repair center for the components with an objective of minimizing the total cost.

8.5.2 Analytical Modeling Approach

We first provide in Table 8.1, the notation used throughout the paper.

The cost components in the model are (a) the expected net cost of using new components when there is no repaired component inventory at service center i (i.e., when the number of components from service center i in the repair queueing system is K_i)—this is the difference between cost of the new component and the expected revenue from disposing of the faulty components (when it is not repaired), (b) the expected cost of repairing the faulty components, (c) the expected inventory holding cost of repaired component at service center i, (d) the expected inventory holding cost of faulty components (to be repaired) from repair center i, and (e) the cost to operate the server at the repair center.

1. New components are used for demand arrivals (of faulty components) from service center i when the number of faulty components in the repair facility belonging to service center i is equal to K_i (i.e., $I_{Fi} = K_i$). Here, the expected cost per unit time of new components at service center i is $C_n \lambda_i p_{K_i}$. Correspondingly, the faulty components do not enter the queue for repair in order to ensure stability of repair queues. The expected revenue per unit time from disposing of faulty components at service center i is $R_s \lambda_i p_{K_i}$. Therefore, the expected net cost of using a new component for a demand from service center i is

Table 8.1 Notation

Model parameters	
λ_i	Arrival rate of faulty components at service center i
C_n	Unit cost for manufacturing a new component
C_r	Unit cost of repairing a faulty component ($C_r < C_n$)
C_μ	Cost per unit time per unit rise in production rate of the server at the repair center
R_s	Revenue from selling a faulty component without repairing ($C_n - R_s > C_r$)
h_{Ri}	Holding cost rate per unit per unit time of repaired component at service center i
h_F	Holding cost rate per unit per unit time of component in the repair system ($h_F < h_{Ri}$)
Decision variables	
K_i	Maximum inventory of repaired components and faulty components at service center i
μ	Processing rate of server at the repair center
Other Variables	
I_{Fi}	Number of faulty components in the repair system (in queue or under repair) from service center i
I_F	Number of faulty components in the repair center $\left(I_F = \sum_i I_{Fi}\right)$
I_{Ri}	Number of repaired components at service center i.
ρ	$=(\lambda_1 + \lambda_2)/\mu$
p_{ni}	Steady-state probability that the number of faulty components from service center i in the repair system is n [$P(I_{Fi} = n)$]

$$Z_{ni} = (C_n - R_s)\lambda_i \, p_{K_i}. \tag{8.1}$$

2. The arriving faulty components from service center i are chosen to be repaired (and demand is satisfied with a repaired component) only when the number of repaired component inventory at service center i is nonnegative (i.e., $I_{Ri} > 0$ or $I_{Fi} < K_i$). Therefore, the expected cost per unit time of repairing the faulty components from service center i is

$$Z_{ri} = C_r \, \lambda_i \left(1 - p_{K_i}\right). \tag{8.2}$$

3. The expected cost per unit time of holding repaired component inventory at service center i is

$$Z_{h1i} = h_{Ri} \, E\left[I_{Ri}\right] = h_{Ri} \left(K_i - E\left[I_{Fi}\right]\right). \tag{8.3}$$

4. The expected cost per unit time of holding inventory of faulty components from service center i to be repaired is

$$Z_{h2i} = h_F \, E\left[I_{Fi}\right]. \tag{8.4}$$

5. The cost per unit time to operate the server at the repair center is

$$Z_s(\mu) = C_\mu \, \mu. \tag{8.5}$$

The expression for the expected total cost, $Z(K, c)$, for a fixed value of K and c is

$$Z(K_1, K_2, \mu) = \sum_{i=1}^{2} \left[Z_{ni} + Z_{ri} + Z_{h1i} + Z_{h2i}\right] + Z_s(\mu). \tag{8.6}$$

Substituting the expressions (8.1)–(8.5) in Eq. (8.6), we get

$$Z(K_1, K_2, \mu) = \sum_{i=1}^{2} \begin{array}{l} \left\{C_r \lambda_i + (C_n - C_r - R_s)\lambda_i p_{K_i} + h_{Ri}\left(K_i - E\left[I_{Fi}\right]\right)\right\} \\ + h_F E\left[I_{Fi}\right] + C_\mu \mu. \end{array} \tag{8.7}$$

In this model, we assume that the production capacity of the server allocated to service center i is $\mu_i = \frac{\lambda_i}{\sum_{j=1}^{2} \lambda_j} \mu$. The allocation of capacity to the individual service centers makes the problem separable, as the multi-echelon repaired component inventory system can be viewed as two independent service centers with their respective repair centers. This problem now reduces to the problem studied by Srivathsan and Viswanathan (2017) in a single server setting. Since the sum of inventory of repaired components at service center i and faulty components from service center i waiting to be repaired at the repair center is fixed at a value of K_i, the repair center for service center i can be modeled as an M/M/1/K_i queueing system. Note that the utilization of the repair center corresponding to repair orders from service center i is given by $\rho_i = \lambda_i/\mu_i$.

As the repair center is modeled as an M/M/1/K_i queueing system (Shortle et al. 2018), the probability that the repair queuing system is full, p_{K_i}, can be obtained as

$$p_{K_i} = \begin{cases} \dfrac{(1-\rho_i)\rho_i^{K_i}}{1-\rho_i^{K_i+1}}; & \rho_i \neq 1 \\ \dfrac{1}{K_i+1}; & \rho_i = 1 \end{cases}. \tag{8.8}$$

The expected inventory of components (in queue or in process) to be repaired from service center i, $E[I_{Fi}]$, is given as

$$E[I_{Fi}] = \begin{cases} \left(\dfrac{\rho_i}{1-\rho_i}\right)\left[\dfrac{1-(K_i-1)\rho_i^{K_i}+K_i\rho_i^{K_i+1}}{1-\rho_i^{K_i+1}}\right]; & \rho_i \neq 1 \\ \dfrac{K_i}{2}; & \rho_i = 1 \end{cases}. \tag{8.9}$$

Substituting Eqs. (8.8) and (8.9) into Eq. (8.7), we obtain the expression for the expected cost $Z(K, c)$ as a function of K and c as follows.

8.5.2.1 Case 1: $\rho \neq 1$

$$\begin{aligned} Z(K_i,\mu) = {} & \frac{(C_n - C_r - R_s)\lambda_i(1-\rho_i)\rho_i^{K_i}}{1-\rho_i^{K_i+1}} + K_i h_{Ri} \\ & -(h_{Ri}-h_F)\left(\frac{\rho_i}{1-\rho_i}\right)\left[\frac{1-(K_i-1)\rho_i^{K_i}+K_i\rho_i^{K_i+1}}{1-\rho_i^{K_i+1}}\right] \\ & + C_r\lambda_i + C_\mu \mu. \end{aligned}$$

8.5.2.2 Case 2: $\rho = 1$

$$Z(K_i,\mu) = \frac{(C_n - C_r - R_s)\lambda_i}{K_i+1} + \frac{K_i}{2}(h_{Ri}+h_F) + C_r\lambda_i + C_\mu \mu.$$

$$Z(K_1,K_2,\mu) = Z(K_1,\mu) + Z(K_2,\mu). \tag{8.10}$$

In the case of a single-echelon after-sales service network, Srivathsan and Viswanathan (2017) showed that total cost function for a multi-server case is convex with respect to K_i when the utilization of the repair center does not exceed one (i.e., $\rho \leq 1$) and that the problem of minimizing the total cost function has a global optimal solution. Since Eq. (8.10) is the sum of two such terms, we can extend the result and show that the function $Z(K_1, K_2, \mu)$ is also convex when $\rho \leq 1$. Subsequently, we can employ the bisection search to obtain the optimal values of K_i for a fixed production rate (μ), and perform an exhaustive search over the processing rate to obtain optimal combination of K_i's and μ.

In the case of $\rho > 1$, Srivathsan and Viswanathan (2017) were not able to show convexity of the total cost function with respect of K. However, it was shown that there exists a value of K, $\hat{K}$, such that the total cost function always increases when $K > \hat{K}$. Since our total cost function $Z(K_1, K_2, \mu)$ is the sum of two cost functions, $Z(K_1, \mu)$ and $Z(K_2, \mu)$, it has to be shown if the results from Srivathsan and Viswanathan (2017) can be extended to obtain the optimal inventory levels. This is the topic of our current research.

8.5.3 Future Research Directions

The above model considers allocation based on the proportion of arrival rates. However, there is a need to consider other advanced allocation schemes. Especially with the development in information technology, the information about pending orders from the individual service centers can aid in better planning of the transfer of the repaired components to the service centers. In this case, if only one service center has pending orders, then the repaired component will have to be sent to that facility. This availability of information means that the model cannot be separated as in the case of allocation based on the proportion of arrival rates. In this scenario, there is a need to solve an underlying continuous time Markov chain model.

Also, one has to keep in mind that the repair facility may be in a different location compared to the service center as discussed in Tan et al. (2003). Note that the optimal values of K_i obtained in our preliminary model would represent a lower bound on the capacity as our model (at present) does not consider transit operations. If these are considered in the model, the capacity needs to increase. Transit operations can be considered in our model as an infinite server queue (Srivathsan and Kamath 2012, 2017). Also, the parametric decomposition-based approach (Whitt 1983) could be employed in our modeling approach. The expected inventory levels obtained from these queueing models can then be used as inputs into the total cost function. However, obtaining closed-form expressions for these performance measures may be difficult, and this may warrant a need for black-box optimization models to obtain the optimal processing rates and the optimal capacities.

Acknowledgments The author thanks Dr. Viswanathan (Professor, Nanyang Business School, Nanyang Technological University) for his valuable research inputs at various stages of this study.

The author would also like to thank Dr. Sailatha Ravi of Medlish Communications for her inputs in writing and development of this chapter.

References

Ahiska, S. S., Gocer, F., & King, R. E. (2017). Heuristic inventory policies for a hybrid manufacturing/remanufacturing system with product substitution. *Computers & Industrial Engineering, 114*, 206–222. https://doi.org/10.1016/j.cie.2017.10.014.

Albright, S. C. (1989). An approximation to the stationary distribution of a multiechelon repairable-item inventory system with finite sources and repair channels. *Naval Research Logistics, 36*, 179–195.

Albright, S. C., & Soni, A. (1988). Markovian multiechelon repairable inventory system. *Naval Research Logistics Quarterly, 35*, 49–61.

Alinovi, A., Bottani, E., & Montanari, R. (2012). Reverse logistics: A stochastic EOQ-based inventory control model for mixed manufacturing/remanufacturing systems with return policies. *International Journal of Production Research, 50*, 1243–1264. https://doi.org/10.1080/00207543.2011.571921.

Assid, M., Gharbi, A., & Hajji, A. (2019a). Production planning of an unreliable hybrid manufacturing–remanufacturing system under uncertainties and supply constraints. *Computers & Industrial Engineering, 136*, 31–45. https://doi.org/10.1016/j.cie.2019.06.061.

Assid, M., Gharbi, A., & Hajji, A. (2019b). Production and setup control policy for unreliable hybrid manufacturing-remanufacturing systems. *Journal of Manufacturing Systems, 50*, 103–118. https://doi.org/10.1016/j.jmsy.2018.12.004.

Avsar, Z. M., & Zijm, W. H. (2003). Capacitated two-echelon inventory models for repairable item systems. In S. B. Gershwin, Y. Dallery, C. T. Papadopoulos, & J. M. Smith (Eds.), *Analysis and modeling of manufacturing systems, International series in operations research & management science* (Vol. 60, pp. 1–36). New York: Springer. https://doi.org/10.1007/978-1-4615-1019-2_1.

Bazan, E., Jaber, M. Y., & Zanoni, S. (2016). A review of mathematical inventory models for reverse logistics and the future of its modeling: An environmental perspective. *Applied Mathematical Modelling, 40*, 4151–4178. https://doi.org/10.1016/j.apm.2015.11.027.

Bijesse, J., McClusky, M., & Sodano, L. (2002). *Service lifecycle management (part 1): The approaches and technologies to build sustainable competitive advantage for service* (AMR Research Report). Boston, MA.

Ching, W.-K., Yuen, W.-O., & Loh, A. W. (2003). An inventory model with returns and lateral transshipments. *Journal of the Operational Research Society, 54*, 636–641. https://doi.org/10.1057/palgrave.jors.2601521.

Cohen, M. A., Agrawal, N., & Agrawal, V. (2006). Winning in the aftermarket. *Harvard Business Review, 84*, 129–138.

Cohen, I., Cohen, M. A., & Landau, E. (2017). On sourcing and stocking policies in a two-echelon, multiple location, repairable parts supply chain. *Journal of the Operational Research Society, 68*, 617–629. https://doi.org/10.1057/s41274-016-0108-0.

De Brito, M. P., & Dekker, R. (2003). Modelling product returns in inventory control—Exploring the validity of general assumptions. *International Journal of Production Economics, 81–82*, 225–241. https://doi.org/10.1016/s0925-5273(02)00275-x.

De Brito, M. P., & van der Laan, E. A. (2009). Inventory control with product returns: The impact of imperfect information. *European Journal of Operational Research, 194*, 85–101. https://doi.org/10.1016/j.ejor.2007.11.063.

DeCroix, G. A., & Zipkin, P. H. (2005). Inventory Management for an Assembly System with product or component returns. *Management Science, 51*, 1250–1265. https://doi.org/10.1287/mnsc.1050.0394.

DeCroix, G. A., Song, J.-S., & Zipkin, P. H. (2005). A series system with returns: Stationary analysis. *Operations Research, 53*, 350–362. https://doi.org/10.1287/opre.1040.0184.

Díaz, A., & Fu, M. C. (1997). Models for multi-echelon repairable item inventory systems with limited repair capacity. *European Journal of Operational Research, 97*, 480–492. https://doi.org/10.1016/s0377-2217(96)00279-2.

Esenduran, G., Kemahlioğlu-Ziya, E., & Swaminathan, J. M. (2012). Product take-back legislation and its impact on recycling and remanufacturing industries. In *Sustainable supply chains* (pp. 129–148). New York: Springer. https://doi.org/10.1007/978-1-4419-6105-1_9.

Feng, Y., & Viswanathan, S. (2014). Heuristics with guaranteed performance bounds for a manufacturing system with product recovery. *European Journal of Operational Research, 232*, 322–329. https://doi.org/10.1016/j.ejor.2013.07.006.

Fisher-Vanden, K. A., Shukla, P. R., Edmonds, J. A., Kim, S. H., & Pitcher, H. M. (1997). Carbon taxes and India. *Energy Economics, 19*, 289–325. https://doi.org/10.1016/s0140-9883(96)01020-1.

Fleischmann, M., & Kuik, R. (2003). On optimal inventory control with independent stochastic item returns. *European Journal of Operational Research, 151*, 25–37. https://doi.org/10.1016/s0377-2217(02)00592-1.

Fleischmann, M., Bloemhof-Ruwaard, J. M., Dekker, R., van der Laan, E. A., van Nunen, J. A. E. E., & Wassenhove, L. N. V. (1997). Quantitative models for reverse logistics: A review. *European Journal of Operational Research, 103*, 1–17. https://doi.org/10.1016/s0377-2217(97)00230-0.

Fleischmann, M., Kuik, R., & Dekker, R. (2002). Controlling inventories with stochastic item returns: A basic model. *European Journal of Operational Research, 138*, 63–75. https://doi.org/10.1016/s0377-2217(01)00100-x.

Fu, K., Gong, X., & Liang, G. (2019). Managing perishable inventory systems with product returns and remanufacturing. *Production and Operations Management, 28*, 1366–1386. https://doi.org/10.1111/poms.12987.

Govindan, K., Soleimani, H., & Kannan, D. (2015). Reverse logistics and closed-loop supply chain: A comprehensive review to explore the future. *European Journal of Operational Research, 240*, 603–626. https://doi.org/10.1016/j.ejor.2014.07.012.

Graves, S. C. (1985). A multi-echelon inventory model for a repairable item with one-for-one replenishment. *Management Science, 31*, 1247–1256. https://doi.org/10.1287/mnsc.31.10.1247.

Gross, D., Gu, B., & Soland, R. M. (1993). Iterative solution methods for obtaining the steady-state probability distributions of Markovian multi-echelon repairable item inventory systems. *Computers & Operations Research, 20*, 817–828. https://doi.org/10.1016/0305-0548(93)90103-p.

Guide, V. D. R., Jayaraman, V., Srivastava, R., & Benton, W. C. (2000). Supply-chain management for recoverable manufacturing systems. *Interfaces, 30*, 125–142. https://doi.org/10.1287/inte.30.3.125.11656.

Gupta, A. (1993). Approximate solution of a single-base multi-indentured repairable-item inventory system. *Journal of the Operational Research Society, 44*, 701–710. https://doi.org/10.1057/jors.1993.119.

Helmrich, M., Retel, J., Jans, R., Heuvel, W., & Wagelmans, A. P. M. (2014). Economic lot-sizing with remanufacturing: Complexity and efficient formulations. *IIE Transactions, 46*, 67–86. https://doi.org/10.1080/0740817x.2013.802842.

Heyman, D. P. (1977). Optimal disposal policies for a single-item inventory system with returns. *Naval Research Logistics Quarterly, 24*, 385–405. https://doi.org/10.1002/nav.3800240302.

Jalil, M. (2011). *Customer information driven after sales service management: Lessons from spare parts logistics*. Ph.D. dissertation, Erasmus Research Institute of Management, Rotterdam, Netherlands.

Jayaraman, V., Guide, V. D. R., & Srivastava, R. (1999). A closed-loop logistics model for remanufacturing. *The Journal of the Operational Research Society, 50*, 497. https://doi.org/10.2307/3009998.

Klausner, M., & Hendrickson, C. T. (2000). Reverse-logistics strategy for product take-back. *Interfaces, 30*, 156–165. https://doi.org/10.1287/inte.30.3.156.11657.

Kleindorfer, P. R., Singhal, K., & Wassenhove, L. N. (2009). Sustainable operations management. *Production and Operations Management, 14*, 482–492. https://doi.org/10.1111/j.1937-5956.2005.tb00235.x.

Korugan, A., & Gupta, S. M. (1998). A multi-echelon inventory system with returns. *Computers & Industrial Engineering, 35*, 145–148. https://doi.org/10.1016/s0360-8352(98)00041-2.

Kruse, W. K. (1979). *An exact N echelon inventory model: The simple Simon method*. Fort Belvoir: Defense Technical Information Center.

Lau, H. C., & Song, H. (2008). Multi-echelon repairable item inventory system with limited repair capacity under nonstationary demands. *International Journal of Inventory Research, 1*, 67. https://doi.org/10.1504/ijir.2008.019209.

Lee, H. L. (1987). A multi-echelon inventory model for repairable items with emergency lateral transshipments. *Management Science, 33*, 1302–1316. https://doi.org/10.1287/mnsc.33.10.1302.

Linton, J. D., Klassen, R., & Jayaraman, V. (2007). Sustainable supply chains: An introduction. *Journal of Operations Management, 25*, 1075–1082. https://doi.org/10.1016/j.jom.2007.01.012.

MacCracken, C. N., Edmonds, J. A., Kim, S. H., & Sands, R. D. (1999). The economics of the Kyoto protocol. *The Energy Journal, 20*, 25–71.

Mirzahosseinian, H., & Piplani, R. (2011). A study of repairable parts inventory system operating under performance-based contract. *European Journal of Operational Research, 214*, 256–261. https://doi.org/10.1016/j.ejor.2011.04.035.

Mirzahosseinian, H., & Piplani, R. (2013). Impact of operational systems on suppliers response under performance-based contracts. *Procedia Computer Science, 16*, 639–647. https://doi.org/10.1016/j.procs.2013.01.067.

Mitra, S. (2009). Analysis of a two-echelon inventory system with returns. *Omega, 37*, 106–115. https://doi.org/10.1016/j.omega.2006.10.002.

Mitra, S. (2012). Inventory management in a two-echelon closed-loop supply chain with correlated demands and returns. *Computers & Industrial Engineering, 62*, 870–879. https://doi.org/10.1016/j.cie.2011.12.008.

Muckstadt, J. A. (1973). A model for a multi-item, multi-echelon, multi-indenture inventory system. *Management Science, 20*, 472–481. https://doi.org/10.1287/mnsc.20.4.472.

Muckstadt, J. A. (1978). Some approximations in multi-item, multi-echelon inventory systems for recoverable items. *Naval Research Logistics Quarterly, 25*, 377–394. https://doi.org/10.1002/nav.3800250302.

Muckstadt, J. A. (1979). A three-echelon, multi-item model for recoverable items. *Naval Research Logistics Quarterly, 26*, 199–221. https://doi.org/10.1002/nav.3800260203.

Muckstadt, J. A., & Isaac, M. H. (1981). An analysis of single item inventory systems with returns. *Naval Research Logistics Quarterly, 28*, 237–254. https://doi.org/10.1002/nav.3800280207.

Nahmias, S., & Rivera, H. (1979). A deterministic model for a repairable item inventory system with a finite repair rate. *International Journal of Production Research, 17*, 215–221. https://doi.org/10.1080/00207547908919609.

Pan, J., Tao, Y., Lee, L. H., & Chew, E. P. (2015). Production planning and inventory control for a two-product recovery system. *IIE Transactions, 47*, 1342–1362. https://doi.org/10.1080/0740817x.2015.1056389.

Perlman, Y., Mehrez, A., & Kaspi, M. (2001). Setting expediting repair policy in a multi-echelon repairable-item inventory system with limited repair capacity. *Journal of the Operational Research Society, 52*, 198–209. https://doi.org/10.1057/palgrave.jors.2601079.

Polotski, V., Kenne, J.-P., & Gharbi, A. (2017). Production and setup policy optimization for hybrid manufacturing–remanufacturing systems. *International Journal of Production Economics, 183*, 322–333. https://doi.org/10.1016/j.ijpe.2016.06.026.

Polotski, V., Kenné, J.-P., & Gharbi, A. (2019). Production control of hybrid manufacturing–remanufacturing systems under demand and return variations. *International Journal of Production Research, 57*, 100–123. https://doi.org/10.1080/00207543.2018.1461272.

Rappold, J. A., & Roo, B. D. V. (2009). Designing multi-echelon service parts networks with finite repair capacity. *European Journal of Operational Research, 199*, 781–792. https://doi.org/10.1016/j.ejor.2008.08.006.

Rubio, S., & Corominas, A. (2008). Optimal manufacturing–remanufacturing policies in a lean production environment. *Computers & Industrial Engineering, 55*, 234–242. https://doi.org/10.1016/j.cie.2007.12.009.

Schrady, D. A. (1967). A deterministic inventory model for reparable items. *Naval Research Logistics Quarterly, 14*, 391–398. https://doi.org/10.1002/nav.3800140310.

Shanker, K. (1981). Exact analysis of a two-echelon inventory system for recoverable items under batch inspection policy. *Naval Research Logistics Quarterly, 28*, 579–601. https://doi.org/10.1002/nav.3800280407.

Sherbrooke, C. C. (1968). Metric: A multi-echelon technique for recoverable item control. *Operations Research, 16*, 122–141. https://doi.org/10.1287/opre.16.1.122.

Sherbrooke, C. C. (1971). An evaluator for the number of operationally ready aircraft in a multi-level supply system. *Operations Research, 19*, 618–635. https://doi.org/10.1287/opre.19.3.618.

Sherbrooke, C. C. (1986). VARI-METRIC: Improved approximations for multi-indenture, multi-echelon availability models. *Operations Research, 34*, 311–319. https://doi.org/10.1287/opre.34.2.311.

Shortle, J. F., Thompson, J. M., Gross, D., & Harris, C. M. (2018). *Fundamentals of queueing theory*. Hoboken: Wiley. https://doi.org/10.1002/9781119453765.

Simon, R. M. (1971). Stationary properties of a two-echelon inventory model for low demand items. *Operations Research, 19*, 761–773. https://doi.org/10.1287/opre.19.3.761.

Slay, F. M. (1984). *VARI-METRIC: An approach to modelling multi-echelon resupply when the demand process is Poisson with a gamma prior*. Tech. Rep. (Report AF501-2). Logistics Management Institute, Washington, DC.

Sleptchenko, A., Heijden, M. C., & Harten, A. (2002). Effects of finite repair capacity in multi-echelon, multi-indenture service part supply systems. *International Journal of Production Economics, 79*, 209–230. https://doi.org/10.1016/s0925-5273(02)00155-x.

Sleptchenko, A., Heijden, M. C., & Harten, A. (2003). Trade-off between inventory and repair capacity in spare part networks. *Journal of the Operational Research Society, 54*, 263–272. https://doi.org/10.1057/palgrave.jors.2601511.

Srivathsan, S., & Kamath, M. (2012). An analytical performance modeling approach for supply chain networks. *IEEE Transactions on Automation Science and Engineering, 9*, 265–275. https://doi.org/10.1109/tase.2012.2189564.

Srivathsan, S., & Kamath, M. (2017). Performance modeling of a two-echelon supply chain under different levels of upstream inventory information sharing. *Computers & Operations Research, 77*, 210–225. https://doi.org/10.1016/j.cor.2016.07.011.

Srivathsan, S., & Viswanathan, S. (2017). A queueing-based optimization model for planning inventory of repaired components in a service center. *Computers & Industrial Engineering, 106*, 373–385. https://doi.org/10.1016/j.cie.2017.01.020.

Sun, H., Chen, W., Liu, B., & Chen, X. (2018). Economic lot scheduling problem in a remanufacturing system with returns at different quality grades. *Journal of Cleaner Production, 170*, 559–569. https://doi.org/10.1016/j.jclepro.2017.09.184.

Tan, A. W. K., Yu, W. S., & Arun, K. (2003). Improving the performance of a computer company in supporting its reverse logistics operations in the Asia-Pacific region. *International Journal of Physical Distribution & Logistics Management, 33*, 59–74. https://doi.org/10.1108/09600030310461007.

Teunter, R. H. (2001). Economic ordering quantities for recoverable item inventory systems. *Naval Research Logistics, 48*, 484–495. https://doi.org/10.1002/nav.1030.

Teunter, R. H. (2004). Lot-sizing for inventory systems with product recovery. *Computers & Industrial Engineering, 46*, 431–441. https://doi.org/10.1016/j.cie.2004.01.006.

Turan, H. H., Sleptchenko, A., Pokharel, S., & ElMekkawy, T. Y. (2018). A clustering-based repair shop design for repairable spare part supply systems. *Computers & Industrial Engineering, 125*, 232–244. https://doi.org/10.1016/j.cie.2018.08.032.

Turan, H. H., Sleptchenko, A., Pokharel, S., & ElMekkawy, T. Y. (2020). A sorting based efficient heuristic for pooled repair shop designs. *Computers & Operations Research, 117*, 104887. https://doi.org/10.1016/j.cor.2020.104887.

Turki, S., Sauvey, C., & Rezg, N. (2018). Modelling and optimization of a manufacturing/remanufacturing system with storage facility under carbon cap and trade policy. *Journal of Cleaner Production, 193*, 441–458. https://doi.org/10.1016/j.jclepro.2018.05.057.

van der Laan, E. A., & Teunter, R. H. (2006). Simple heuristics for push and pull remanufacturing policies. *European Journal of Operational Research, 175*, 1084–1102. https://doi.org/10.1016/j.ejor.2005.06.030.

van der Laan, E. A., Salomon, M., Dekker, R., & Wassenhove, L. V. (1999). Inventory control in hybrid systems with remanufacturing. *Management Science, 45*, 733–747. https://doi.org/10.1287/mnsc.45.5.733.

van Jaarsveld, W., Dollevoet, T., & Dekker, R. (2015). Improving spare parts inventory control at a repair shop. *Omega, 57*, 217–229. https://doi.org/10.1016/j.omega.2015.05.002.

Wei, C., Li, Y., & Cai, X. (2011). Robust optimal policies of production and inventory with uncertain returns and demand. *International Journal of Production Economics, 134*, 357–367. https://doi.org/10.1016/j.ijpe.2009.11.008.

Whitt, W. (1983). The queueing network analyzer. *Bell System Technical Journal, 62*, 2779–2815. https://doi.org/10.1002/j.1538-7305.1983.tb03204.x.

Chapter 9
Towards a Unified Understanding and Management of Closed Loop Operations

Santosh Mahapatra, Dwayne Cole, Raktim Pal, and Scott Webster

Abstract In recent years, companies have been emphasizing several initiatives to address environmental concerns that have given rise to regulatory risks, natural resource constraints and climate change which have negatively affected the business prospects. Among the initiatives, closed loop operation is the most comprehensive one because of its potential to accomplish the lofty goal of "cradle to cradle" in which material resources used in a business cycle are returned back to their original states for reuse in the next cycle. In light of four interconnected studies, this chapter identifies the key elements of closed loop operations, analyzes their significance in business and environmental performance and suggests models and considerations for managing the operations cost-effectively. Specifically, the chapter articulates the significance of remanufacturing and reuse of used products as the corner stone of closed loop operations, recognizes the criticality of acquisition of high quality used products in adequate quantity for cost-effective remanufacturing, establishes the need for appropriate incentives to motivate consumers for timely return products, and highlights the significance of intense communication with internal and external stakeholders for effective integration for environmental operations. The exposition aims to aid in developing a unified understanding of closed loop operations and suggests directions for future research to address additional challenging issues.

Keywords Cradle-to-cradle operations · Returns acquisition · Remanufacturing Consumer valuation · Communication · Alignment

---

S. Mahapatra (✉)
Clarkson University, Potsdam, NY, USA
e-mail: smahapat@clarkson.edu

D. Cole
Florida A&M University, Tallahassee, FL, USA

R. Pal
James Madson University, Harrisonburg, VA, USA

S. Webster
Arizona State University, Tempe, AZ, USA

C. Chen et al. (eds.), *Pursuing Sustainability*, International Series in Operations Research & Management Science 301,
https://doi.org/10.1007/978-3-030-58023-0_9

9.1 Introduction

9.1.1 Significance of Closed Loop Supply Chains for Sustainability

Over the last two decades, environmental challenges leading to regulatory risks, resource constraints, and climate change have motivated organizations to emphasize environmentally sustainable operations (Roy et al. 2018). Many global companies (e.g., Kimberly-Clark, Best Buy, and LEGO)[1] have had to abandon some of their products and practices because of environmental concerns even though they were profitable. Organizational approaches to environmental efforts have been be characterized in a variety of ways—proactive and reactive, preventative and end-of-the-pipe, eco-efficient and eco-effective by scholars (Porter and van der Linde 1995; Unruh 2008; Orsato 2006). Proactive and reactive approaches differ in terms of the triggering influence—voluntary initiatives without any regulatory pressure are proactive whereas initiatives undertaken as a response to regulation are reactive. Preventative are those that aim to prevent occurrence of waste; end-of-the-pipe in turn remediate or clean up the waste. Eco-efficiency aims to minimize waste for a specified level of output, whereas eco-effectiveness focuses on eliminating the concept of waste by emphasizing complete reutilization of materials by disallowing any material going to waste. Among these classifications, the eco-efficient versus eco-effective characterization seems most comprehensive and inclusive of all three types of approaches. Eco-efficient approach advocates efficiency in utilization of depletable resources; in contrast, eco-effective approach relies on "waste = food" or "cradle-to-cradle" principles.

Typically, eco-efficiency involves "reduce, reuse and recycle" (i.e., 3R) actions in a decreasing order of priority (Lee and Lionel 2007). 'Reduce' refers to lesser waste creation, "reuse" advocates extended use of product after one cycle of use and "recycle" refers to recovering useful materials and components from a used product for alternative applications. However, in the context of resource constraints, 3R actions entail limited level of demand satisfaction and/or value recovery. For example, "reduce" would emphasize lesser usage of resources even at the expense of lower demand fulfilment to slow down the environmental damage; "reuse" would imply multiple use/sell of products in which repeat usages fetch lesser revenue, and "recycle" would involve recovery of constituents for creating goods of lower economic value. In contrast, eco-effectiveness is about creating a production system that promotes the concept of "virtuous closed loop" in which all resources used to create a product or services can be reused many times for producing similar or higher value products, thereby preventing the depletion of natural resources. In this system, materials are not "recycled"; they are "upcycled" (Unruh 2008—ingredients are recovered to the original state of purity). Fundamentally, eco-effectiveness

[1] https://www.theguardian.com/sustainable-business/2015/feb/09/corporate-ngo-campaign-environment-climate-change.

is most robust to address the environmental challenges of current times because it aims to accomplish the "cradle-to-cradle" goal in the material flow system.

9.2 Elements of "Closed Loop" Supply Chain

"Eco-effectiveness" is realized through closing the material flow loop while maximizing economic and environmental payoffs. It is based upon three fundamental principles that mimic the biosphere's working (Unruh 2008). The principle are (a) use of a few types of constituents, (b) modular architecture, and (c) virtuous closed loop (cycle up). Just four elements, carbon, hydrogen, nitrogen, and oxygen that make most of the weight of the living system of planet earth reflect the principle of use of a few types of constituents. The almost identical cellular structure configured in various patterns across all living bodies indicates the principle of modular architecture. Finally, the natural decomposition of any dead organism to its basic elements that again transforms into the physical substance of a living being illustrates the principle of virtuous closed loop.

Translation of the three biosphere principles to industrial context has distinct design and operational implications. Use of a few types of components implies rethinking the material chemistry in order to simplify the number and types of materials used in a company's production. Further, these materials should be usable across a range of products and applications utilizing the principle of a common architecture to provide the benefits of scale and scope economies. Finally, consistent with principle of virtuous closed loop, the material chemistry should enable production of quality products that can be cost-effectively recycled back into the basic elements for producing the next generation of products. This necessitates an easy-to-disassemble product design/architecture and reconfiguration of the conventional buyer–supplier relationship in the supply chain to ensure that products that are in use with the customers are timely returned and processed for cost-effective value recovery and use in manufacturing new products.

Effective replication of biosphere rules in an industrial context requires a significant amount of rethinking to ensure that the ingredients used to manufacture a product are separable economically to their original states so that the recycled materials maintain the quality and the purity of the virgin materials. Such environmentally benign ingredients are few and difficult to find. Moreover, implementation of the necessary closed loop system requires a significant amount of operational, technological, and financial cooperation and coordination in the supply chains. The aforesaid features at the highest level of excellence would represent the "cradle-to-cradle" supply chain. Figure 9.1 presents the broad elements of the "cradle-to-cradle" oriented closed loop operations in a supply chain. One of the greatest challenges in realizing the "cradle-to-cradle" objective pertains to the need for reusing the materials across multiple technological cycles without being affected by the issue of technological obsolescence. Thus realizing a truly "cradle-to-cradle" supply chain is utopian. Nevertheless, the sustainability efforts of companies such as Herman Miller

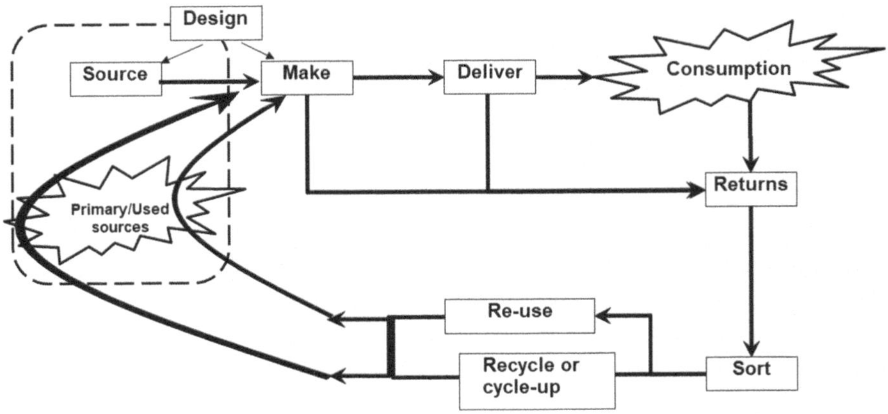

Fig. 9.1 A schematic view of virtuous closed loop (i.e., cradle-to-cradle) supply chain

(Lee and Lionel 2007), Patagonia (O'Rourke and Strand 2017), Shaw Industries (Unruh 2008), Recreational Equipment Inc. (REI) (Beil and Hopp 2013), and Hewlett-Packard (Cole et al. 2017) are striving to accomplish this goal. In these supply chains while getting back products from consumers and reusing those for manufacturing newer version of the products or upcycling to the original material state is an ideal target, due to economic and technological constraints there are always some used products that are disposed out as waste or are used for lesser applications in the supply chain.

9.3 Remanufacturing and Waste Reduction in Closed Loop Supply Chains: Empirical Evidence

There can be several environmental practices for a company to consider in the closed loop context.

While "cradle-to-cradle" oriented closed loop supply chain seems to be the most robust approach to address the issues of environmental damage, natural resource constraints and climate change, the costs of conducting operations in an environmentally sensitive way has been concerning to managers (Mahapatra et al. 2015). Accomplishing a good balance between environmental and business performance is often difficult because of the potential *trade-offs* between the two (Epstein 1996). The costs and benefits of different environmental initiatives can vary significantly; and, companies making investments in environmental programs usually expect the payback in a short timeframe (Cook 2014). As a result most of the environmental initiatives of firms follow the 3R (i.e., Reduce, Reuse, and Recycle) framework (Sarkis 2001) although there might be an aim to get closer to "cradle-to-cradle" over time. "Reduce" involves minimization of waste (i.e., resource consumption that does not create value) through product and process (re)designs, "reuse" involves refurbishing/remanufacturing for repeat use of materials/components until they are unusable, and "recycle" focuses on recovery of ingredients from process wastes such as scrap/effluents. Recycled materials are often of inferior quality with lesser economic value compared to virgin materials. Since "reduce" minimizes wasteful resource consumption and "reuse" restricts amount of virgin resource usage, the ecological and economic benefits of these two types of environmental practices are usually higher than "recycle." In general, preventative initiatives put more emphasis on "reduce" and "reuse" than on "recycle." However, "upcycle" replacing "recycle" in the "cradle to-cradle" system, "upcycle" might get higher priority than "reduce" and "reuse" in a "cradle to-cradle" system.

Appropriate evaluation and selection of environmental practices is critical to their successful implementation. Firms pursue different strategies even in the same industry due to the unique dynamic relationship between external constraints and a firm's internal considerations (Sousa and Voss 2008). As a result, the cost–benefit criteria for evaluating the usefulness of various practices are firm specific. Accordingly, it is difficult to assess and theorize the usefulness of different environmental practices for firm. Making a distinction between more and less effective practices is important because the beneficial associations between various environmental initiatives and financial and environmental performance are not universal. One approach to investigate this issue is to compare the intensity of implementation of various environmental practices and link those to the financial and business performance of companies. The aim is to identify a set of practices that is more effective than another set with respect to specified performance outcomes when firm-specific priorities for those practices and performance outcomes are considered (Link and Naveh 2006; Nga 2009). However, the issue is complicated, because, the various environmental practices being highly nonuniform with varied levels of implementation. This makes is difficult to analyze and compare especially when the direct information from companies are available. Such information are hard to obtain. To navigate the challenges, Mahapatra et al. (2015) analyzed the information on environmental practices of firms in the sustainability reports published by www.corporateregister.com. The study reviewed closed-loop environmental practices of 30 exemplar environmental companies and classified the practices into five categories. The five categories are remanufacturing, *proactive* waste/packaging reduction, material substitution (in favor of benign substitutes), material/packaging reuse, and energy reduction.

A key challenge in analyzing the effectiveness of different environmental practices is the difficulty in measuring their costs and benefits on a common scale. Closed-loop initiatives (e.g., waste elimination, reduction in material and energy consumption, substitution of ecologically sensitive materials with less sensitive ones, collection of used products to facilitate remanufacturing and reuse) in general involve product and processes improvement practices. Intensity of these practices is difficult to measure and compare objectively. However, subjective perceptions about the emphasis on these practices are measurable. Consequently, the authors utilized a 1–5 point Likert scale (1 representing a very low level of existence and 5 representing a very high level of existence) perceptual ratings for measuring the environmental practices and performance. To evaluate the cost-effectiveness in terms of business and environmental performance, the authors considered return on assets (ROA), compound annual growth of sales (CAGR) and certifications in environmental standards such as ISO 14000, Eco-Management and Audit Scheme (EMAS), EPA 33/50. The financial impacts of the operational initiatives are likely to be experienced with some time lag. Firms reportedly expect a payback in less than 18 months on their environmental investments (Cook 2014). Accordingly, the study used 1–2 year lags in the analysis.

Mahapatra et al. (2015) utilized a modified version of data envelopment analysis (DEA) technique proposed by Dehnokhalaji et al. (2010) to identify the set of environmental practices that is more cost-effective than others. The technique is amenable to using both interval and ratio scaled data simultaneously. Their analysis obtains the efficiency and uniqueness scores for each decision unit (i.e., organization) for comparing the cost-effectiveness of different environmental practices of firms. In DEA, managerially relevant antecedent and consequent variables are considered as inputs and outputs to derive efficiency score that define how efficient a decision unit is compared to others in a group. The inputs and outputs in a DEA need not be related in a strict physical sense.

While analyzing cost-effectiveness of practices, the DEA technique has some unique advantages over comparable techniques such as regression, structural equation modeling (SEM), analytical hierarchy process (AHP), multi-criteria decision making (MCDM), multi-objective programming, and stochastic frontier analysis (SFA). Regression-based models are useful in evaluating performance when just one output is considered (Chen and Zhu 2003). In addition, while regression and SEM estimate the average relationship across all units in the sample, DEA compares the optimal performance of each decision-making unit (DMU) in the sample with the best combination of units in the same sample (Charnes et al. 1994). Thus, DEA facilitates accounting for firm-specific considerations, provides deeper understanding of DMU's performance, and generates specific information to make recommendations regarding inputs and outputs for enhancing efficiency (Thanassoulis et al. 2011). Unlike parametric statistical and econometric methods, DEA does not hold any distributional assumptions and is less demanding on sample size (Premachandra et al. 2009). Unlike an econometric regression model, DEA does not require the analyst to specify the production function to link inputs and outputs (Bussofiane et al. 1997). There is no need to assign an a priori weight to each input and output in DEA, which is a requirement for several widely applied multi-criteria decision-making methods. Another methodology for measuring efficiency of organizational units is stochastic frontier analysis (SFA) (Aigner et al. 1977). While DEA assumes that data is essentially noise free, SFA allows for noise or random shock in the data and can make stochastic inferences. On the other hand, unlike SFA, DEA does not hold assumptions regarding distribution of the error terms and the functional form linking the outputs and inputs. Thus DEA has several advantages over other techniques.

The DEA based approach for interval data used in Mahapatra et al. (2015) is explained below. Let there be n decision making units (DMUs) and each of the DMUs needs m inputs (denoted by vector $\mathbf{x} \in R^m$) to generate p outputs (denoted by vector $\mathbf{y} \in R^p$). The corresponding input and output information of all DMUs (i.e., n number of DMUs) can be presented by $\mathbf{X}$($m \times n$ matrix) and $\mathbf{Y}$($p \times n$ matrix), respectively. For the unit (i.e., organization) under consideration (DMU_0), the inputs and outputs are denoted by vectors $\mathbf{x}_0$ and $\mathbf{y}_0$, respectively.

Let $\mathbf{U} = \begin{pmatrix} \mathbf{Y} \\ -\mathbf{X} \end{pmatrix}$ and $\mathbf{u}_0 = \begin{pmatrix} \mathbf{y}_0 \\ -\mathbf{x}_0 \end{pmatrix}$. Using an input oriented model, the input–output vector for the unit under consideration is: $w = \begin{pmatrix} \mathbf{0} \\ \mathbf{x}_0 \end{pmatrix}$.

The interval scaled model of Dehnokhalaji et al. (2010) that extends the general DEA model (Joro et al. 1998) is:

$$
\begin{aligned}
&\min \theta = -\boldsymbol{\rho}^T \mathbf{u}_0 + \eta \\
&s.t. \\
&-\left(\boldsymbol{\rho}^T \mathbf{U}\right)^T + \eta \mathbf{1} \geq \mathbf{0} \\
&-\boldsymbol{\rho}^T w = 1 \\
&\boldsymbol{\rho} \geq \varepsilon \mathbf{1} \quad \left(\varepsilon > 0 \text{ is Non Archimedian}\right) \\
&\eta \text{ is free}
\end{aligned}
\tag{9.1}
$$

The weights are defined by vector $\boldsymbol{\rho}$. The parameter η determines the returns to scale property. For example, $\eta = 0$ enforces constant returns to scale. The objective of the formulation above is to minimize the difference between weighted inputs and outputs. This is equivalent to maximizing the difference between weighted outputs and inputs in a typical input-oriented model. The first set of constraints ensure all efficiency measures must be less than or equal to one. In order to avoid infinite number of solutions the second constraint is imposed. The third set of constraints impose strictly positive lower limit to the weights used in the model so that all inputs and outputs can take some positive values and weakly efficient units are not diagnosed to be efficient.

The model finds a maximal subset of n units in which DMU_0 is efficient. If DMU_0 is not efficient, then some of the inequalities in $-\boldsymbol{\rho}^T\mathbf{U} + \eta \mathbf{1} \geq 0$ are not satisfied. Each of the inequalities is associated with one unit. Thus, violation of an inequality implies that the corresponding unit will need to be omitted from the set to make DMU_0 efficient. The Big M technique is employed to address the violations. Accordingly, model (9.1) is modified as follows:

$$
\begin{aligned}
&\min \ \mathbf{1}^T z \\
&s.t. \\
&-\boldsymbol{\rho}^T \mathbf{u}_0 + \eta = 0 \\
&-\left(\boldsymbol{\rho}^T \mathbf{U}\right)^T + \eta \mathbf{1} + M\mathbf{z} \geq \mathbf{0} \quad \left(M \gg 0\right) \\
&-\boldsymbol{\rho}^T \mathbf{w} = 1 \\
&\boldsymbol{\rho} \geq \varepsilon \mathbf{1} \quad \left(\varepsilon > 0 \text{ is Non Archimedian}\right) \\
&\eta \text{ is free} \\
&z_j \in \left\{0, 1\right\}, \ j = 1, \ldots, n
\end{aligned}
\tag{9.2}
$$

For each unit under consideration (DMU_0), the model (9.2) finds the optimal solution. If z_j has value of 1, the second constraint in (9.2) becomes redundant indicating the corresponding unit (DMU_j) is excluded from the set in which DMU_0 is efficient. If no units are excluded from the set then z_j will have value of zero for all of the units indicating not a single unit is better than DMU_0. This implies DMU_0 is efficient with respect to all units. In such scenario DMU_0 will be considered to be fully efficient with an efficiency score of 1. On the other hand, if many units are excluded from the set (i.e., these units will have z_j value of 1) to make DMU_0 efficient then the efficiency score of DMU_0 will be much lower than one. With respect to the above model, the efficiency score of DMU_0 can be obtained as:

$$\text{Efficiency score} = \frac{n - \sum_{j=1}^{n} z_j}{n} \tag{9.3}$$

Note that each of the n units takes a turn to become the unit under consideration (i.e., DMU_0). Thus, the above optimization model is solved n times. In these runs, it is of interest to see how many times a specific unit is excluded (i.e., has $z_j = 1$) from the sets of firms in which the other $n - 1$ DMU_0 are efficient.

Based on the above efficiency scores, Mahapatra et al. (2015) proposed: the more frequently a unit is excluded, the more distinctly superior the unit is compared to others. They referred to this characteristic as an index of uniqueness. Formally, they characterized the uniqueness measure of a unit as:

$$\text{Uniqueness measure of unit } j = \frac{\sum_{i=1}^{n} z_{ji}}{n} \tag{9.4}$$

$$\text{where, } z_{ji} \text{ is value of } z_j \text{ in the "}i\text{" th optimization run}$$

Proactive waste reduction and remanufacturing were found to be most impactful in contributing to stronger business and environmental performance. Remanufacturing enables the closed loop value recovery and reuse strategies and is an outcome "of" proactive design of products such that they can be easily disassembled to recover components that can be reassembled and upgraded to "like new" condition after their usage. Proactive waste reduction involves significant (re)thinking to improve products and processes for reducing the generation of waste. Further, these practices vouchsafe strong processes leading to higher environmental certifications. In light of the efficiency and uniqueness scores Mahapatra et al. (2015) grouped the companies into high performers (top 20 percentile of the units), medium performers (units between 40th percentile and 60th percentile), and low performers (bottom 20th percentile of the units). There was no evidence of bias for any specific industry across the three performance groups. The authors noted that the high performance groups do not deemphasize closed-loop initiatives for deriving higher financial performance suggesting a *complementary* relationship between environmental practices and economic and environmental performance. The above complementary relationship is somewhat less noticeable with the medium performers highlighting the fact that excellent closed loop operations delivers strong business advantages.

9.4 Decision Models for Cost-Effective Remanufacturing and Reuse

Companies have traditionally focused on and invested in the forward supply chains (i.e., raw materials to finished products). Since it is difficult to quickly and completely switch to upcyclable ingredients for a particular product, transforming forward supply chains to cradle-to-cradle closed loop supply chains is strategically complex, operationally challenging and financially risky. As a result, as noted by Mahapatra et al. (2015), most of the environmental firms currently emphasize remanufacturing and reuse to the maximum possible extent. Yet design of supply chains for collecting of products back from the customers is difficult. These products vary widely in quality and value—bringing those back into the supply chain to recover value or to upgrade for the next business cycle without sacrificing the business prospects of the new products requires a tremendous insight into the future technological cycles, customer preference, profitability, and so on. In light of the significance of remanufacturing and reuse in closed-loop supply chains, it is critical to develop appropriate decision models for facilitating cost-effective returns and remanufacturing.

Mahapatra et al. (2012) investigated the issues with respect to integrated manufacturing and remanufacturing (i.e., hybrid manufacturing) when heterogeneous quality and nonuniform quantity of returns influence the optimal production rates and inventory levels. They used a mixed integer linear programming (MILP) model to obtain the optimal production plan for a specified planning horizon. The model was applied to the illustrative operational data of an office equipment manufacturer to evaluate the impacts of quality of returns, quality based segregation of returns, and capacity readjustment costs. It was noted that (a) relative to simple manufacturing, the hybrid manufacturing system generates higher profit; (b) higher proportion of good quality returns and effective sorting of returns lead to higher profitability, higher amount of remanufacturing, more responsive production, and lesser inventory; (c) the benefits of hybrid manufacturing decline and greater operational instability is experienced with deterioration of quality of returns and lack of sorting of returns. Additionally, profitability and stability of operations are impacted negatively by higher production readjustment costs.

Mahapatra et al.'s (2012) analysis highlights the significance of *integrated* planning and operational decision-making in the closed-loop supply chain context. It suggests that organizations should deploy practices to ensure acquiring higher proportion of good quality returns, achieving greater flexibility (i.e., lower cost) to adjust the volume of operations, and sorting and segregating of returns by quality levels. However, motivating the consumers to return the used products is difficult; therefore, the results highlight the significance of accurate estimation of quality of returns, incentivizing consumers to return quality used products for cost-effective remanufacturing and good relationship with virgin component suppliers for higher flexibility to accommodate the uncertainties induced due to hybrid manufacturing environment.

In light of above constraints and opportunities, manufacturers need to consider two possible business scenarios while planning for cost-effective returns acquisition and remanufacturing in closed loop supply chains: (a) the consumers own the products and (b) the consumers lease or rent the products. We discuss approaches to analyze those scenarios next.

9.4.1 Consumer Owned Products Scenario

In this scenario, consumers need to be motivated to replace the currently owned product when the product has enough value to make returns and remanufacturing/reuse rewarding to consumers and manufacturers respectively. Towards that end, manufacturing companies focusing on closed-loop operations need to develop appropriate acquisition policies for used products. Cole et al. (2017) investigated this issue while accounting for consumer's relative preference for new and used products purchase when sales evolve over time. They utilized the classic Bass diffusion model that supports the concept of product life cycle (PLC) to represent the evolution of sales. A critical aspect of their analysis is the consideration of time dynamics—how customer choice influences the new product sales, the quantity and condition of used products, customer willingness-to-return and the demand for remanufactured products. The analysis considers the used product acquisition cost, cost of remanufacturing, used product supply constraints, and profit margins of new and remanufactured products in a dynamic context. The study included two types of used product acquisition methodologies—buyback and trade-in. In a buyback, the customer returns the product for a price. In a trade-in, the customer returns the product to buy a new product and at a discounted price. The key aspects of their analysis of both contexts are as follows.

9.4.1.1 Consumer Valuation, New Product Sales and Remanufactured Product Demand

Cole et al. (2017) considered two categories of consumers—innovators and imitators. The demand for new product sales is modeled such that each consumer group's valuation of the product decreases over time due to obsolescence and the valuation is higher than the price paid at a specific time during the product life cycle [refer to Eq. (9.1)].

$$d_n(t) = [\theta(1-p_n)](M - D_n(t-1)) + [(1-\theta)\iota M]\left(\frac{D_n(t-1)}{M}\right)(M - D_n(t-1)) \quad (9.5)$$

where, $d_n(t)$ = demand for new products at time t, $D_n(t)$ = cumulative new product sales through period t, M = upper limit on total sales over the life-cycle, θ = fraction of innovator segment of new product consumers; p_n = new product price, and ι = fraction of whose valuation greater than the purchase price.

The demand for remanufactured product demand over time closely follows the pattern of new product demand (Östlin et al. 2009). Cole et al. (2017) captured the essence of this linkage in terms of a relative-size parameter α and a time-lag parameter τ. The value of α defines the size of the remanufactured product market relative to the new product market. The value of τ is the number of periods that the remanufactured product demand is behind the new product sales. Thus they model the remanufactured product demand in period t as:

$$d_r(t) = \alpha d_n(t-\tau) \text{ for } \tau < t, d_r(t) = 0 \text{ for } \tau \geq t \quad (9.6)$$

and the total demand in periods 1 through t is

$$D_r(t) = \alpha D_n(t-\tau) = \sum_{j=1}^{t} d_r(j) \text{ for } \tau < t. \quad (9.7)$$

The model assumes that demand for new product in any period occurs at the beginning of the period, whereas the demand for remanufactured product occurs at the end of the period. Accordingly, demand for a remanufactured product can be satisfied using the buyback (or trade-in) returns in the period. If the duration of a period is long, then it is possible that $\tau = 0$.

9.4.1.2 Buyback Acquisition, Volume and Profitability

The cost-effectiveness of the buy-back alternative depends on the acquisition cost of returns, remanufacturing cost and replacement purchase of consumers. Total number of products with the consumers, consumer's valuation for the product (i.e., propensity to return), buyback price, age of the product (i.e., time duration of usage with the consumer) influence the quality and quantity of returns. Accounting for these considerations Cole et al. (2017) formulated the profitability maximization objective of buyback alternative as a dynamic function as follows.

$$\Pi_b(t) = p_r(t)x(t) - hI(t) - \sum_{i=1}^{t}\left(c_b(t,i) - \delta(t) + c_m(i) - \gamma(m - \varepsilon(t))\right)s_b(t,i) \quad (9.8)$$

$$\Pi_b^p = \max_{c_b(t,i)} \left\{ \sum_{t=1}^{T_r} (1-r)^t \, \Pi_b(t) | , x(t) \le d_r(t) | , I(t) \ge 0 \right\} \tag{9.9}$$

where,

m: the profit margin from new product.
$\varepsilon(t)$: change in variable cost in period t which is same as difference in margin in period $t = 1$ and period t; $\varepsilon(1) = 0$.
γ: the fraction of buybacks that translate into a purchase of a new product.
$c_b(t, i)$: the *gross buyback price* of a product of age i in period t without considering product obsolescence.
$\delta(t)$: loss in product value in period t due to obsolescence that reduces the attractiveness of the product over its life-cycle.
$c_m(i)$: transforming a returned unit of age i into a remanufactured product.
$s_{bj}(t, i)$ = number of products of age i returned via a buyback program in period t from customer segment j (i.e., imitator or innovator).
$p_r(t)$ = remanufactured product price in period t.
h = inventory holding cost per unit-period for remanufactured product.
r = net discount rate (e.g., cost of capital less inflation).
$x(t)$ = sales of remanufactured product in period t.
$I(t)$ = inventory of remanufactured product at end of period t, $I(t) = I(t-1) + \sum_i s_b(t, i) - x(t)$.
T_r = last period in the remanufactured product life-cycle.

In the formulation, Expression (9.8) represents the profit in period t and expression (9.9) represents the profit over the remanufactured product life cycle. In the formulation, the age dependent *gross buyback prices* and the volume of returns from the imitator and innovator customer segments are dynamically determined keeping in view consumer's assessment of the market value of a worn-out product vis-à-vis the buy-back price. In their formulation, some remanufactured product demand might not be satisfied in the solution to (9.9) because of the shortfall in returns volume.

9.4.1.3 Trade-in Acquisition, Volume, and Profitability

The trade-in alternative has similar rationale as that of the buyback alternative except a key difference—a trade-in customer must purchase the new product from the firm while returning the used product. This constraint will require the trade-in price to be higher than the buyback price for motivating the customer to go for the lower flexible option. The corresponding trade-in price premium will depend upon consumer's perceived cost of reduced flexibility in the trade-in program vis-à-vis the buyback program. Thus if ϕ is the *trade-in-to-buyback disutility* then the trade-in price, $c_t(t, i)$ must be $(\phi - 1)\%$ higher than the buyback price, $c_b(t, i)$ in order to match the chances of returns in the buyback policy. Consistent with above rationale,

Cole et al. (2017) formulated the profitability maximization objective of trade-in alternative as a dynamic function as follows.

$$\Pi_t(t) = p_r(t)x(t) - hI(t) - \sum_{i=1}^{t}\left(c_t(t,i) - \delta(t) + c_m(i) - \left(m - \varepsilon(t)\right)\right)s_t(t,i), \quad (9.10)$$

$$\Pi_t^p = \max_{c_t(t,i)}\left\{\sum_{t=1}^{T_r}(1-r)^t\,\Pi_t(t) | , x(t) \leq d_r(t) | , I(t) \geq 0\right\}. \quad (9.11)$$

The notations in expressions (9.10) and (9.11) follow from the buyback formulation. It is apparent from (9.8) and (9.10), in trade-in alternative there is an additional fraction $(1 - \gamma)$ of returns which translate into a new product purchase, thereby increasing profit through the profit margin term.

9.4.1.4 Relative Usefulness of Trade-in and Buyback Alternatives

Cole et al.'s (2017) numerical analysis for the formulations (9.5)–(9.11) yielded the following interesting insights. It helped understand how supply–demand dynamics influence the profitability of remanufacturing activities. The study showed how the new product sales, consumers' valuations of the owned products, and age-dependent quality of returns shape the product acquisition and remanufacturing costs. It established that the optimal acquisition policy is to target used products of an age when the acquisition and remanufacturing costs are minimum. Further, it noted that the optimal age for trade-in is longer compared to that of the buyback. The difference in the ideal acquisition age in the two policies can be significant in settings where a firm must offer a large percentage increase in the buyback price to realize the same return volume under a trade-in program as that under a buyback program. The difference in the ideal age can affect the degree of supply/demand mismatch between buyback and trade-in programs and may play a role in their relative profitability. Management may wish to account for this difference when choosing between a buyback and trade-in program (e.g., a large reduction in supply/demand mismatch when moving from a buyback program to a trade-in program increases the likelihood that a trade-in program will be more profitable than a buyback program, and vice-versa). The study clarified why and how a firm should identify how old the used products should be while adopting the buyback or trade-in policies, and use this information while implementing the take-back policy in closed loop operations.

Cole et al.'s (2017) numerical analysis also indicated that when the cost of holding inventory is substantial, the profit over the remanufactured product life cycle is maximum if the age for minimal acquisition and remanufacturing costs matches with the time lag between the beginning of new product life cycle and remanufactured product life cycle. Under these conditions, optimal acquisition pricing can be derived myopically with respect to the potential supplies from customers in each period. Since product design influences the acquisition and remanufacturing costs, and the time lag between the beginning of new product life cycle and remanufactured product life cycle is influenced by consumer awareness and interest in a remanufactured product, companies must ensure that the supply of used product aligns with the remanufactured product demand. When it is difficult to match the optimal used product age (age at which acquisition and remanufacturing costs are minimum) and remanufactured product demand time lag, proactive pricing that accounts for the sales, returns and demand over the life has the potential to increase the profit significantly.

Cole et al.'s (2017) study also informs managers about the merits of considering a mixed strategy of using a buyback program during certain periods and a trade-in program during other periods. The new product margin is a key driver of the relative profitability of buyback and trade-in programs in an upcoming period, and typically the margin tends to be highest near the beginning of the product life-cycle and lowest near the end of the product life-cycle. Thus, when a mixed strategy is used, it is likely to be more profitable to begin with a trade-in program and end with a buyback program in the product life cycle. Furthermore, when a product life-cycle is over and the next generation of the new product begins, it is in the firm's interest to revert to a trade-in program (for acquiring the old generation product for remanufacturing) because the margin on the new generation of the product at the beginning of its life-cycle is likely to be higher than the margin on the old generation of the product at the end of its life-cycle.

9.4.2 Consumer Leasing and Renting Scenario

Although buyback and trade-in are two important approaches for getting the products returned from customers, their economic and environmental usefulness depends upon consumer's willingness to return a product and buy a new one from the same company and the reusability of the returned products by the manufacturer. In a market context when the technology life cycle is short or there are companies who introduce upgraded products frequently then the valuation and reusability of older generation products drops drastically. Under such circumstances the remanufacturing and reuse of options are difficult to implement especially when consumers are not sensitive about environmental benefit of closed loop operations and therefore do not have the motivations to return the owned, used products.

In this context, rental or leasing as opposed to selling can serve as an effective alternative mechanism for closed loop operations to realize the cradle-to-cradle "objective." The key requirement for successful renting/leasing vis-à-vis selling is to have renting/leasing economically attractive to both producers and consumers. Several industries such as cable services, satellite TV, telecom, seasonal sports goods, and carpets have adopted leasing and rental business model successfully utilizing the principles of closed loop operations.

Companies such as Recreational Equipment Inc. (REI), Interface illustrate how leasing and rental based business could offer both financial and environmental payoffs. In leasing and renttal business the ownership of the products lies with the companies and consumers use the products for a monthly or annual fee. As a result, while the companies have better control on accessing the used products from consumers, the consumers benefit from for not having to make a lump sum payment as in buying. In these supply chains, reverse (returns) supply chain is not just important; it is an integral part of every company's customer relationship. In addition to regular returns of used goods at the end of contractual leasing/rental period, products may need to be collected prematurely because a customer may switch to another company or product/service, cancel the service, experience technical problems, default on the agreement, upgrade to a different product, and so on. Thus successful reverse supply chain management involves deciding the cost-effective level of pricing, and centralization/decentralization of logistics, remanufacturing, repair, refurbishing, and recycling operations for different level of quality and volume of returns over time. The main operational objectives are maximizing efficiency at varied scales of operations by balancing facility and infrastructure costs, transportation costs and inventory costs, reducing returned product turnaround times through reduced operational variability, and minimizing the cost of (poor) quality by minimizing the trade-offs (e.g., cost of avoiding defects versus customer service costs). Accordingly, the supply chain and operational implications are to increase the inventory (asset) turnover ratio, reducing operational complexity and designing the product for easy disassembly and serviceability.

In the steady state of a leasing and rental business cycle in a closed-loop supply chain, efficiency can be computed as follows.

Efficiency = Number of units leased or rented / Total number units owned by the business,

where, the total number of units owned is the sum of units leased/rented, the number of returned units that are being processed, transported, remanufactured, refurbished, repaired, or are in inventory in the reverse supply chain.

In the above equation, using the Little's law, the number of units at any stage in the reverse supply chain can be computed as the product of throughput rate and time spent in the processing stage. Companies need to (re)design products and processes in terms of modular architecture, simpler processes, standardized components, environmentally benign ingredients, and so on to minimize the returns management times for the target throughput rates.

9.5 Role of Strategic and Organizational Alignment

Operations in closed loop supply chain necessitates integration of forward and reverse supply chain. Managing operations in the reverse supply chain is especially complex because there are numerous moving parts with so many organizations to coordinate with. Accessing used products with varied levels of quality in different volumes from ubiquitously dispersed customers with varied types of preference for used and remanufactured products makes is incredibly challenging. Integrating the returns flows with the virgin product flows present additional challenges because of the mismatch between the scales, quality levels, and complexity of operations in both supply chains. It requires a significant of coordination and collaboration among the business partners and stakeholders across design, procurement, manufacturing, retailing, and logistics stages to achieve the necessary alignment for smooth closed loop operations. However, such coordination and collaboration are difficult to occur when organizations in a supply chain consider one another as competitors. Organizations need to overcome this competitive attitude of winning against the exchange partners while aiming for efficient closed-loop operations. Communication about environmental challenges and opportunities with all stakeholders including industry peers, environmental interest groups, regulatory bodies, investors, and customers can play a significant role in accomplishing the necessary alignment for integrated the closed-loop operations.

Mahapatra et al. (2020) highlighted the role of alignment based on a content analytic assessment of operations of exemplar environmental companies that are part of the Global Environmental Management Initiative (GEMI) consortium. They noted that strategic environmental orientation, adoption of proactive environmental initiatives such as waste reduction, remanufacturing, substitution with environmentally benign materials, and implementation of environmental innovation infrastructure (i.e., decision support systems and design and manufacturing standards) are not enough to accomplish significant environmental product and process improvements. Deep and engaged communications with internal and external stakeholders about the environmental missions, practices, and infrastructure are germane to align the organizations and translate the environmental efforts into environmentally superior products and processes, which in turn could yield strong business performance. Transparent communication of environmental efforts, challenges and accomplishments with stakeholders can be useful in obtaining feedback, managing expectations, and informing about the operational challenges and opportunities in finding cost-effective environmentally benign solutions jointly. A counterintuitive insight from their study is that firms engaging in proactive environmental efforts should be open (and not secretive for competitiveness) about their innovations and operations to foster stakeholder collaboration, which, in turn, could contribute to stronger reputation leading to environmental competitive advantage.

9.6 Concluding Remarks

Closed loop operations presents the potential to realize the lofty ideal of "cradle-to-cradle" business. In light of the research publications discussed in this chapter, three foundational principles need to hold true to realize the lofty idea—selection and use of environmentally benign materials and processes that reduce waste generation and promote "upcyclability," cost-effective remanufacturing, and collaborative and mutually rewarding engagement among internal and external exchange partners in the supply chain of organizations. Traditionally, identification and selection of raw materials and ingredients has been the responsibility of R&D and engineering personnel in an organization. The need for cost-effective collecting back the used products from the consumers and upcycling, reusing or remanufacturing those in the closed loop context, necessitates the design, engineering, sourcing, manufacturing, and logistics personnel jointly share the responsibility of all operations in the closed loop supply chain.

Given the complexity of issues, it is important for organizations aspiring to advance closed loop environmental goals to implement necessary decision-making frameworks, tools, and operational mechanisms. With respect to that perspective, the key takeaways are as follows:

- Efforts towards preventative waste reduction and remanufacturing should be emphasized.
- To make remanufacturing profitable, the supply demand dynamics of the new products, used products and remanufactured products, consumer preference for different types of products at different prices and cost of remanufacturing should be considered.
- Over the life cycle of a product, remanufacturing is likely to be most profitable when the remanufactured product demand start time matches with the age (i.e., time since the beginning of new product life cycle) at which the remanufacturing cost is the minimum.
- Since the cost of remanufacturing is influenced by product design (e.g., reusable components and ingredients, modular architecture, ease of disassembly and serviceability, etc.) and remanufactured product demand start time is influenced by marketing efforts which influence consumer awareness and interest, the marketing, remanufacturing and used product acquisition operations should align with one another.
- Since the new product margin is a driver of relative profitability of buyback and trade-in programs in an upcoming period, and typically margin tends to be highest near the beginning of the product life cycle and lowest near the end of the product life cycle, it is likely to begin remanufacturing with a trade-in program and end with a buyback program in the product life cycle.
- Finally, organizations must emphasize open communication and collaboration with internal and external stakeholders for successful implementation of closed-loop initiatives. Inadequate communication could hinder cost-effective environmental product and process innovation.

Although, the studies discussed in this chapter offer several theoretical and practical insights into cost-effective closed loop operations, there is scope to improve the analyses for deeper understanding of the challenges and opportunities. For example, appropriate measurement of efficiency and effectiveness of different environmental alternatives, consumer valuation of products at any time to decide the purchase of a new or remanufactured product, and so on should be investigated with more nuanced considerations. Towards that end, future studies may focus on a more comprehensive understanding of the factors influencing the cost-effectiveness of "cradle-to-cradle" closed-loop operations.

References

Aigner, D. J., Lovell, C. A. K., & Schmidt, P. (1977). Formulation and estimation of stochastic frontier production function models. *Journal of Econometrics, 6*(1), 21–37.

Beil, D. R., & Hopp, W. (2013). WDI Publishing, Product ID#: 1-429-292.

Bussofiane, A., Martin, S., & Parker, D. (1997). The impact of technical efficiency of the UK privatization programme. *Applied Economics, 29*(3), 297–310.

Charnes, A., Cooper, W., Lewin, A., & Seiford, L. (Eds.). (1994). *Data envelopment analysis, theory, methodology and applications*. Boston: Kluwer Academic Publishers.

Chen, Y., & Zhu, J. (2003). DEA models for identifying critical performance measures. *Annals of Operations Research, 124*(1), 225–244.

Cole, D., Mahapatra, S., & Webster, S. (2017). A comparison of buyback and trade-in policies to acquire used products for remanufacturing. *Journal of Business Logistics, 38*(3), 217–232.

Cook, J. A. (2014, Quarter 1). Green supply chains are still too costly. *CSCM's Supply Chain Quarterly*. Retrieved from www.SupplyChainQuarterly.com

Dehnokhalaji, A., Korhonen, P. J., Köksalan, M., Nasrabadi, N., & Wallenius, J. (2010). Efficiency analysis to incorporate interval-scale data. *European Journal of Operational Research, 207*(2), 1116–1121.

Epstein, M. J. (1996). Improving environmental management with full environmental cost accounting. *Environmental Quality Management, 6*(1), 11–22.

Joro, T., Korhonen, P., & Wallenius, J. (1998). Structural comparison of data envelopment analysis and multiple objective linear programming. *Management Science, 44*, 962–970.

Lee, D., & Lionel, B. (2007). *Cradle-to-Cradle design at Herman Miller: Moving toward environmental sustainability*. Boston: Harvard Business School Publications.

Link, S., & Naveh, E. (2006). Standardization and discretion: Does the environmental standard ISO 14001 lead to performance benefits? *IEEE Transactions on Engineering Management, 53*(4), 508–519.

Nga, J. K. H. (2009). The influence of ISO 14000 on firm performance. *Social Responsibility Journal, 5*(3), 408–422.

O'Rourke, D., & Strand, R. (2017). Patagonia: Driving sustainable innovation by embracing tensions. *California Management Review, 60*(1), 102–125.

Orsato, R. J. (2006). When does it pay to be green? *California Management Review, 48*(2), 127–143.

Östlin, J., Sundin, E., & Björkman, M. (2009). Product life-cycle implications for remanufacturing strategies. *Journal of cleaner production, 17*(11), 999–1009.

Mahapatra, S., Pal, R., & Narasimhan, R. (2012). Hybrid (re)manufacturing: Manufacturing and operational implications. *International Journal of Production Research, 50*(14), 3786–3808.

Mahapatra, S., Pal, R., Hult, G. T. M., & Talluri, S. (2015). Assessment of proactive environmental initiatives: Evaluation of efficiency based on interval scale data. *IEEE Transactions in Engineering Management, 62*(2), 280–293.

Mahapatra, S., Barbieri, P., & Rahimian, S. (2020). A content-analytic understanding of factors contributing to the success of proactive environmental strategy. *The International Journal of Environmental Sustainability, 16*(1), 1–28.

Porter, M. E., & van der Linde, C. (1995). Green and competitive: Ending the stalemate. *Harvard Business Review, 73*(5), 120–134.

Premachandra, I. M., Bhabra, G. S., & Sueyoshi, T. (2009). DEA as a tool for bankruptcy assessment: A comparative study with logistic regression technique. *European Journal of Operational Research, 193*(2), 412–424.

Roy, V., Schoenherr, T., & Charan, P. (2018). The thematic landscape of literature in sustainable supply chain management (SSCM): A review of the principal facets in SSCM development. *International Journal of Operations & Production Management, 38*(4), 1091–1124.

Sarkis, J. (2001). Manufacturing's role in corporate environmental sustainability—Concerns for the new millennium. *International Journal of Operations & Production Management, 21*(5/6), 666–686.

Sousa, R., & Voss, C. A. (2008). Contingency research in operations management practices. *Journal of Operations Management, 26*(6), 697–713.

Thanassoulis, E., Kortelainen, M., Johnes, G., & Johnes, J. (2011). Costs and efficiency of higher education institutions in England: A DEA analysis. *Journal of Operational Research, 62*(7), 1282–1297.

Unruh, G. C. (2008). The biosphere rules. *Harvard Business Review, 86*(2), 111–117.

Chapter 10
Orchestrating Sustainable Stakeholder Value Creation: A Product Life Cycle Extension Perspective

Suresh Srinivasan and Vaidyanathan Jayaraman

Abstract The primary role of managers at the helm of companies is to create superior "shared value" to customers, shareholders and other key stakeholders. Orchestrating a business model that identifies "sweet spots" creating value to all stakeholders is the essence of sustainable competitive advantage. Product life cycle extension and the circular economy is one such "sweet spot" that creates competitive advantage. This chapter builds a conceptual model bringing together value creation framework grounded in the resource based theory, sustainability theories including triple bottom line and corporate social responsibility frameworks and the product life cycle extension, circular economy, and reverse supply chain principles to postulate that product life cycle extensions are an important paradigm to achieve shared and sustainable competitive advantage for firms. The chapter ends with a set of challenges in implementing circular economy and raises relevant research questions that could deepen scholarly research in this critical area of shared value and product life cycle extension strategies.

Keywords Value creation · Shared value · Circular economy · Product life cycle extensions

10.1 Introduction

The primary role of managers at the helm of companies is to create shareholder value. Investors and markets closely watch managers' strategies and actions. Such shareholder value is manifested in the share price and market capitalization of companies. The CEOs at the helm of companies like Microsoft, Alphabet, and Apple Inc. persistently focus on superior value creation within their firms which in effect

---

S. Srinivasan (✉) · V. Jayaraman
Great Lakes Institute of Management, Chennai, India
e-mail: suresh.s@greatlakes.edu.in

C. Chen et al. (eds.), *Pursuing Sustainability*, International Series in Operations Research & Management Science 301,
https://doi.org/10.1007/978-3-030-58023-0_10

resulted in a growing share price and a larger market capitalization. Each of these companies enjoying a market capitalization of more than a trillion dollars, in their pursuit to create superior firm value, relentlessly strives to deliver unmatched value to their customers in comparison to the competition, and in parallel, strive to ensure such value creation is at the optimal cost (Bowman and Ambrosini 2000). Such firm level value creation is achieved through dynamic strategies that build superior capabilities that are idiosyncratic in terms of people, skill sets, technologies, innovation, and an effective business model (Wernerfelt 1984; Barney 1991).

Beyond the doctrine of shareholder value creation, the voice of "stakeholder" activism is gaining prominence. Environment and society are other major stakeholders, whose well-being is now becoming a key for sustainability. Shareholder value at the cost of other key stakeholders is no more considered to be a sustainable proposition. The concepts of triple bottom line (Elkington 1998) and corporate social responsibility (Carroll 1991) will document such stakeholder value creation and the stakeholder model (Freeman 2008). However, Friedman's shareholder model (2007) presents a strong view that the primary role of managers in companies is to create shareholder value; and any effort to create value for other stakeholders like the environment and the society could dilute shareholder value creation and thereby destroy shareholder value.

The "shared value" concept proposed by Porter and Kramer (2011) argues that if the business models of companies are carefully orchestrated, it is indeed possible to create stakeholder value in terms of the well-being of the environment and society, in parallel without diluting shareholder value in terms of firm level profits. In essence the concept of shared value fuses the shareholder and stakeholder model. The shared value doctrine underlines the principle that when the business model of the company if carefully chosen such that investments in a unique set of capabilities are made that not only creates superior firm level value in terms of meeting the customer and shareholder value propositions but also ensures other stakeholder well-being. Porter and Kramer (2011) argue that such "sweet spots" and capabilities that create "shared value" further strengthen the competitive advantage of companies and deepen firm level value creation.

Product life cycle extension and the circular economy is one such "sweet spot" that creates competitive advantage to companies not only in terms of firm level value creation, but also does good for other key stakeholders including the environment. In response to the continuous growth of waste production which is threatening the environment, regulations and societal pressures are driving companies and consumers to reduce, reuse, and recycle in an effort to diminish the bulk waste that is dispatched to junkyards. Companies are thus battling to achieve economic profits with minimal environmental foot print and at an optimal cost of such product recovery (Dwivedi and Madaan 2020). In the healthcare industry, for example, solid waste generation in the form of disposable medical devices and electronics, plastics, bandages, and furniture is a serious concern. A large part of such waste contain radiological, biological, and chemical hazards; with a population of over 60 years expected to increase to more than two billion by 2050, such municipal waste not only exposes the world to major environmental issues but also constitutes a financial drain on the economy (Ertz and Patrick 2020; United Nations 2017). The goal of product life cycle extensions, circular economy, and product recovery options are to retain the identity and functionality of the used products as much as possible (Dwivedi and Madaan 2020).

This chapter has built a conceptual model (see Fig. 10.1) based on literature review by bringing together value creation framework (Bowman and Ambrosini 2000; Brandenburger and Stuart 1996) grounded in the resource based theory (Wernerfelt 1984; Barney 1991), sustainability theories including triple bottom line (Elkington 1998) and corporate social responsibility frameworks (Carroll 1991) and the product life cycle extension, circular economy, and reverse supply chain principles to postulate that product life cycle extensions are an important paradigm to achieve shared value in terms of sustainable competitive advantage for firms. Linking product life cycle extension strategies to the principle of Industrial ecology (Goods and Ehrenfeld 1995), industrial symbiosis or by-product synergy (BPS) firms can gain unique competitive advantages. This chapter specifically dwells upon few product recovery and lifecycle extension strategies including remanufacture, recycle, recovery, and refurbishment and explores the challenges for companies in implementing the same to achieve sustainable competitive advantage.

While the concept of circular economy and sustainable development has become an area of interest for both scholars and practitioners, it has been described to be too vague to be implement and thus losing some momentum (Kirchherr et al. 2017).

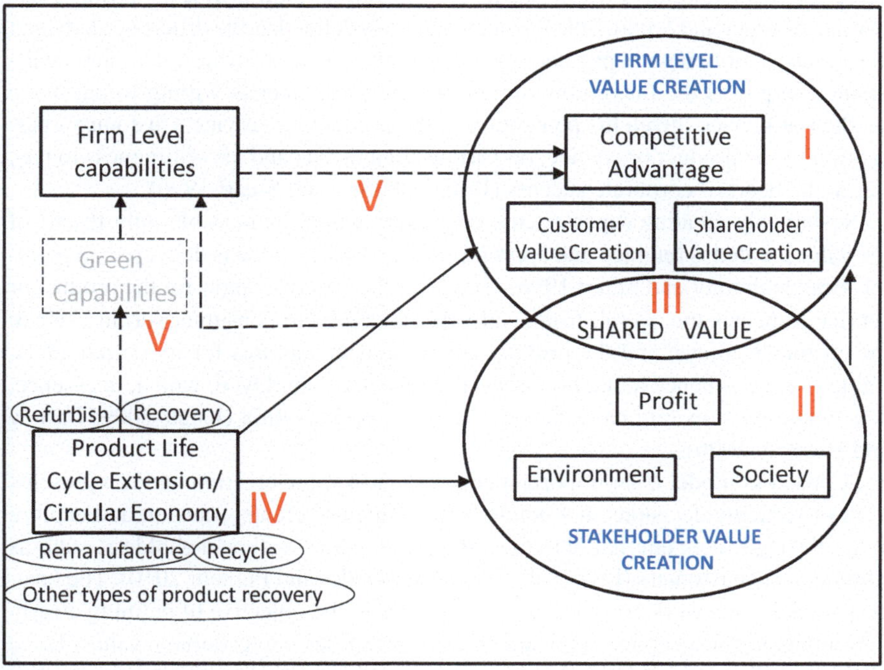

Fig. 10.1 Conceptual model

The chapter also ends with a set of challenges in implementing circular economy and raises relevant research questions that could deepen scholarly research in this critical area of shared value and product life cycle extension strategies.

10.2 Firm Level Value Creation

Resource based theorists argue that a company's competitive advantage stems from idiosyncratic organizational resources and capabilities (Wernerfelt 1984; Barney 1991). Resources and capabilities are considered unique when they are valuable, rare, and inimitable by the competition, non-substitutable and organization wide (Wernerfelt 1984; Barney 1991; Barney and Griffin 1992). The resource-based view advocate that competitive advantage and superior capabilities of a firm derived through such unique capabilities manifest through superior and sustained firm level profitability and firm level value creation in comparison to industry peers (Grant 1996; Rumelt 1991).

The value creation framework argues that the ability of a firm to create value is a combination of superior value created to customers (Sirmon et al. 2007; Kraaijenbrink et al. 2010; Webb et al. 2011) and the captured value within the firm (Bowman and Ambrosini 2000). Competitive advantage, in essence is the ability of firms to create value to customers, superior to competitors, and in parallel capture value for its shareholders within the firm, achieved through firm level asymmetries (Brandenburger and Stuart 1996). Value creation is defined as the difference between customer's willingness to pay and the opportunity cost to the firm. Customer's willingness to pay is the maximum amount which a customer is willing to pay for a product or service, given his perception of the product, or service. Customer's perception of the product or service, on various dimensions and his willingness to pay, lies at the heart of corporate success (Brandenburger and Stuart 1996).

Sustainably creating superior customer value would be possible only if part of the value created is retained and appropriate shareholder value is delivered, in parallel (Brandenburger and Stuart 1996). The price the customer pays for the product or service truncates the overall firm level value created into consumer surplus, which the customer enjoys, and the profits, which the firm captures for its shareholders. Value capture is hence a function of cost of resources employed, which, in essence, is a function of operational efficiency and cost optimization (Knott 2009; Barney and Hesterly 2006).

A business model is an important component in value creation and it represents a set of strategic decisions that define how companies create, transfer, and capture value through their internal activities and partnerships with stakeholders, such as suppliers and customers (Ertz et al. 2019; Osterwalder and Pigneur 2010). The business model framework contributes compellingly to the objective of defining clearly how an organization converts resources and capabilities into economic value (Teece 2010) including but not limited to positioning advantageously in the market against the competition (Urbinati et al. 2017), by designing an organizational structure such that the value created to the customer and the shareholder is superior to the competition (Ertz et al. 2019; Osterwalder and Pigneur 2010).

Extending the resource based theory, corporate diversification is the means by which firms leverage unique firm specific capabilities into related and unrelated markets to achieve competitive advantage (Piskorski 2005). Unique combination of related but different business results in economies of scope where the cost of producing two goods jointly is less than the combined cost of producing each product separately (Piskorski 2005). Integrating corporate strategy and related diversification with the principle of Industrial ecology (Goods and Ehrenfeld 1995), industrial symbiosis or by-product synergy (BPS) suggests that companies from traditionally separate industries can be engaged in some form of network involving physical exchange of material, energy, water, and by-products, in order to exploit unutilized resources or by-products and thereby increase resource utilization and competitive advantage (Chertow 2000).

10.3 Stakeholder Value Creation

Creating superior firm value and merely creating shareholder value is not holistic; moving beyond shareholders and creating stake holder value is becoming the key for corporate sustainability. Superior value created at the firm level translates into satisfied customers willing to pay a higher price for the product or service than the competition and in parallel the firm achieving superior profits. The "triple bottom line" framework integrates the environmental and social dimensions into the traditional value creation framework and argues that the definition of value creation needs to be broadened by bringing in value created to other key stakeholders that include the environment and the society (Elkington 1998). Extending the "triple bottom line" framework the concept of corporate social responsibility (CSR) argues that companies need a broader perspective of "stakeholder" well-being for sustainable shareholder value creation. The doctrine of CSR argues that the value created at a firm level and to the shareholders of a company does not necessarily mean that value is created to the key stakeholders; in fact, superior shareholder value creation can also result in value destruction for other stakeholders like the environment. CSR is a company's sense of responsibility towards the community and environment in which it operates (Carroll 1991); and can be achieved through investments in waste and pollution reduction processes in parallel to earning profits to shareholders by earning adequate returns on the resources employed in the business (Sheehy 2015; Malhotra and Dann 2009).

Contrary views and perspectives argue that a company's sole purpose is to maximize returns to its shareholders and the obeying the laws of the jurisdictions within which it operates (Friedman 2007). Investments in CSR can erode shareholder value creation; for example, providing societal and environmental benefits imposes a financial constraint on the company, increasing costs and reducing profits and thereby shareholder value destruction (Friedman 2007). As a result, companies have largely excluded social and environmental considerations from their economic thinking and in their pursuit to create shareholder value (Porter and Kramer 2011).

10.4 Shared Value

Porter and Kramer's (2011) concept of "shared value" argues that expenditure and investments for the well-being of the other stakeholders does not necessarily mean it would be at the cost of shareholder value creation. Shared value is about designing a unique business model that enhances the competitive advantage of the company while in parallel advancing the environmental and societal conditions in the communities in which it operates, thereby creating superior stakeholder value. Shared value is all about choosing that "sweet spot," an investment into which will create not only shareholder value but also value to other key stakeholders of the company.

Let us consider the increasing technology requirements like big data and high-power computing and the "on premises" servers required to support such needs. Data storage and server operations consume high amounts of energy and have a direct impact on the environment. With more and more companies increasing the number and capacity of their servers, fragmented servers end up consuming higher energy, lower server utilization rates and thus operate inefficiently in comparison to a "large-scale" cloud service provider. Although such companies using fragmented servers could be creating shareholder value by achieving superior profits, they are in fact destroying stakeholder value by negatively impacting the environment. A large-scale cloud operator achieved approximately 65% server utilization rates versus 15% on premises, which means migrating data to the cloud reduces the hardware requirement by around 75%. Further energy consumption reduces by 85% and more importantly the large-scale data center operators use a more favorable power mix that has a 28% lesser carbon intensive component as compared to the global average (Amazon 2020). In effect by companies carefully moving data and computing to the cloud and by creating a unique capability to seamlessly outsource data storage and computing, they not only create better shareholder value but also create superior stakeholder value in terms of lesser environmental footprint. This is a classic case of shared value creation.

10.5 Product Life Cycle Extensions and Circular Economy

Resource scarcity, population growth and climate change is forcing the economic players to move away from the "make–use–dispose" model to "make–use–return" model (Alamerew and Brissaud 2019). The objective of the circular economy or end of life product circularity is economic prosperity while maintaining environmental quality and social equity (Kirchherr et al. 2017).

Circular economy is defined as an economic system that replaces the "end-of-life" concept with reducing, reusing, recycling, and recovering materials in production, distribution and consumption processes (Kirchherr et al. 2017; Jayaraman et al. 1999). End-of-life product recovery encompass collection of used products, reprocessing of a recovered product and redistribution of the processed product; these include remanufacture, repair, recondition, repurpose, cannibalization, redesign, refurbish, upgrading, and recycle. Companies like BMW and IBM target to reuse more than 80% of all plastics (Thierry et al. 1995). Dell Computers for example developed a product life extension plan to increase the useful life of their products through maintenance, repair, reuse, and recycling, thereby reducing the energy requirements for providing a given function (Shokohyar et al. 2014). Circular economy facilitates an effective flow of resources keeping products, components and materials at their highest value at all times through extension of product life times by repair, reconditioning, remanufacturing, and recycling (Bocken et al. 2017).

Remanufacture, recycle, recover, and refurbishment are some of the product extension strategies that have gained importance in recent years (Jayaraman et al. 2003). Remanufacture is about using parts of discarded products in a new product with the same function. Reuse deals with reusing by another consumer of discarded product which is still in good condition and fulfills its original function. Recycling is about processing materials to obtain the same (high grade) or lower (low grade) quality. Recovery is again gaining prominence in terms of reverse supply chain strategies. The original equipment manufacturers (OEM) are becoming more and more responsible for their products that have been discarded by the consumer, especially given the regulatory enforcements and the growing environmental and economic concerns (Hosseinzadeh and Roghanian 2012).

Economic players need to carefully consider the holistic impact of their actions on all the stakeholders. For example, researchers are looking at the benefits of technology developments like the Industry 4.0 from an economic perspective, which is an imperative. However, understanding of the ecological and social impact of such developments is considerably less. Investigating for downstream aspects especially towards the End-of-Life (EOL) and recycling of products is much less understood (Rahman et al. 2020). Digital and emerging technologies consume large amounts of raw materials, extraction of which produces acidic sludge, radioactive substances and other toxic waste all of which pose threat to the environment (Birkel et al. 2019). Such technologies demand new machineries and servers and also lead to high energy consumption and will progressively become dependent on energy producers and suppliers and thus on energy prices. Further such new generation machinery replace many of the existing legacy machinery, majority of which have to be discarded and will end up in landfills. All of these have a major negative impact on the environment decomposition and their degradation takes long time (Birkel et al. 2019) severely impacting the environment. Again, "Lot size one," mass customization and individualization of products offers benefit in terms of meeting precise requirements of customers, but such products become more difficult to for reuse or to resell the products again for a different customer. Such products result in increased waste and recycling might become more difficult with such non standardized products (Birkel et al. 2019).

The principle of by-product synergy (BPS) can offer true business opportunities beyond cost reduction, if wastes are viewed as raw materials for other industries (Zhu et al. 2020). As BPS networks develop, industry goals may shift from reducing waste generation towards producing near-zero waste and finally to producing 100% product, while emissions are lowered and energy use is minimized (Mangan and Olivetti 2010). The economic activity orchestrated in a BPS network creates new businesses and jobs, under the premise that turning biomass output from one organization into a product stream for another organization can generate revenue while reducing emissions and the need for virgin materials.

The Guitang Group (GG) operating one of China's largest sugar refineries uniquely stands apart in implementing industrial symbiosis strategy by developing a collection of downstream companies that link production of sugar, alcohol, cement, compound fertilizer, and paper by recycling and reuse. Guitang's industrial symbiosis and the "sugar chain" model strives to utilize nearly all by-products of sugar production, thereby generating new streams of revenue across closely synergized businesses and reduced environmental emissions and disposal costs, while simultaneously improving the quality of sugar (Zhu et al. 2007).

10.6 Challenges to Circular Economy and Future Research Agenda

The shift to circular economy demand far reaching changes across the organization from business models to technical aspects including systems and work flows within companies, and the relationship with the customers, distributors, suppliers, production systems, and so on (Ritzén and Sandström 2017).

There are a number of barriers to achieving circular economy and product life cycle extensions (Ritzén and Sandström 2017). There are three levels at which the circular economy operates, with the primary objective to achieving sustainable development: first, at the company, product, and consumer level; second, at the eco-industrial park level; and third, at the city, region, or national level (Kirchherr et al. 2017). There are implementation challenges at each such level. Existing contributions on reverse supply chain design and configuration are limited almost totally to quantitative research and modeling, and they fail to provide business rules and a comprehensive framework for designing and implementing product recovery and reverse value chain principles (Gobbi 2011). Key challenges in implementing circular economy and relevant research questions that could deepen scholarly research in this critical area are discussed in the rest of this chapter.

10.6.1 Integration Across Functions and Green Supply Chain Capabilities

The implementation of circular economy needs to be integrated across functional departments with distributed responsibilities across departments and functions within a company; which currently is highly centralized. Further, the understanding on circular economy is shallow as we move down the organizational structure. This results in low risk-taking ability to move towards disruptive changes that are required in implementing circular economy. The overall understanding of the concept and insights need to be enhanced for circular economy to be successfully implemented (Ritzén and Sandström 2017).

Green capabilities of supply chains relate to integration of internal functional flows (finance, logistics and information) with components of green manufacturing that include packing, supply, marketing, eco-design, and environmental participation; these contribute significantly to firm performance. These have triggered the notion that green capabilities can serve as a catalyst for the adoption of green practices such as product return management and circular economy in the organizations (Shaharudin et al. 2019). A firm's ability to handle returns and its external integration capabilities is important for an effective reverse logistics system which is a key to motivating product returns from customers. The breaking down of departmental silos is thus key to facilitating the flow of product returns in the reverse supply chain (Shaharudin et al. 2019). IBM for example introduced a process for repair or upgrade in order to extend the product's useful life and can be taken back to the manufacturer with ease. Toshiba financial services offer to consumers products with different warranty arrangements and also provide product take back programs (Shokohyar et al. 2014). IBM has been practicing product recovery management since the 1990s which is set up as an in-house operation. The company has been recovering used main frames and personal computers that were discarded by itself and by its customers (Thierry et al. 1995). Success of such initiatives requires integration of functions at the organizational level, which companies like IBM have organized.

Research also shows that value creation in closed loop supply chains is limited by too many constraints. The strategic success factors may relax constraints, but they themselves are also constrained, as multiple stakeholders are involved, each having different interests, and hence breaking these cycles proves to be difficult and requires integral thinking particularly among internal stakeholders. Further research is needed to differentiate between different types of cycles, for example in a taxonomy, and different stakeholder viewpoints, both quantitative and qualitative (Schenkel et al. 2019)

10.6.2 Value Chain Considerations

The original equipment manufacturers (OEM) are becoming more and more responsible for their products that have been discarded by the consumer. However, the challenge is that the role certain companies and OEMs play in the value chain could be serious impediments; for example companies selling their products through retailers lose control on its products at the point of sale. In such cases up time of their products is also not a critical efficiency factor. A possible solution to overcome this challenge could be to move towards a "product-service-systems" model that gives the company a higher control on its products in the hands of the consumers (Ritzén and Sandström 2017). Companies that lease their products are generally in a more favorable position than companies that only sell products. Leasing companies usually have more information on the quality and return of used products (Thierry et al. 1995).

10.6.3 Product Design and Product Recovery Design

Manufacturing industries account for a significant part of the world's consumption of resources and generation of waste. They must design and implement integrated sustainable practices and develop products and services for long-term use and ease of repair, remanufacturing, and recycling of their products at the "end of life" phase for achieving sustainable development (Linton and Jayaraman 2005; Shokohyar et al. 2014). The solution to the environmental degradation due to landfill could either be lifetime extension business models that involve either "nature" strategies that improve product durability through improved design or "nurture" strategies that prolong product lifetime as long as the product remains functional (Ertz and Patrick 2020). Nature strategies facilitate longer life spans such that it lasts for a longer duration achieved through product design, self repair, refurbishing, and upgrading including mid-life efficiency checks, replacements, standardization, and reassembly. It also includes products designed for durability, increased performance and efficiency, products compatible for sharing, reuse and for a second hand ownership (guaranteed buy back). Nature strategies also encompass product oriented service system (PSS), maintenance and extended warrantee, use-oriented product service system including rental lease share, pay per use, remanufacturing, next life sales, take-back management, and upgrading (Ertz and Patrick 2020).

Nurture strategies include allowing temporary access to products while retaining ownership (access model). Remanufacturing, refurbishment, offer product access and retain ownership to internalize benefits of circular resource productivity. Product lifecycle extensions through repair upgrade and resell, reverse logistics, product collection and reintegration into the value chain are vital components of nurture strategies (Ertz and Patrick 2020). The organization structure, the geographical location of return acceptance and especially the process for product returns plays a crucial role in the success of the circular economy (Gobbi 2011; Bhatia et al. 2019).

Developing a modular architecture by considering the manufacturing and recovery process distinctively in the early design stage can improve the efficiency of disassembly and product recovery at its end of life stage. Such eco-modular architecture needs to consider interface complexity, material similarity and lifespan similarity for effective and efficient product recovery (Kim and Moon 2019). Sony for example developed products with consideration for their upgradability. At the end of life phase products are taken back to remanufacture by global take-back enters (Shokohyar et al. 2014).

The BMW-Z1 for example is a specifically designed product for disassembly and recycling. It had an all-plastic skin that was designed to be disassembled from its metal chassis in 20 minutes. The design architecture provided for improved product recovery including reducing the number of materials used; avoiding composite components; marking parts and components; and using two-way fasteners instead of screws and glue (Thierry et al. 1995).

10.6.4 Integration Across Network Partners

External and localized community level collaborations in the form of eco-industrial parks are vital for the circular economy. Eco-industrial parks are groups of business which cooperate with each other and the local community in an attempt to reduce waste and pollution and efficiently share resources such as information materials water energy infrastructure and natural resources and help achieve sustainable development with an intention of improving economic gains and improving environmental quality. They could be located on a common property and have collaborative strategies that includes "by-product synergy," "waste to feed" exchanges, waste water cascading, shared logistics, shipping, and receiving facilities where designs and processes are integrated to address multiple objectives (Hein et al. 2015).

Firms must move away from their traditional strategies of mutual undermining to new and innovative forms of symbiosis helping each partner perform traditional tasks more efficiently that will foster economic, social, and environmental partnerships leading to synergistic stakeholder value creation required for a sustainability transition (Elkington 1998). Corporate sustainability's triple bottom line is broadening the environmental agenda given the key role that business can play in forging workable solutions (Elkington 1998). Business is not merely war, anymore; gone are the days of "I win–you lose"; outsmarting competition, capturing market share, beating up suppliers, locking up customers, making a killing; its already moved towards listen to customers, work with suppliers, create teams, entering into strategic partnerships, even with competitors (Elkington 1998). Competition is more to do with sustainability partnerships ending up in "win–win–win" when it comes to triple bottom line (Elkington 1998). Competitors like IBM and DEC in Europe have set up joint programs to recycle personal computers (Thierry et al. 1995).

A vital and interesting research agenda would also be how companies can bring together a unique combination of closely related business resulting in industrial symbiosis enhancing resource utilization, reducing costs and thereby establishing sustainable competitive advantage. How such unique corporate strategies can be linked with industrial symbiosis and circular economy will form a valuable research question in the coming years.

10.6.5 Green Human Capital and Organizational Culture

Availability of green human capital and a sustainability culture within the organization is important for a successful reverse supply chain without which product life time extensions will not be possible (Bag and Gupta 2019). Organizational culture is a key influencing behavioral factor along with the commitment from the senior management of the company, for the success of sustainable supply chain management (Kumar et al. 2020).

10.6.6 Networked Architecture

Integrated information backed decision making within companies is vital for product recovery strategies (Dwivedi and Madaan 2020). Introducing new generic network architecture and networked IT systems including barcode readers and radio frequency identification devices (RFID) need to seamlessly integrated across the value chain for an efficient product recovery (Dwivedi and Madaan 2020).

10.6.7 Success of a Product "Return" Model

The acquisition price a firm will pay to the customer for a returned product is crucial for the closed supply chain model to successfully work as well has implication on its original product's continued success (Gaur et al. 2017). How does a firm's product stand in relation to a used product? Especially from a customer value proposition perspective, pricing perspective (Subramanian and Subramanyam 2012; Ray et al. 2005; Govindan et al. 2019), and given the different revenue models and cost structures. In what circumstances is a refurbished product profitable from an economic perspective as compared to a firm's new product? Research has also not yet focused much on the customer side and legislative side of the end of life product circularity (Alamerew and Brissaud 2019). Companies like BMW have remanufactured high value components such as engines, starter motors, and alternators for a number of years. Each component is disassembled, tested, repaired, and reassembled according to strict quality standards and have the same warranty as a new part and is sold at 50–70% of the new product price (Guide et al. 2000; Guide and Jayaraman 2000; Thierry et al. 1995).

Remanufacturing is an alternative to mitigate e-waste such as mobile phones and laptops that are getting obsolete faster (Guide et al. 2003; Linton et al. 2007). Pricing is a significant aspect to ensure successful remanufacturing implementation. Several studies show the existence of green consumers. An integrated scenario is better than independent-scenario, with higher profit. The existence of green segment increases the profits of manufacturer and retailer, but it is limited. Price-sensitivity of new and remanufactured products, speed of change, and remanufacturing cost influence the optimum prices (Gan et al. 2019).

There is a clear difference in the customer's behavior and willingness to pay between remanufactured products and new products and hence the risk of cannibalization is low (Govindan et al. 2019; Guide and Li 2010). More research is required to explicate the conditions that sets out such differences.

10.7 Reinforcing Capabilities Loop

By ignoring the efficient return and refurbishment or disposal of products, many companies miss out a significant return on investment. If appropriately managed, product life cycle extension strategies can provide a competitive advantage by consolidating the market position with the overall benefit of improving company image (Gobbi 2011). This results in reverse flow where closed loop supply chain and product lifecycle extension capability, or green capabilities themselves strengthen the overall capabilities of the firm, further deepening its competitive advantage and in achieving stakeholder value creation.

Overall, product life cycle extension and the circular economy is one such "sweet spot" that creates competitive advantage to companies in terms of firm level value creation, but also does good for other key stakeholders including the environment. It, however, comes with a number of challenges and implementation bottlenecks for a viable product recovery at the end of life stage. This subject, however, provides enough scope for further research.

References

Alamerew, Y. A., & Brissaud, D. (2019). Circular economy assessment tool for end of life product recovery strategies. *Remanufacturing, 9*(3), 169–185.

Amazon. (2020). On-premises vs. in the cloud. It's greener in the cloud. Retrieved February 25, 2020, from https://aws.amazon.com/about-aws/sustainability/.

Bag, S., & Gupta, S. (2019). Examining the effect of green human capital availability in adoption of reverse logistics and remanufacturing operations performance. *International Journal of Manpower*. Vol. ahead-of-print No. ahead-of-print. https://doi.org/10.1108/IJM-07-2019-0349

Barney, J. (1991). Firm resources and sustained competitive advantage. *Journal of Management, 17*(1), 99–120.

Barney, J. B., & Griffin, R. W. (1992). The management of organizations: Strategy, structure, behavior. Houghton Mifflin College Div.

Barney, J., & Hesterly, W. (2006). *Strategic management and competitive advantage: Concepts and cases*. Hoboken, NJ: Pearson Education.

Bhatia, M. S., Dora, M., & Jakhar, S. K. (2019). Appropriate location for remanufacturing plant towards sustainable supply chain. *Annals of Operations Research*, 1–22.

Birkel, H. S., Veile, J. W., Müller, J. M., Hartmann, E., & Voigt, K. I. (2019). Development of a risk framework for Industry 4.0 in the context of sustainability for established manufacturers. *Sustainability, 11*(2), 384.

Bocken, N. M., Olivetti, E. A., Cullen, J. M., Potting, J., & Lifset, R. (2017). Taking the circularity to the next level: A special issue on the circular economy. *Journal of Industrial Ecology, 21*(3), 476–482.

Bowman, C., & Ambrosini, V. (2000). Value creation versus value capture: Towards a coherent definition of value in strategy. *British Journal of Management, 11*(1), 1–15.

Brandenburger, A. M., & Stuart, H. W. (1996). Value-based business strategy. *Journal of Economics and Management Strategy, 5*(1), 5–24.

Carroll, A. B. (1991). The pyramid of corporate social responsibility: Toward the moral management of organizational stakeholders. *Business Horizons, 34*(4), 39–48.

Chertow, M. R. (2000). Industrial symbiosis: Literature and taxonomy. *Annual Review of Energy and the Environment, 25*(1), 313–337.

Dwivedi, A., & Madaan, J. (2020). A hybrid approach for modeling the key performance indicators of information facilitated product recovery system. *Journal of Modeling in Management*, 15(3), pp. 933–965. https://doi.org/10.1108/JM2-01-2019-0003

Goods, G., & Ehrenfeld, J. R. (1995). Industrial Ecology: A Strategic Framework for Product Policy and Other Sustainable Practices. Available at: http://citeseerx.ist.psu.edu/viewdoc/download?doi=10.1.1.202.507&rep=rep1&type=pdf (accessed on 06/10/2020)

Elkington, J. (1998). Partnerships from cannibals with forks: The triple bottom line of 21st-century business. *Environmental Quality Management, 8*(1), 37–51.

Ertz, M., Leblanc-Proulx, S., Sarigöllü, E., & Morin, V. (2019). Made to break? A taxonomy of business models on product lifetime extension. *Journal of Cleaner Production, 234*, 867–880.

Ertz, M., & Patrick, K. (2020). The future of sustainable healthcare: Extending product lifecycles. *Resources, Conservation and Recycling, 153*, 104589.

Freeman, E. (2008). Ending the so-called Friedman-Freeman debate. *Business Ethics Quarterly, 18*(2), 162–165.

Friedman, M. (2007). The social responsibility of business is to increase its profits. In D. R. McKay, R. Nitsch, & D. A. Peters (Eds.), *Corporate ethics and corporate governance* (pp. 173–178). Berlin: Springer.

Gan, S. S., Pujawan, I. N., Wahjudi, D., & Tanoto, Y. Y. (2019). Pricing decision model for new and remanufactured short life-cycle products with green consumers. *Journal of Revenue and Pricing Management, 18*(5), 376–392.

Gaur, J., Amini, M., & Rao, A. K. (2017). Closed-loop supply chain configuration for new and reconditioned products: An integrated optimization model. *Omega, 66*, 212–223.

Gobbi, C. (2011). Designing the reverse supply chain: The impact of the product residual value. *International Journal of Physical Distribution and Logistics Management, 41*(8), 29.

Govindan, K., Jiménez-Parra, B., Rubio, S., & Vicente-Molina, M. A. (2019). Marketing issues for remanufactured products. *Journal of Cleaner Production, 227*, 890–899.

Grant, R. M. (1996). Prospering in dynamically-competitive environments: Organizational capability as knowledge integration. *Organization Science, 7*(4), 375–387.

Guide, V. D. R., Jr., Jayaraman, V., Srivastava, R., & Benton, W. C. (2000). Supply chain management for recoverable manufacturing systems. *Interfaces, 30*, 125–143.

Guide, V. D. R., Jr., & Jayaraman, V. (2000). Product acquisition management: A framework and current industry practice. *International Journal of Production Research, 2000*(38), 3779–3801.

Guide, V. D. R., Jr., Jayaraman, V., & Linton, J. D. (2003). Building contingency planning for closed loop supply chains with product recovery. *Journal of Operations Management, 21*(3), 259–279.

Guide, V. D. R., Jr., & Li, J. (2010). The potential for cannibalization of new products sales by remanufactured products. *Decision Sciences, 41*(3), 547–572.

Hein, A. M., Jankovic, M., Farel, R., & Yannou, B. (2015). A conceptual framework for eco-industrial parks. In *ASME 2015 international design engineering technical conferences and computers and information in engineering conference*. American Society of Mechanical Engineers Digital Collection.

Hosseinzadeh, M., & Roghanian, E. (2012). An optimization model for reverse logistics network under stochastic environment using genetic algorithm. *International Journal of Business and Social Science, 3*(12), 1–17.

Jayaraman, V., Guide, V. D. R., Jr., & Srivastava, R. (1999). A closed-loop logistics model for remanufacturing. *Journal of Operational Research Society, 50*, 497–509.

Jayaraman, V., Patterson, R., & Rolland, E. (2003). The design of reverse distribution networks: Model and heuristic solution procedure. *European Journal of Operational Research, 150*, 128–149.

Kim, S., & Moon, S. K. (2019). Eco-modular product architecture identification and assessment for product recovery. *Journal of Intelligent Manufacturing, 30*(1), 383–403.

Kirchherr, J., Reike, D., & Hekkert, M. (2017). Conceptualizing the circular economy: An analysis of 114 definitions. *Resources, Conservation and Recycling, 127*, 221–232.

Knott, P. (2009). Integrating resource-based theory in a practice-relevant form. *Journal of Strategy and Management, 2*(2), 163–174.

Kraaijenbrink, J., Spender, J. C., & Groen, A. J. (2010). The resource-based view: A review and assessment of its critiques. *Journal of Management, 36*(1), 349–372.

Kumar, A., Moktadir, M. A., Khan, S. A. R., Garza-Reyes, J. A., Tyagi, M., & Kazançoğlu, Y. (2020). Behavioural factors on the adoption of sustainable supply chain practices. *Resources, Conservation and Recycling*: X, 100031.

Linton, J. D., & Jayaraman, V. (2005). A framework for identifying differences and similarities in the managerial competencies associated with different modes of product life extension. *International Journal of Production Research, 43*(9), 1807–1829.

Linton, J. D., Klassen, R., & Jayaraman, V. (2007). Sustainable supply chains: An introduction. *Journal of Operations Management, 25*(6), 1075–1082.

Malhotra, N., & Dann, J. (2009). Business ethics integral to corporate strategy, says Stanford's Malhotra. cbsnews.com. Archived from the original on March 18, 2017. Retrieved from August 14, 2018.

Mangan, A., & Olivetti, E. (2010). By-product synergy networks: Driving innovation through waste reduction and carbon mitigation. In J. Harmsen & J. B. Powell (Eds.), *Sustainable development in the process industries: Cases and impact* (pp. 81–108). Hoboken, NJ: John Wiley & Sons.

Osterwalder, A., & Pigneur, Y. (2010). *Business model generation: A handbook for visionaries, game changers, and challengers*. Hoboken: John Wiley & Sons.

Piskorski, M. (2005). *Note on corporate strategy* (pp. 1–10). Boston: Harvard Business School Publication. 9-705-449.

Porter, M. E., & Kramer, M. R. (2011). The big idea: Creating shared value. How to reinvent capitalism—and unleash a wave of innovation and growth. *Harvard Business Review, 89*(1–2), 62–77.

Rahman, S. M., Perry, N., Müller, J. M., Kim, J., & Laratte, B. (2020). End-of-Life in industry 4.0: Ignored as before? *Resources, Conservation and Recycling, 154*, 104539.

Ray, S., Boyaci, T., & Aras, N. (2005). Optimal prices and trade-in rebates for durable, remanufacturable products. *Manufacturing & Service Operations Management, 7*(3), 208–228.

Ritzén, S., & Sandström, G. Ö. (2017). Barriers to the circular economy—Integration of perspectives and domains. *Procedia Cirp, 64*, 7–12.

Rumelt, R. P. (1991). How much does industry matter? *Strategic Management Journal, 12*(3), 167–185.

Schenkel, M., Krikke, H., Caniëls, M. C., & Lambrechts, W. (2019). Vicious cycles that hinder value creation in closed loop supply chains: Experiences from the field. *Journal of Cleaner Production, 223*, 278–288.

Shaharudin, M. R., Tan, K. C., Kannan, V., & Zailani, S. (2019). The mediating effects of product returns on the relationship between green capabilities and closed-loop supply chain adoption. *Journal of Cleaner Production, 211*, 233–246.

Sheehy, B. (2015). Defining CSR: Problems and solutions. *Journal of Business Ethics, 131*(3), 625–648.

Shokohyar, S., Mansour, S., & Karimi, B. (2014). A model for integrating services and product EOL management in sustainable product service system (S-PSS). *Journal of Intelligent Manufacturing, 25*(3), 427–440.

Sirmon, D. G., Hitt, M. A., & Ireland, R. D. (2007). Managing firm resources in dynamic environments to create value: Looking inside the black box. *Academy of Management Review, 32*(1), 273–292.

Subramanian, R., & Subramanyam, R. (2012). Key factors in the market for remanufactured products. *Manufacturing & Service Operations Management, 14*(2), 315–326.

Teece, D. J. (2010). Business models, business strategy and innovation. *Long Range Planning, 43*(2–3), 172–194.

Thierry, M., Salomon, M., Van Nunen, J., & Van Wassenhove, L. (1995). Strategic issues in product recovery management. *California Management Review, 37*(2), 114–136.

United Nations (2017). World Population Ageing. Available at: https://www.un.org/en/development/desa/population/publications/pdf/ageing/WPA2017_Highlights.pdf (accessed on 06/10/2020).

Urbinati, A., Chiaroni, D., & Chiesa, V. (2017). Towards a new taxonomy of circular economy business models. *Journal of Cleaner Production, 168*, 487–498.

Webb, J. W., Ireland, R. D., Hitt, M. A., Kistruck, G. M., & Tihanyi, L. (2011). Where is the opportunity without the customer? An integration of marketing activities, the entrepreneurship process, and institutional theory. *Journal of the Academy of Marketing Science, 39*(4), 537–554.

Wernerfelt, B. (1984). A resource-based view of the firm. *Strategic Management Journal, 5*(2), 171–180.

Zhu, Q., Lowe, E. A., Wei, Y. A., & Barnes, D. (2007). Industrial symbiosis in China: A case study of the Guitang Group. *Journal of Industrial Ecology, 11*(1), 31–42.

Zhu, Y., Dawande, M., Gavirneni, N., & Jayaraman, V. (2020). Industrial Symbiosis: Impact of Competition on Firms' Willingness to Implement, IISE Transactions, https://doi.org/10.1080/24725854.2020.1781305.

Chapter 11
Exercise Your Rs! You Never Know When You May Need Them: Revisiting and Extending Modes of Product Life for the Future

Jonathan D. Linton and Vaidyanathan Jayaraman

Abstract Do you remember the slogans regarding the *3 Rs*; well, there is actually 12. Each R can provide opportunity, but can also create peril if a management team is unacquainted with it at an inopportune time. Hence, we revisit the urgent need for companies—and society in general—to rethink supply chain strategy with respect to Rs. The expansion of the 3 Rs to 12 is inherent with the view that consumers and markets are focusing on product life extension—not product obsolescence. In an obsolescence economy, there are only three Rs and the R most revered and reported is Recycling. However, if the focus is on product life extension—getting all the value possible from a product and the by-products that result from the production, use and disposal of the product—recycling is considered the least attractive of the Rs. This is increasingly important as we face millennial challenges associated to greater glocal sustainability. The pandemic of 2020 shows us that even with tremendous shuttering of the world economy, greenhouse gas emissions and other measures of pollution are still substantial. In fact, the level of economic activity during the height of global lockdowns is described as being consistent with what is needed to keep the global average temperature to the target of 1.5 °C above baseline. Consequently, marshalling and extending our technical knowledge and related managerial skills is needed to meet the challenges of reduced greenhouse gas emissions and ensuring sustainability more broadly. Firms that are unaware of their position and that of their supply chains in relation to the 12 Rs, will have difficulties at multiple points in the future. Supply chains that engage the 12 Rs in an appropriate

J. D. Linton (✉)
School of Management, University of Sheffield, Sheffield, UK

Science Technology Studies Laboratory, Higher School of Economics, Moscow, Russian Federation

V. Jayaraman
Great Lakes Institute of Management, Chennai, India

C. Chen et al. (eds.), *Pursuing Sustainability*, International Series in Operations Research & Management Science 301,
https://doi.org/10.1007/978-3-030-58023-0_11

manner not only can avoid future difficulties but reduce their cost basis at a time when the growth in many markets is flat at best.

11.1 Introduction[1]

The 12 R framework is an enabler for supporting and efficiently extending the adoption of different modes of product life extension (PLE). Each mode presents unique challenges, opportunities, and economic benefits. However, certain modes offer the challenges and competencies needed for success with other modes. By understanding this and managing accordingly, the uncertainty and challenge associated with doing something new can be reduced greatly. The outcome is a reduced likelihood of failure and a reduction in the resources needed to learn and master skills that are already present within one or more parts of the organization. The overall goal in all such modes is to decrease the environmental impact while increasing the societal and economic value of the product. Having introduced the challenge in general, the modes and challenges associated with each mode are introduced and then described in greater detail. The Rs of Product Life Extension are: (1) Recall, (2–5) Repair—multiple modes, (6) Restore, (7) Reuse, (8) Remanufacture, (9) Recycle, (10) Recover, (11) Reduce, and (12) Redirect. The challenges associated with these modes of life extension are: (a) Uncertain timing and quantity of returns; (b) Need to balance demand with returns; (c) Uncertainty in subassemblies, parts, and materials recovered from returned items; (d) Requirement for a reverse logistics network; (e) Complication of material matching requirements; (f) Need to disassemble the returned products; and (g) Problem of stochastic routings and highly variable processing times (Guide et al. 2000). Having listed the 12 modes and seven complicating characteristics, each mode of PLE will be briefly described.

11.2 Product Life Extension Modes

11.2.1 Recall

This is a directive by a manufacturer or an agent for the return of a product that has already been shipped to the market or has been sold to consumers but discovered to be contaminated, unsafe or defective (Flexner 1987). Product recalls incur direct cost of collecting, replacing, or repairing the defective products in the short term and could lead to damaging the brand name and lost sales in the long run (Borah and

[1]This study was conducted within the framework of the Basic Research Program at the National Research University Higher School of Economics (HSE) and supported within the framework of implementation of the HSE 5-100 Program Roadmap.

Tellis 2016). The end customer is frequently required to deliver the product to a specific location either to correct any problems or return for a refund or exchange. Jayaraman et al. (2003) indicate that channel design for retro-movement requires careful consideration of the product and demand characteristics and the nature of the existing distribution channel. Product recalls are often caused by quality and safety issues (He et al. 2020). Recalls are quite common in a variety of industry such as automotive and electronics (Rupp 2004; Nystedt 2007) and often are associated with significant present and future costs to a company. These recalls result from a lack of quality assurance in the manufacturing or design processes of its supply chain partners and can impact a large number of products manufactured over an extended period of time. Some of the recent studies in the literature debate the operational drivers of product recalls such as outsourcing (Stevens et al. 2014) and product variety (Shah et al. 2016).

11.2.2 Repair: Maintain Post-failure

Is to impart new life and promote recuperation through some adjustments made on a component of a product to bring it back to working order, but not necessarily to an "as good as new" state. This repair may be carried out by the consumer, manufacturer or a third party. The key question for any consumer or organization to pose is whether the cost of repair is less than the cost of replacement. This repair involves a moderate unit transformation and is usually carried out using materials provided either by the manufacturer, or salvaged from another piece of equipment or procured from a third-party supplier.

11.2.3 Repair: Maintain Pre-failure pReventative Maintenance

Preventative Maintenance (PM) includes inspection and servicing activities that have been preplanned to occur at specific time intervals to retain the functional capabilities of a product. PM is done to restrict the timing of product downtime that includes unanticipated failure avoidance and avoiding failures that will result in damage to numerous parts of a product rather that only the part that failed (Gertsbakh 2000). Preventative maintenance may be provided by the customer, manufacturer or a third-party provider and involves moderate transformation of a unit.

11.2.4 Repair: Maintain Pre-failure pRedictive Maintenance

Predictive maintenance is a condition driven program that uses direct monitoring of the system efficiency, mechanical condition and other key indicators to determine the actual average time to loss of efficiency or failure for each system (Mobley 1990). By adopting predictive maintenance strategies, replacements of parts need only be done on an as-is required basis such as repair.

11.2.5 Repair: Improve Existing Product

An accounting practice is to gradually depreciate an asset over time to reflect a decrease in both the quality of output and the amount of time that the product will still be "fit for use." Japanese industry demonstrated this assumption to be a limitation of the treatment of assets and products through their practice of continuous improvement. Hence, leading Japanese manufacturing firms demonstrate that a product could be modified over time to improve its performance and extend the anticipated life. This practice is now used by leading manufacturers as part of the maintenance/upgrade process and can involve both hardware and software changes to both the product and the processes it is associated with.

11.2.6 Restore (Refurbish)

Involves large aesthetic improvements on a product which may bring it to a "as good as new" state, but with limited functionality improvements. In the case of refurbishing (Aneesh and Kumar 2020), the process involves renovating an older or damaged product to a workable or better condition and typically is sold at a reduced price in the market.

11.2.7 Reuse

A product is used without any change in its form, shape or content. This mode includes product leasing or a product being transferred (sold) by the current owner to a new owner. The product might be used for the same or a different application. For example, this could involve rental cars being sold to a private owner or a cell-phone being repurposed from being a means of communication to being a means of contact in the case of emergency. Typically, the cost of reuse (collection + testing) is lower than the cost of manufacturing. Reuse can be applied to an entire product or to some subset of its parts.

11.2.8 Remanufacturing

The transformation of used units that consist of components and parts into units which satisfy the same quality and other standards as new units (Lund 1984; Moosmayer et al. 2020). For a remanufacturing operation to function in an effective and efficient manner, an organization must be focused not only on the (traditional) flow of products to the customer from the manufacturer but also on the reverse flow from the customer (end-user) back to the manufacturer or a third-party remanufacturer. Remanufacturing involves a major transformation of a component, unit or part where the labor value-added could be quite high since many parts may be tested and repaired (Chen and Ulya 2019).

11.2.9 Recycle

A transformation process of a product's material into raw materials for use in new products. In this process, materials that are destined for disposal are collected, processed, and used to make new products. The only value retained is that of the material. The value of the material is often much lower as it is extracted from a used product. In the case of materials such as copper and aluminum, the adverse impact on the value of the material is low. However, for most polymers the adverse impact on the value is high as the length of polymer chains is reduced during reprocessing resulting in a degradation in mechanical properties.

11.2.10 Recover

Under Extended Producer Responsibility (EPR) there is a shift of financial and/or physical responsibility for end-of-life products from society to beneficiaries of products. Product or material recovery plays a huge role in a sustainable environment. Some examples of EPR includes packaging materials recovery such as take-back of electronic goods (Walls 2006) and EU's green dot program (Tojo 2001). Product recovery generally refers to collection activities that include end-of-life products collection and appropriate treatment (Jacobs and Subramanian 2011). Recovery programs typically include targets for both collection and salvage of used products such as the EU's WEEE directive. This is an area of tremendous concern currently, as it was recently established that many recovery operations were unable to extract value from plastics and other post-consumer waste. In many cases, this has led to the return of mixed post-consumer waste to the place of origin and a cessation of future material transfers. Recent developments have necessitated a combination of reconsideration of product design, managing, disposal pricing and storing of solid waste and application of the other 11 Rs listed here.

11.2.11 Reduce

In a sustainable environment, the goal is to reduce the input required in a product. This could include reducing: energy use during manufacturing and distribution, power consumption during product use, and the quantity of raw materials that go into producing the product. Eco-efficiency—adoption of an environmental lens to look for opportunities to reduce and eliminate waste and inefficiency—continues to gain attention in lifecycle consideration in the downstream supply chain. It is clear that lower energy consumption, scrap reduction and increased efficiency reduces cost while protecting the environment. This involve any environmental improvements that directly improve the company's bottom line.

11.2.12 Redirect

During the production, use and disposal of a product, by-products result due to the inefficiency of transformation of energy and materials from raw to intermediate to finished state. These inefficiencies often pollute some medium—water, air, land—other times they represent a missed opportunity—generation of waste heat. This is the domain of the field of industrial ecology (Frosch 1992; Graedel and Allenby 1995). Industrial ecology likens industrial systems to that of nature and seeks to keep systems in balance by ensuring that by-products and outputs are consumed by some other part of the system. For example, slag—a by-product of steel production—was considered for centuries to be a waste suitable only for dumping in landfill. However, it was recently determined to be a suitable input for Portland cement. Hence a liability that had to be dumped at an economic and social cost, is now an income generating by-product. Similarly, traditional propulsion and braking systems of automobiles were modified to produce a hybrid car—a propulsion system that can recapture energy wasted during braking. The result is in an automobile that is more fuel efficient. *Redirect* belongs with modes of Product Life Extension, as it is part of the supply chain and systems approach pursued in this chapter. To tackle traditional and extended forms of Product Life Extension, we must move away from consideration of only the firm and take the perspective of the entire supply chain. Finally, there is one conspicuous absence in the discussion of redirect. While material and energy have been discussed, information has so far been overlooked. Failure to consider the utilization of information not only can lead to missed opportunities in terms of the duration of product life extension and determination of most appropriate customer group and application but also the value of information as a by-product. Information is a highly valuable and underutilized by-product of product systems not only demonstrated by the many firms focused on the provision and processing of information, but also on OEMs making physical products—see Wise and Baumgartner (2000) for examples. Having considered each of the twelve Rs, the unique operational challenges are presented as a series of seven complicating characteristics.

11.2.12.1 Unique Operational Challenges to Rs

Each challenge is considered briefly for a detailed assessment see Linton and Jayaraman (2005). In most cases, the operational challenges can be divided into low, medium, and high categories in relation to each R. However, in some cases an operational challenge is not applicable. The seven complicating characteristics are listed below and their impact on each of the 12 Rs are summarized in Table 11.1.

(a) *Uncertain timing and quantity of returns*—reflects the uncertain nature of the use and durability of a product.
(b) *Need to balance demand with returns*—imperfect balance between demand and return can result in either shortages or stockpiling of inventory. The result is that the policy for the acquisition, management, and control of inventory is critical.
(c) *Need to disassemble the returned products*—Need to know the degree and method of disassembly prior to the arrival of a product. This is a challenge for production planning and control.
(d) *Uncertainty in subassemblies, parts, and materials recovered from returned items*—If the condition of a part or product influences its recoverability then this is an important issue. In cases of flow uncertainty, traditional materials planning approaches, such as material requirements planning systems, may be used.
(e) *Requirement for a reverse logistics network*—addresses how products are collected from users and returned to a facility for processing. Once the requirement for a reverse logistics network is identified, a series of decisions must be made on the nature of the network, since these networks can be very different from each other (Fleischmann et al. 2001).
(f) *Complication of material matching requirements*—in some cases it may be critical for the matching of a specific product to a specific customer, specific components to a specific product or both. In such cases, special care must be taken to ensure that proper records are kept linking the components, core, and customer so the finished product is returned with its original components to its original customer. There are alternative repair systems that may be used. For example, a firm may maintain a spares pool and allow customers to trade a nonfunctional item for a functional unit. The nonfunctional item is then repaired and returned to the spares pool.
(g) *Problem of stochastic routings and highly variable processing times*—routings can be stochastic, since it may not be apparent which processing steps are required, due to the uncertain condition of the returned products. This uncertainty can also lead to great variability in the processing time required for each step in the manufacturing process.

Table 11.1 Complicating characteristics and the 12 Rs

	Uncertain **tim**ing	**Bal**ance demand and supply	Need to **dis**assemble returns	Uncertainty in materials **rec**overed	Need for reverse **log**istics network	Material **match**ing requirements	Stochastic **rout**ings for material repair and remanufacture
Recall	Moderate	NA	Moderate	Low	Yes	High	NA
Repair—post-failure	Moderate	NA	Moderate	High	No	High	High
Repair—preventative maintenance	Low	NA	Moderate	Low	No	High	Low
Repair—predictive maintenance	Low	NA	Moderate	Low	No	High	Low
Repair—upgrade	Moderate	NA	Moderate	Low	Yes/No	NA	Low
Restore—refurbish	High	High	High	High	Yes	High	High
Reuse	Low	High	None	Low	Yes/No	NA	NA
Remanufacture	High	High	High	High	Yes	High	High
Recycle	High	Moderate	Low	Low	Yes	NA	NA
Recover	High	High	High	High	Yes	Low	High
Reduce	NA	NA	NA	NA	NA	NA	NA
Redirect	High	High	NA	NA	NA	NA	NA

Not only do the complicating characteristics add challenges to the management of manufacturing, supply chains, and product systems, but these challenges require and heighten the competence of the management team in working with and responding to uncertainty.

11.3 Need to Consider the Challenge that Rs Present to Functional Areas

Specific functional implications have been addressed in detail in Linton and Jayaraman (2005). Consequently, the relationship between different Rs and functions are not reviewed here. However, the functional areas that interact heavily with a number of the different Rs are briefly listed below.

1. *Focus* is the major reason or driver of the life extension modes. There is no clear relationship between the *foci* of the modes of product life extension and their impact on different parts of a firm's operations.
2. *Forecasting,* in a typical manufacturing environment, involves estimating the future demand for products/services and the resources necessary to produce them. In some of our Rs one must forecast both supply and demand and to attempt to coordinate the two.
3. *Purchasing* in a typical environment is based on the demand forecasts. The automation of purchasing through system like Materials Requirements Planning (MRP) is limited since most Rs offer greater uncertainty regarding what will need to be purchased.
4. For most manufacturing processes *inventory control and management* assumes that parts and finished products are interchangeable and that raw materials and components can be obtained on an as required basis; this is not the case for many Rs.
5. *Production planning and control* in conventional manufacturing is typified by certainty in routing and processing times. This is also the case in some of the Rs.
6. In conventional manufacturing, *logistics* consider forward flows along the supply chain based on relation between demand forecasting, inventory and production policies. Rs have greater complexity due to uncertainties associated with forecasting, inventory and production already mentioned. In other cases, reverse flows are also a consideration.

Having indicated the functional areas within operations and their relation to the Rs, we now provide some managerial implications of the Rs are considered.

11.4 Managerial Implications

A shift in perspective is required for management teams to successfully navigate a twelve R economy. A supply side focus on production, product ownership and obsolescence is replaced with a focus on the delivery of value to the customer. This move to a customer value focus is consistent with a supply chain perspective (as opposed to more inward strategic considerations of the OEM as a focal firm). Hence, forward looking OEMs are increasing their profitability and strengthening their position, by understanding how to extract value in additional and nontraditional ways from their product and their supply chain partners. This involves an increase in consideration of the product—both upstream (process) and downstream (usage). Such an approach also invites consideration of the product in terms of multiple applications, multiple lives, and post-user disposal. The question being considered by the 12 Rs and the rhetoric of Product Life Extension is *how we can extract value at as many possible points as possible while minimizing complications and the possibility of future liabilities*.

For many managers and management teams, the suggestion that the future of competition and operation lie heavily on mastery of 12 Rs will sound too daunting, different and difficult to approach. Few organizations lack competence with all the modes of Product Life Extension. However, a close inspection of manufacturers will show how many of the Rs are already resident within each firm. Consequently, a first step for management teams is to review internal operations to determine which of the 12 Rs are currently present. As indicated earlier, the move to a supply chain perspective means that access to and understanding Rs is no longer limited to the legal boundaries of a firm. Any supply chain member's expertise in each R is also relevant. In some cases, firms have made conscious strategic decisions to have R activities addressed by a trusted strategic partner. In other cases, firms have not made conscious decisions for Rs to be handled by partners, but these relationships have gradually emerged over time. Consequently, a six step-process is proposed to take stock of a firm's position in regards to the 12 Rs. (1) A review is conducted to determine the firm's position in terms of the 12 Rs. (2) A review is conducted to determine the supply chain's position in terms of the 12 Rs—regarding satisfaction of the firm's needs and also the needs of other parts of the supply chain. (3) The question is then asked of whether there is any evidence that firms outside of the supply chain have an active involvement in the supply chains Rs. (4) A SWOT analysis is conducted for each of the Rs. (5) The implications of the document are summarized and any decisions to modify the current situations—changes in future strategic direction—are recorded. (6) Periodic reassessment occurs in response to changes in firm or supply chain strategy, changes in customer expectations, or changes in the policy environment. Having summarized the six-step process, each step is considered in more detail.

11.4.1 Reviewing Firm Activity with Each of the 12 Rs

Most firms are currently engaged in one or more of the Rs, but awareness of this involvement is often limited. Consequently, the first step is to determine for which of these activities a firm is actually actively pursuing.

11.4.2 Reviewing Supply Chain Partner Activity with Each of the 12 Rs

The firm may have outsourced certain Rs to supply chain partners—or even have spun-out an R to become an independent entity. In such a case the firm is not actively involved with an R, but to become actively involved in the future with the R requires the displacement of an existing supply chain partner. Hence, the existence of this type of relationship is important to understanding the firm's positioning in relation to the Rs.

Even if a supply chain partner is not managing one or more Rs on your behalf, they should be included in the firm audit of Rs. The two most notable findings with supply chain partners will be the presence/absence of Rs differing. If a partner lacks an R for which your firm has competence, there may be an opportunity to provide the R-related services that are currently absent. There are multiple possibilities; including: managing the R for your partner or assisting the partner in developing internal capabilities. Such arrangements are far from unusual, many OEMs offer monitoring, maintenance and repair services for their products. These activities were pioneered by sector leaders including IBM and Xerox. The high profitability and strategic advisability of this tactic has been considered by Wise and Baumgartner (2000). Hence the presence of an R in your firm that is absent for a supply chain partner could be an attractive opportunity financially and strategically.

The presence of an R in a supply chain partner that your firm lacks capabilities with provides different opportunities. Your supply chain partner may be well positioned to provide your firm with the missing R. Prior to considering such an option it is important to recognize that as the two firms have a relationship there is also an opportunity to learn how a firm that you are accustomed to working with deals with something that you have not yet mastered.

If the supply chain lacks capabilities with an R, then either an external organization can view this gap as an opportunity or this R, currently an orphan, could become problematic at any time. There are a substantial number of cases in which third-parties are providing an R, but are clearly outside of the supply chain. Examples include firms that refill toner cartridges once the ink empties and firms that capture reusable components from products that are disposed. In both of these cases, the OEM is impacted by the operation of the third parties in an economically unattractive manner due to the lack of consideration of these operations in their strategic and operational model (Linton 2008). In the case of toner cartridges, the OEM's business model was originally based on the assumption that empty ink cartridges are replaced with new toner cartridges produced by the OEM. In some cases, modifications have been made to the business model in an apparent response to the rise of third-party refillers. Changes to business models in response to third parties include: (1) a pre-bate program in which the customer pays a lower price for a toner cartridge in exchange for a promise not to refill the cartridge and (2) an offer to take back cartridges via a prepaid courier for recycling. At best, such programs lead to low value recycling. At worst, cartridges that still retain economic and functional value to house and deliver ink become part of the landfill.

In the case of third-parties recovering components or products that have been disposed of by their original owners, the challenges to the OEM's supply chain are somewhat different. As the disposed of product carries the name of the OEM, the OEM and supply chain partners take reputational damage if the product is sold through grey market channels and fails while being used by a customer. When customers are advised that they should not expect warranty and service for product purchased in the grey market, the typical customer response is that if a product carries the OEM's name, the logo of the OEM represents the OEM's commitment (or lack of commitment) to quality and customer satisfaction. A further complication for the OEM is that a lack of knowledge regarding the number and disposition of third-party grey market products complicates the planning and operations of the OEM. The automotive supply chain offers an exemplar of this challenge due to the large number of parts, the long life of the typical automobile and a lack of OEM awareness of the number and location of all the product they have produced during the many years in which a car model is produced. Historically, automotive supply chains are designed to support spare parts requirements for each car model many years after the discontinuation of the automobile. The cost of producing and inventorying small numbers of spare parts is substantial. Low cost/low priced competition from third-parties is provided by scrap yards that remove parts from discarded vehicles. This complicates the management of manufacturing and inventorying spare parts. Some form of cooperation with third-party suppliers would save the automotive supply chain tremendous cost as it would allow for supply planning to focus on the market that cannot be satisfied through removal of parts from discarded vehicles. In other words, at the time of the introduction of a new model, the automotive supply chain would provide all spare parts. However, as time progresses the automotive supply chain would cede larger parts of the spare part market to the third-party suppliers that could provide parts from recently disposed of vehicles. This would result in a coordinated exit by OEMs from the spare parts market. It would also eliminate the expenses of maintaining readiness to prepare small numbers of spare parts on demand—an expensive option historically written into the contracts of all automotive suppliers. While many focal firms and supply chain partners are displeased with the presence of firms that profit from under-serviced Rs, to ignore the presence of these firms can be costly in ways and to an extent neither easily anticipated.

The most common example of orphan Rs relate to packaging material that are designated as recyclable. This *recyclable* packaging has been an issue that appears and reappears for decades. Earlier complaints were about recyclable German green dot packages that were dumped in neighboring (French) landfills instead of being reprocessed as advertised and paid for. In recent years, developing countries have refused to unload ships laden with *recyclable* waste. The waste then gets returned to its originating point in the developed world and eventually gets designated as waste due to the lack of recycling facilities in either the developing or developed world.

11.4.3 Are Firms Outside Your Supply Chain Involved with Your Rs

The possibility and repercussions of firms outside of your supply chain being involved with your Rs was mentioned in the previous section. For each R, the possibility of the presence of unrecognized third parties in your supply chain should be investigated. While the identity of such firms may be well understood, consideration of the interaction of downstream partners and customers is helpful in this regard. For example due to heightened environmental sensitivity, an OEM asked a local government of what they intended to do with the firm's waste water. Due to this inquiry, the OEM discovered that the waste water was going to be used for watering parks and agricultural land. The OEM recognized that either they would need to treat their water in a different—more thorough manner—or convince the local government to utilize their water in a different way.

In some cases, conversations with supply chain partners and customers are sufficient. At the very least, this is a good starting point. Conducting a waste flow audit is the best approach to ensuring an understanding of the disposition of products and by-products at their end-of-life. One food processor reputed as a leader in environmental management found surprising value in conducting a waste flow audit. Initially, management anticipated that the audit would just verify the excellence of their approach and processes. While most by-products of production did not end up in landfill, the audit found there was much greater potential for value generation. Examples of findings include: (a) Pig feed to product—food waste was diverted from landfill to pig feed changing a disposal cost into a small revenue stream. However, further consideration of the waste stream indicated that the pig feed included hundreds of pounds of nut dust. Utilizing existing equipment, the nut dust waste could be converted into a nut butter product. This recognition redirected waste from a product in which value is measured by the ton to a product for which value is measured by the kilo; (b) Carrying fruit instead of worms—berries for the production of pie fillings were provided in large plastic buckets. These buckets were redirected away from landfills by selling them to an individual that distributed them as bait buckets for fisherman. While this redirection was imaginative, the firm found that by returning the buckets (reuse) clean and empty to their original supplier much larger savings in product cost were possible, and (c) Reducing casual employee theft—during audit it was determined that the very large plastic garbage bags that were typically utilized for waste disposal were being purchased at a rate that far exceeded the supply of waste disposed. A multifunctional team was assembled to discuss the possible mysterious alternative uses that could explain the large number of bags that were being consumed. It was mentioned that these bags made very good tarpaulins due to their size and thickness. A decision was made to make bags that were received from suppliers accessible for the personal use of employees for their renovation projects. Employees were advised as specific bags were being made available; it was expected that the preferred bag would no longer be *borrowed.* Having given some examples of the sort of findings that a waste audit can provide in terms of understanding what happens to the non-supply disposition of one's waste, the assessment of the suitability of a firm's interaction with each R is addressed.

11.4.4 SWOT for Each R

By considering the Strengths, Weaknesses, Opportunities, and Threats offered by each R; it is possible to better assess whether the current approach is appropriate. To a large extent strengths and weaknesses represent the availability of the appropriate skills (complicating characteristics) in-house or easily accessible within the supply chain.

For better understanding of strengths, weaknesses and opportunities related to developing skills with a new R, we provide Tables 11.2 and 11.3.

Many of the complicating characteristics associated with Product Life Extension Rs are shared amongst multiple Rs. Consequently, skills inherent with one R can be applied to assist in developing another R. In fact, developing skills with complicating characteristics by pursuing new Rs may be attractive for other reasons. While the management of manufacturing and services typically focuses on smooth integration and elimination of uncertainty, the complicating characteristics require the management of different types of uncertainty. Management of uncertainty skills are valuable to an organization under a variety of circumstances. For example, external disruptions to supply chains from trade wars, national disaster or pandemics creates tremendous disruption and uncertainty. A firm that is skillful in mastering the complicating characteristics of product life extension has experience and capabilities with managing and responding to uncertainty that a manufacturer with level loading and tightly integrated supply chains lacks. Consideration of new Rs will identify opportunities for a firm to extend their value proposition while reducing future risks associated with waste generation and resource underutilization. Such considerations should also identify where threats exist—including the potential for third-parties to position themselves into your supply chain in an unwelcome manner filling the gaps left by an overlooked R (e.g., grey market recovery of product mentioned earlier) or the possibility of external changes in regulation that might impact the firm (e.g., sudden refusal of traditional waste receivers to accept post-consumer plastic). When

Table 11.2 Location of capabilities and experience for each R and accompanying SWOT analysis from the perspective of the focal firm

R	Firm	Supply chain	3rd party	Orphan	Strength	Weakness	Opportunity	Threat
Recall								
Repair—post failure								
Repair—preventative maintenance								
Repair—predictive maintenance								
Repair—upgrade								
Reuse								
Restore—refurbish								
Remanufacture								
Recycle								
Recover								
Reduce								
Redirect								

presented with a SWOT summary for each of the 12 Rs, appropriate strategic redirection should be apparent.

11.4.5 Strategic Redirection

The implications of the 12 R Review document are summarized. Any decisions to modify the current situations are recorded. In some cases, these changes are limited by budget. In other cases, the changes will be self-funding. In the case of self-funding opportunities, the limitation on new projects is not budget but available management bandwidth.

Attention is drawn to the use of already existing internal knowledge on managing the complicating characteristics (Table 11.3)—this is important as firm failure to move into new specialties is often linked to trying to bridge too many competency and capability gaps at the same time (Walsh and Linton 2011). It is also important to consider how the skills inherent in the supply chain are to be utilized to strengthen intended future positioning.

Table 11.3 Effect of presence of one mode of PLE (row) on the gap between the firm's current capabilities and required capabilities to adopt a second mode of PLE(column)

	Recall	Repair	Preventative maintenance	Predictive maintenance	Upgrade	Product reuse	Restore	Remanufacture	Recycle	Recover	Reduce	Redirect
Recall	XXX	rec **rout**	**rout**	**rout**	**rout**	**bal**	tim **bal** dis rec **rout**	tim **bal** dis rec **rout**	tim **bal**	tim **bal** dis rec **rout**	*N.A.*	*N.A.*
Repair—post failure	**log**	XXX	None	None	rec **log**	**bal log**	tim **bal** dis **log**	tim **bal** dis **log**	tim bal. log	tim **bal** dis **log**	*N.A.*	*N.A.*
Repair—preventative maintenance	tim **log** rout	tim rec rout	XXX	None	tim **log**	**bal log**	tim bal. dis rec **log** rout	tim bal. dis rec **log** rout	tim bal. log	tim bal. dis rec **log** rout	*N.A.*	*N.A.*
Repair—predictive maintenance	tim **log**	tim rec rout	None	XXX	tim **log**	**bal log**	tim **bal** dis rec **log** rout	tim **bal** dis rec **log** rout	tim bal. log	tim **bal** dis rec **log** rout	*N.A.*	*N.A.*
Repair—upgrade	**log match**	**rec match** rout	**match**	**match**	XXX	**bal log**	tim **bal** dis rec **log match** rout	tim **bal** dis rec **log match** rout	tim bal. **log**	tim **bal** dis rec **log match** rout	*N.A.*	*N.A.*
Reuse	tim **dis log match**	tim **dis** rec **match rout**	**dis match rout**	**dis match rout**	tim **dis log rout**	XXX	tim **dis** rec **log match rout**	tim **dis** rec **log match rout**	tim dis log	tim **dis** rec **log match rout**	*N.A.*	*N.A.*
Restore—refurbish	None	None	None	None	None	None	XXX	None	None	None	*N.A.*	*N.A.*
Remanufacture	None	None	None	None	None	None	None	XXX	None	None	*N.A.*	*N.A.*
Recycle	dis **match**	dis rec **match rout**	dis **match rout**	dis **match rout**	dis **rout**	bal	bal dis rec **match rout**	bal dis rec **match rout**	XXX	bal dis rec **match rout**	*N.A.*	*N.A.*

	Recall	Repair	Preventative maintenance	Predictive maintenance	Upgrade	Product reuse	Restore	Remanufacture	Recycle	Recover	Reduce	Redirect
Recover	match	match	match	match			match	match		XXX	*N.A.*	*N.A.*
Reduce	*N.A.*	*N.A.*	*N.A.*	*N.A.*	*N.A.*	*N.A.*	*N.A.*	*N.A.*	*N.A.*	*N.A.*	XXX	*N.A.*
Redirect	*N.A.*	*N.A.*	*N.A.*	*N.A.*	*N.A.*	*N.A.*	*N.A.*	*N.A.*	*N.A.*	*N.A.*	*N.A.*	XXX

Acronyms in bold indicates the absence of skills associated with a complicating characteristics. Acronyms in normal font indicate the presence of some experience with a complicating characteristic, but recognizes the need for additional skills in dealing with the complicating characteristic. Acronyms are as follows: (1) Uncertain **tim**ing, (2) **Bal**ance demand and supply, (3) Need to **dis**assemble returns, (4) Uncertainty in materials **rec**overed, (5), Need for reverse **logi**stics network (6) Material **match**ing requirements, and (7) Stochastic **rout**ings for material repair and remanufacture. *N.A.* is used to indicate that the R does not have strong interactions with the complicating characteristics

11.4.6 Periodic Reassessment

Reassessment occurs in response to changes in firm or supply chain strategy, changes in customer expectations, or changes in the policy environment.

The ideal time for reassessment is unclear as it is a function of change in customer, supply chain dynamics, environment, and technology. Hence period does not refer to the elapse of a specific amount of time, but an amount of change that a firm has experienced. This varies as a function of the earlier mentioned factors at the discretion of the management team.

In summary a firm should consider: (1) which Rs they have, (2) How the capabilities gained with these Rs can be helpful in introducing new Rs, and (3) the direct and indirect advantages to mastering a wider range of Rs. Having examined the managerial implication of the 12 Rs, policy implications are now considered.

11.5 Implications for Policy

Past practice of policy makers had been to advocate a simple message: 3 Rs with a focus on Recycling. While simple it has proven to be insufficient. Policy makers need to first understand the full gamut of options for reducing environmental impact and increasing the economic and social value that production offers. Policy makers need to encourage the development of appropriate skills so that organizations and society as a whole have the receptor capacity to engage successfully with all Rs. Such abilities to deal with the higher levels of uncertainty associated with the complicating characteristics have other positive impacts. By increasing understanding and receptor capacity, the long existing challenge of policy and regulation creating unintended blockages can be lessened. For example, European Electronic Take Back Legislation aims to divert electronic waste from landfill by requiring an increasing percentage of waste to be made into product. However new design legislation enacted at roughly the same time, bans the use of some categories of the recovered material to be used in new products. In summary one regulation makes a resource available for use, while the other regulation makes the resource unusable. This difficulty becomes more pronounced when multiple levels or parts of government have jurisdiction simultaneously.

11.6 Conclusion and Future Direction

As traditional models of product and manufacturing management have focused on integration and risk reduction, the opportunities for future research relating to Product Life Extension relate to the flexibility and agility for firms and supply chains through the management of uncertainty.

The nature of Product Life Extension and Sustainability has changed over the last few decades. The number of modes or Rs has increased (12 considered here). Consideration has moved from an individual firm (or focal firm) to the supply chain of the firm. In line with supply chain practices there is an increasing need to consider upstream (processes and raw materials) as well as downstream (customer and use). Furthermore, the move away from obsolescence and domestic market to product servitization and different applications, uses and markets results in a focus on the possible multiple lives of a product and even includes the product's eventual disposal. Hence, we are moving from the earlier perspective of selling product and encouraging the fast replacement of the product to extracting value at all possible points in time and minimizing the complications and future liabilities at all points in time.

Many business leaders have come to recognize the importance of climate change, water air pollution, waste management, food safety and other environmental issues. Firms are facing regulatory requirements, natural resource scarcity, shifting customer expectations and increasing demand from business customers who demand green products—all of this will continue to shape the nature of competition. Eco-risk mitigation strategies to deal with pollution and waste is the most immediate environmental challenge for many companies. Stories of eco-risk management gone wrong have become a regular fodder for the media. This involves contamination in products such as milk, chocolate, peanuts, butter, eggs, spinach, and cough syrup. Hence, effective risk management can spot the full spectrum of impending threats to the business and maps out what liabilities might emerge.

Through this chapter, we have revisited the need for companies to understand how to manage their Rs. By exploring opportunities to engage in one or more forms of Product Life Extension, companies can better employ risk management systems that consider various scenarios related to potential threats. This exploration can happen not only within a company's own operations but also at their upstream and downstream touch points leading to the adoption of various Rs and Product Life Extension strategies. The important message that is offered in this chapter is adopting one or more of the Rs should be looked on as a potentially profitable venture—not a costly or burdensome activity. Innovative companies can modify their existing products to reduce wastes and costs while also meeting customer demand for environmentally friendly products. Systemic innovations within a firm can also be used to break free of old ways of thinking and allow them to reposition to enter new markets and sectors. It is encouraging to note that business interest in the environment has grown rapidly and continues to do so over the past couple of decades as companies go through a multistage tactical evolution from eco-resistance to eco-compliance to eco-efficiency and finally to eco-advantage.

References

Aneesh, S., & Kumar, N. A. (2020). Reverse Logistics Network for used refrigerators. *International Journal of Research in Engineering, Science and Management, 2*(6), 280–284.

Borah, A., & Tellis, G. J. (2016). Halo (spillover) effects in social media: Do product recalls of one brand hurt or help rival brands? *Journal of Marketing Research, 53*(2), 143–160.

Chen, C.-K., & Ulya, M. A. (2019). Analyses of the reward-penalty mechanism in green closed-loop supply chains with product remanufacturing. *International Journal of Production Economics, 210*, 211–223.

Fleischmann, M., Beullens, P., Bloemhof-Ruwaard, J. M., & Van Wassenhove, L. (2001). The impact of product recovery on logistics network design, Production and Operations Management, 10(2): 156–173.

Flexner, S. B. (1987). *The random house dictionary of the English language* (2nd ed.). New York: Random House.

Frosch, R. A. (1992). Industrial ecology: A philosophical introduction. *Proceedings of the National Academy of Sciences, 89*, 800–803.

Gertsbakh, I. (2000). *Reliability theory*. Berlin: Springer.

Graedel, T. E., & Allenby, B. R. (1995). *Industrial ecology*. Englewood Cliffs, NJ: Prentice-Hall.

Guide, V. D. R., Jr., Jayaraman, V., Srivastava, R., & Benton, W. C. (2000). Supply chain management for recoverable manufacturing systems. *Interfaces, 30*, 125–143.

Guide, V. D. R., Jr., & Jayaraman, V. (2000). Product acquisition management: a framework and current industry practice. *International Journal of Production Research, 2000*(38), 3779–3801.

He, L., Rong, Y., & Shen, Z.-J. M. (2020). Product sourcing and distribution strategies under supply disruption and recall risks. *Production and Operations Management, 29*(1), 9–23.

Jacobs, B. W., & Subramanian, R. (2011). Sharing responsibility for product recovery across the supply chain. *Production and Operations Management, 21*(1), 85–100.

Jayaraman, V., Guide, V. D. R., Jr., & Srivastava, R. (1999). A closed-loop logistics model for remanufacturing. *Journal of Operational Research Society, 50*, 497–509.

Jayaraman, V., Patterson, R., & Rolland, E. (2003). The design of reverse distribution networks: model and heuristic solution procedure. *European Journal of Operational Research, 150*, 128–149.

Linton, J.D. (2008). Assessing the Economic Rationality of Remanufacturing Product. *Journal of Product Innovation Management. 25*, pages 287–302.

Linton, J. D., & Jayaraman, V. (2005). A framework for identifying differences and similarities in the managerial competencies associated with different modes of product life extension. *International Journal of Production Research, 43*(9), 1807–1829.

Lund, R. (1984). Remanufacturing. *Technology Review, 87*, 18–23.

Mobley, R. K. (1990). *An introduction to predictive maintenance*. New York, NY: Van Nostrand Reinhold.

Moosmayer, D. C., Abdulrahman, M. D., Subramanian, N., & Bergkvist, L. (2020). Strategic and operational remanufacturing mental models: A study on Chinese automotive consumers buying choice. *International Journal of Operations and Production Management, 40*(2), 173–195.

Nystedt, D. (2007). Sanyo, Lenovo share cost of defective battery recall, PC World. Retrieved March 4, from www.pcworld.com/article/129579.

Rupp, N. G. (2004). The attributes of a costly recall: Evidence from the automotive industry. *Review of Industrial Organizations, 25*(5), 21–44.

Shah, R., Ball, G. P., & Netessine, S. (2016). Plant operations and product recalls in the automotive industry: An empirical investigation. *Management Science, 63*(8), 2439–2459.

Stevens, A. B., Dong, Y., & Corsi, R. (2014). Global sourcing and quality recalls: An empirical study of outsourcing-supplier concentration-product recalls linkages. *Journal of Operations Management, 32*(5), 241–253.

Tojo, N. (2001). *Effectiveness of EPR programmes in design change*. Lund: International Institute of Industrial Environmental Economics.

Walls, M. (2006). *Extended producer responsibility and product design. Economic theory and selected case studies*. Washington, DC: Resources for the Future.
Walsh, S. T., & Linton, J. D. (2011). The strategy-technology firm fit audit: a guide to opportunity assessment and selection. *Technology Forecasting and Social Change, 78*, 199–216.
Wise, R., & Baumgartner, P. (2000). Go downstream: The new profit imperative in manufacturing. *IEEE Engineering Management Review, 28*, 89–96.

Part III
Sustainable Natural Resource Management

Chapter 12
Economic Management of Electric Power Systems

Alberto J. Lamadrid L.

Abstract The theory of dynamic optimization and optimal automatic control is rich and continues its development in different areas including engineering, mathematics, and the focus of this chapter, economics. The objective of this chapter is to introduce these concepts for early graduate students, as a toolset in their graduate studies and beyond. Due to space limitations by the editors, in this synthesis we present introductory theory, and the applications we include are focused on electric power systems. For further and more in depth presentation of the theory, we refer the reader to sources along the text. We start with a review of the Lagrangean Method, the intuition behind the formulation, and a sketch of the economic applications as shadow prices. I then review the basic dynamic optimization problem in Sect. 12.1, starting with Deterministic formulations of the discrete and the continuous problem and present an example applied to the replacement of an energy storage system (ESS). Section 12.2 presents the stochastic formulation based on geometric Brownian motion (GBM), and presents an example of an agent optimizing the use of an ESS. In Sect. 12.3 I introduce the formulation of stochastic programs and decision making under uncertainty. The reader familiar with research in electricity systems with uncertainty can easily recognize the underlying generalization presented. I present an example of a unit commitment problem and potential ways to deal with the *curse of dimensionality*. The author is particularly indebted to Yihsu Chen, Jon Conrad and Tim Mount for their inputs and feedback.

12.1 Dynamic Optimization

The first section in this chapter discusses deterministic optimization in discrete time, followed by the theory in continuous time. The applications are illustrated with examples in the electric power sector, e.g., optimal replacement for energy

A. J. Lamadrid L. (✉)
Economics and Industrial and Systems Engineering, Lehigh University, Bethlehem, PA, USA
e-mail: ajlamadrid@ieee.org

C. Chen et al. (eds.), *Pursuing Sustainability*, International Series in Operations Research & Management Science 301,
https://doi.org/10.1007/978-3-030-58023-0_12

storage systems. The second section extends to stochastic optimization, as the intermittency associated with renewable energy sources poses significant challenge to the operations of the power system.

12.1.1 Deterministic Discrete Time

12.1.1.1 The Lagrangean Method

Consider a mathematical optimization problem of the form:

$$\max \quad f_0(x) \tag{12.1a}$$

$$\text{subject to} \quad f_i(x) \le 0, \quad i = 1, \ldots, m. \tag{12.1b}$$

where $x \in \mathbb{R}^n$, $f_0 : \mathbb{R}^n \to \mathbb{R}$ and $f_i : \mathbb{R}^n \to \mathbb{R} \quad \forall i = 1, \ldots, m$.

Maximizing a function subject to constraints implies a number of steps. For example, we can check which constraints are binding, and then from the binding constraints, a substitution of variables can be made to reduce the search space. From this we can e.g., find optimality conditions for the left variables. However, a more general approach is to formulate the Lagrangean function.

The idea behind the formulation of the Lagrangean function, sometimes called Lagrangean relaxation, is to allow for violations of the constraints, augmenting the objective function with a weighted sum of the constraint functions. Therefore we 'relax' the constraints. For generality we focus on the case of inequality constraints. Let $\mathcal{L} : \mathbb{R}^n \times \mathbb{R}^m \to \mathbb{R}$ denote the *Lagrangean* associated to problem (12.1):

$$\mathcal{L}(x, \lambda) \equiv f_0(x) + \lambda^{\mathsf{T}} f(x), \tag{12.2}$$

where $\lambda \in \mathbb{R}^m$ and $\{\cdot\}^{\mathrm{T}}$ denotes the transpose operator and the column vector f(x) is given by:

$$f(x) = \begin{pmatrix} f_1(x) \\ \vdots \\ f_i(x) \\ \vdots \\ f_m(x) \end{pmatrix} \tag{12.3}$$

We denote λ^i as the *Lagrange multiplier* associated to the *i*th inequality constraint $f_i(x) \le 0$. By introducing the Lagrange multiplier λ we reduce the constrained optimization problem with n variables and m constraints to an unconstrained problem with $n+m$ variables. We introduce this concept for the sake of completeness, as we will be using it extensively in the rest of this chapter. The Lagrange multiplier has

Table 12.1 Nomenclature for the discrete problem

Variable	Description
T	Horizon for the problem
$y_t>0$	State variable in period t, $t=0, 1, \ldots, T$
$z_t>0$	Control variable chosen in period t, $t=0, 1, \ldots, T$
$Q(y_t, z_t)=y_{t+1}-y_t$	State equation, expressing the evolution of y_t over time
$N(y_t, z_t)$	Net benefit from y_t and z_t in period t, $t=0, 1, \ldots, T$
$S(y_t)$	Final Function or scrap value function of y_t
$\rho=\dfrac{1}{1+\delta}$	Discount factor, with $\delta>0$ being the discount rate

economic interpretation that we encourage the reader to understand, referring to e.g., Simon and Blume (1994); Sioshansi and Conejo (2017).

Consider a finite horizon, multiple discrete period problem version of (12.1) in which in each time period of time an agent wants to maximize a function by choosing a *control variable* z_t (e.g., representing consumption). The choices of z_t affect the stock or level of another *state variable* y_t, and hence the optimization is subject to those changes over time.

Table 12.1 defines the nomenclature for this problem.

Consider a problem with a fixed horizon T and a free state y_T where we choose values for the control variable z_t for a discrete set of periods $t=0, 1, \ldots, T$, with the objective of:

$$\max \quad N=\sum_{t=0}^{T-1}\rho^t N(y_t,z_t)+\rho^T S(y_T) \tag{12.4a}$$

$$\text{Subject to} \quad y_{t+1}-y_t=Q(y_t,z_t) \tag{12.4b}$$

$$y_0=\bar{y}, \tag{12.4c}$$

where (12.4c) indicates an initial condition for the problem. We can formulate the Lagrangean for this problem as follows.

$$\mathcal{L}\equiv\sum_{t=0}^{T-1}\rho^t\left(N(y_t,z_t)+\rho\lambda_{t+1}\left(Q(y_t,z_t)+y_t-y_{t+1}\right)\right)+\rho^T S(y_T). \tag{12.5}$$

In this Lagrangean, λ_{t+1} is the shadow price of y_{t+1}, giving the marginal value of y_{t+1} from the period $t+1$, and it can be interpreted as the spot price. The economic interpretation of λ_{t+1} is just the willingness to pay in period $t+1$ for an incremental amount of y_{t+1}. The first order conditions (FOCs) for period t for this problem are:

$$\frac{\partial\mathcal{L}}{\partial z_t}=\rho^t\left(\frac{\partial N(\cdot)}{\partial z_t}+\rho\lambda_{t+1}\frac{\partial Q(\cdot)}{\partial z_t}\right)=0, \tag{12.6a}$$

$$\frac{\partial \mathcal{L}}{\partial y_t} = \rho^t \left(\frac{\partial N(\cdot)}{\partial y_t} + \rho\lambda_{t+1} \left(\frac{\partial Q(\cdot)}{\partial y_t} + 1 \right) - \lambda_t \right) = 0, \tag{12.6b}$$

$$\frac{\partial \mathcal{L}}{\partial \rho\lambda_{t+1}} = \rho^t \left(Q(y_t, z_t) + y_t - y_{t+1} \right) = 0. \tag{12.6c}$$

In the FOC for the state variable y_t, (12.6b), the last term $\rho^t\lambda_t$ comes from the derivative with respect to y_t for period $t-1$ of the term:

$$\rho^{t-1} \left(N(y_{t-1}, z_{t-1}) + \rho\lambda_t \left(Q(y_{t-1}, z_{t-1}) + y_{t-1} - y_t \right) \right). \tag{12.7}$$

Assuming that $\rho^t > 0$, and modifying the FOCs,

$$\rho\lambda_{t+1} = -\left(\frac{\frac{\partial N(\cdot)}{\partial z_t}}{\frac{\partial Q(\cdot)}{\partial z_t}} \right), \quad \forall t = 0, 1, \ldots T-1 \tag{12.8a}$$

$$\rho\lambda_{t+1} - \lambda_t = -\left(\frac{\partial N(\cdot)}{\partial y_t} + \frac{\partial Q(\cdot)}{\partial y_t} \right), \quad \forall t = 0, 1, \ldots T-1 \tag{12.8b}$$

$$y_{t+1} - y_t = Q(y_t, z_t). \quad \forall t = 0, 1, \ldots T-1 \tag{12.8c}$$

The terminal first order condition is given by the derivative of $\mathcal{L}$ with respect to y_T:

$$\frac{\partial \mathcal{L}}{\partial y_T} = \rho^T \left(-\lambda_T + \frac{\partial S(y^T)}{\partial y^T} \right) = 0, \tag{12.9}$$

which reduces to $\frac{\partial S(y^T)}{\partial y^T} = \lambda_T$. The initial condition (12.4c), the sequence of first order equations (12.8a)–(12.8c) starting from period $t=0$, and the terminal condition (12.9) forms a system of $3T+2$ equations with $3T+2$ unknowns corresponding to.

- $z_t^* \quad \forall t = 0, 1, \ldots T-1$
- $y_t^* \quad \forall t = 0, 1, \ldots T$
- $\lambda_t^* \quad \forall t = 0, 1, \ldots T$

Even though this problem could be solved as a system of $3T+2$ equations and unknowns, its structure allows for a more efficient *recursive* algorithm. To facilitate notation, let N_z and Q_z denote the derivative of $N(\cdot)$ and $Q(\cdot)$ with respect to the control variable, $\frac{\partial N(\cdot)}{\partial z_t}$, and $\frac{\partial Q(\cdot)}{\partial z_t}$. Also, let N_y and Q_y denote the derivative of $N(\cdot)$ and $Q(\cdot)$ with respect to the system state, $\frac{\partial N(\cdot)}{\partial y_t}$, and $\frac{\partial Q(\cdot)}{\partial y_t}$ respectively. Then:

i. For $t=0$, solve the optimality condition for the control variable, (12.6a) and substitute it into the optimality condition for the state variable, (12.6b):

$$\rho\lambda_1 = -\left(\frac{N_z(\bar{y},z_0^*)}{Q_z(\bar{y},z_0^*)}\right) \rightarrow N_y(\bar{y},z_0^*) - \left(\frac{N_z(\bar{y},z_0^*)}{Q_z(\bar{y},z_0^*)}\right)\left(Q_y(\bar{y},z_0^*)+1\right) - \lambda_0 = 0. \quad (12.10)$$

We can guess or estimate an initial value for λ_0 and solve for z_0^*. Once we have a value for z_0^* we can obtain $y_1^* = \bar{y} + Q(\bar{y},z_0^*)$ from (12.4b). And $\lambda_1^* = -(1+\delta)\left(\frac{N_z(\bar{y},z_0^*)}{Q_z(\bar{y},z_0^*)}\right)$

ii. For $t=1$, $N_y(y_1^*,z_1^*) - \left(\frac{N_z(y_1^*,z_1^*)}{Q_z(y_1^*,z_1^*)}\right)\left(Q_y(y_1^*,z_1^*)+1\right) - \lambda_1^* = 0$. From this equation we can solve for z_1^*. Once the value of z_1^* is determined, we can use this to solve for $y_2^* = y_1^* + Q(y_1^*,z_1^*)$. And $\lambda_2^* = -(1+\delta)\left(\frac{N_z(y_1^*,z_1^*)}{Q_z(y_1^*,z_1^*)}\right)$

iii. We repeat the same procedure iteratively till $t=T-1$. The equilibrium condition then is

$$N_y(y_{T-1}^*,z_{T-1}^*) - \left(\frac{N_z(y_{T-1}^*,z_{T-1}^*)}{Q_z(y_{T-1}^*,z_{T-1}^*)}\right)\left(Q_y(y_{T-1}^*,z_{T-1}^*)+1\right) - \lambda_{T-1}^* = 0.$$

From this equation we can solve for z_{T-1}^*. Once the value of z_1^* is determined, we can use this to solve for the terminal state value $y_T^* = y_{T-1}^* + Q(y_{T-1}^*,z_{T-1}^*)$. The terminal value of λ_T^* is $-(1+\delta)\left(\frac{N_z(y_{T-1}^*,z_{T-1}^*)}{Q_z(y_{T-1}^*,z_{T-1}^*)}\right)$

iv. Compare the value of λ_T^* to $\frac{\partial S(y^T)}{\partial y^T}$, as per (12.9).

(a) If $\lambda_T \approx \frac{\partial S(y^T)}{\partial y^T}$, the procedure gives us the optimal paths for z_t^*, y_t^* and λ_t^* for all $t=0, \ldots, T$

(b) If $\lambda_T \not\approx \frac{\partial S(y^T)}{\partial y^T}$, change the guess for λ_0 in step (i). Then repeat algorithm until $\lambda_T \approx S'(y^T)$, step iv. (a).

Counterintuitively, a finite problem is harder to solve than a formulation with an infinite horizon. Consider the infinite horizon problem

$$\max \quad N = \sum_{t=0}^{\infty} \rho^t N(y_t, z_t) \tag{12.11a}$$

$$\text{Subject to} \quad y_{t+1} - y_t = Q(y_t, z_t) \tag{12.11b}$$

$$y_0 = \overline{y}. \tag{12.11c}$$

In this problem, (12.11b) denotes a condition analogous to (12.4b) in the fixed horizon problem.

The Lagrangean for this problem is as follows.

$$\mathcal{L} \equiv \sum_{t=0}^{\infty} \rho^t \left(N(y_t, z_t) + \rho\lambda_{t+1} \left(Q(y_t, z_t) + y_t - y_{t+1} \right) \right). \tag{12.12}$$

The interpretation of the shadow price λ_{t+1} is as before, (12.5), but note that the discounted value term for the terminal period is not present anymore. The first order conditions (FOCs) for period t for this problem are:

$$\frac{\partial \mathcal{L}}{\partial z_t} = \rho^t \left(N_z(\cdot) + \rho\lambda_{t+1} Q_z(\cdot) \right) \qquad = 0, \tag{12.13a}$$

$$\frac{\partial \mathcal{L}}{\partial y_t} = \rho^t \left(N_y(\cdot) + \rho\lambda_{t+1} \left(Q_y(\cdot) + 1 \right) - \lambda_t \right) = 0, \tag{12.13b}$$

$$\frac{\partial \mathcal{L}}{\partial \rho\lambda_{t+1}} = \rho^t \left(Q(y_t, z_t) + y_t - y_{t+1} \right) \qquad = 0. \tag{12.13c}$$

Assuming that $\rho^t > 0$, the FOCs can be recasted as,

$$\rho\lambda_{t+1} = -\left(\frac{N_z(\cdot)}{Q_z(\cdot)} \right), \qquad \forall t = 0, 1, \ldots \infty \tag{12.14a}$$

$$\rho\lambda_{t+1} - \lambda_t = -\left(N_y(\cdot) + \rho\lambda_{t+1} Q_y(\cdot) \right), \qquad \forall t = 0, 1, \ldots \infty \tag{12.14b}$$

$$y_{t+1} - y_t = Q(y_t, z_t). \qquad \forall t = 0, 1, \ldots \infty \tag{12.14c}$$

The FOCs (12.14a)–(12.14c) are identical to the conditions (12.8a)–(12.8c). For this infinite horizon problem, a concern is whether $\sum_{t=0}^{\infty} \rho^t N(y_t, z_t)$ converges to a finite value. This is called a *Transversality Condition*, requiring

1. A zero limit for the discounted value of the state variable

$$\lim_{t \to \infty} \rho^t \lambda_t y_t = 0. \tag{12.15}$$

2. A finite bound on $N(y_t, z_t)$

Both of these conditions guarantee a convergence of $\sum_{t=0}^{\infty} \rho^t N(y_t, z_t)$. But more importantly, an infinite horizon problem raises the possibility of a steady-state solution with $y_t = y_{t+1} = y^*$, $z_t = z_{\{t+1\}} = z^*$, and $\lambda_t = \lambda_{t+1} = \lambda^*$ $\forall t \geq \tau > 0$. If the optimal value of the control variable z_t^* can drive the system to converge to a steady state, then Eqs. (12.14a)–(12.14c) can be expressed as,

$$\rho\lambda = -\left(\frac{N_z(y,z)}{Q_z(y,z)}\right), \tag{12.16a}$$

$$\rho\lambda - \lambda = -\left(N_y(y,z) + \rho\lambda Q_y(y,z)\right), \tag{12.16b}$$

$$0 = Q(y,z). \tag{12.16c}$$

Since the discount factor is $\rho = \dfrac{1}{1+\delta}$, the factor on the left hand side of (12.16b) can be simplified as,

$$\begin{aligned} \rho\lambda\left(1 - \frac{1}{\rho}\right) &= -\left(N_y(y,z) + \rho\lambda Q_y(y,z)\right), \\ -\rho\lambda\delta &= -\left(N_y(y,z) + \rho\lambda Q_y(y,z)\right). \end{aligned} \tag{12.17}$$

Using the condition (12.16a) to replace $\rho\lambda$ in (12.17), we obtain

$$\delta\left(\frac{N_z(y,z)}{Q_z(y,z)}\right) = -\left(N_y(y,z) - \left(\frac{N_z(y,z)}{Q_z(y,z)}\right) Q_y(y,z)\right), \tag{12.18}$$

and therefore we can obtain a value for δ.

$$\delta = -N_y(y,z)\left(\frac{Q_z(y,z)}{N_z(y,z)}\right) + Q_y(y,z). \tag{12.19}$$

Equations (12.16c) and (12.19) form a two-equation system in steady state for the optimal values of the state y^* and the control z^* variables.

12.1.1.2 The Hamiltonian

Consider the Lagrangeans for the finite horizon problem, (12.5) and for the infinite horizon problem (12.12), (Conrad 2010). We define the *current-value Hamiltonian* as.

Definition 12.1 (Current-Value Hamiltonian, Discrete Formulation)

$$H_t \equiv N(y_t, z_t) + \rho\lambda_{t+1} Q(y_t, z_t). \tag{12.20}$$

The current-value Hamiltonian is embedded in both Lagrangean problems, and therefore, for the infinite horizon case, we can rewrite the Lagrangean as.

$$\mathcal{L} \equiv \sum_{t=0}^{\infty} \rho^t \left(H_t + \rho\lambda_{t+1} \left(y_t - y_{t+1} \right) \right). \tag{12.21}$$

The first order conditions (FOCs) for period t for this problem can be expressed in terms of partial derivatives of the current-value Hamiltonian:

$$\frac{\partial \mathcal{L}}{\partial z_t} = 0 \Rightarrow \frac{\partial H_t}{\partial z_t} = \rho^t \left(N_z(\cdot) + \rho\lambda_{t+1} Q_z(\cdot) \right) = 0, \tag{12.22a}$$

$$\frac{\partial \mathcal{L}}{\partial y_t} = 0 \Rightarrow -\frac{\partial H_t}{\partial y_t} = -\left(N_y(\cdot) + \rho\lambda_{t+1} Q_y(\cdot) \right) = \rho\lambda_{t+1} - \lambda_t, \tag{12.22b}$$

$$\frac{\partial \mathcal{L}}{\partial \rho\lambda_{t+1}} = 0 \Rightarrow \frac{\partial H_t}{\partial \rho\lambda_{t+1}} = Q(y_t, z_t) = y_{t+1} - y_t. \tag{12.22c}$$

Comparing Eqs. (12.22a)–(12.22c) to Eqs. (12.13a)–(12.13c) we see the equivalency. The economic interpretation of the current value Hamiltonian is *the income per period received*, considering the current period net benefit, $N(y_t, z_t)$ and the discounted change in the state variable $\rho\lambda_{t+1}Q(y_t, z_t)$. $N(y_t, z_t)$ can also be considered a utility flow, or a dividend. On the other hand, as $Q(y_t, z_t)$ represents the change in the state variable, it can be interpreted as a discounted capital gain from intertemporal moves from y_t to y_{t+1} given the optimal control management exerted by z_t^*. Therefore, the optimal current value Hamiltonian is given by.

$$H_t^* = N(y_t^*, z_t^*) + \rho\lambda_{t+1} Q(y_t^*, z_t^*). \tag{12.23}$$

In Weitzman (2003, p 102), Weitzman interprets the current value Hamiltonian as a firm's *properly accounted net income at time t*. Assuming a perfect capital market where money can be exchanged at the interest rate ρ, the stock market share is consistent with current earnings, which consist of dividends plus net investment (capital gains or losses). Economically, this represents the firm being at the economic possibility frontier, with the marginal rate of transformation between current gains and the investment in future gains being equal to the current price of investment relative to current gains.

12.1.1.3 Bellman Equations

Consider the infinite horizon problem from Eqs. (12.11a)–(12.11c).

Definition 12.2 (Value Function)

$$V(y_t) = \arg\max_{z_\tau} \sum_{\tau=t}^{\infty} \rho^{\tau-t} N(y_\tau, z_\tau). \tag{12.24}$$

Suppose we are trying to solve this problem starting in period t, with an initial condition y_t. Assuming the optimizer is adopting in every period an optimal policy z_τ^* for $\tau=t$, $t+1$, ..., ∞, the value of the problem will depend on the initial condition y_t

Definition 12.3 (Bellman Equation) The Bellman equation expresses an optimality condition based on an unknown value function. It requires that

$$V(y_t) = \arg\max_{z_\tau} \left(N(y_\tau, z_\tau) + \rho V(y_{t+1}) \right) \tag{12.25a}$$

$$V(y_t) = \arg\max_{z_\tau} \left(N(y_\tau, z_\tau) + \rho V\left(y_t + Q(y_t, z_t)\right) \right) \tag{12.25b}$$

Equation (12.25b) implies finding the optimal value for the control variable in period t that balances the net benefit from y_t and z_t, $N(y_t, z_t)$, considering any potential marginal losses in the discounted future values.

The first order condition for the control variable, similar to (12.13a) and (12.22a), implies

$$\frac{\partial V(\cdot)}{\partial z_t} = N_z(\cdot) + \rho \frac{\partial V}{\partial Q} Q_z(\cdot) = 0, \tag{12.26}$$

and therefore $\frac{\partial V(y_{t+1})}{\partial Q} = \lambda_{t+1}$. For this reason in addition to the interpretation of λ_{t+1} as the shadow price of y_{t+1}, we can interpret this as the *marginal value* of a change $\Delta y_t > 0$.

Thus, the value function $V(y_t)$ measures the wealth derived from owning and optimally managing y_t. In case somebody does not own the resource y_t but wanted to acquire it, the value would be given by $V(y_t)$. Suppose you own the resource y_t and decide to sell it and put this value $V(y_t)$ to earn interest δ per period. In this case.

$$\delta V(y_t) = H_t^* = N(y_t^*, z_t^*) + \rho \lambda_{t+1} Q(y_t^*, z_t^*), \tag{12.27}$$

where $\delta V(y_t)$ denotes the rent income from selling the resource y_t; and $H_t^* = N(y_t^*, z_t^*) + \rho \lambda_{t+1} Q(y_t^*, z_t^*)$, the income from the continued ownership of the resource y_t. Therefore the current value Hamiltonian is directly related to the Bellman equation.

12.1.1.4 The Pontryagin's Maximum Principle

The maximum principle applies to deterministic problems, and provides a mechanism to manage the curse of dimensionality associated to dynamic programming problems. Consider the following problem:

$$\max N = \sum_{t=0}^{T} F(y_t, z_t, t) \tag{12.28a}$$

$$\text{Subject to } y_{t+1} - y_t = Q(y_t, z_t, t) \tag{12.28b}$$

$$G(y_t, z_t, t) \leq 0. \tag{12.28c}$$

The necessary FOCs for the problem (12.28a)–(12.28c) can be expressed in terms of the Hamiltonian, and are given by

i. For each t we have that z_t maximizes the Hamiltonian $H_t(y_t, z_t, \rho\lambda_{t+1}, t)$ subject to (12.28c)
ii. The following difference equations control the changes over time

$$\rho\lambda_{t+1} - \lambda_t = -\frac{\partial H_t^*}{\partial y_t}, \tag{12.29a}$$

$$y_{t+1} - y_t = \frac{\partial H_t^*}{\partial \rho\lambda_{t+1}}. \tag{12.29b}$$

Note the relation to our previously derived (12.22b)–(12.22c) and the relation to the Lagrangean formulation of the problem. The maximum principle is an ordinary differential equation, and can be generalized to model predictive control settings (i.e., a receding horizon optimization).

12.1.2 Deterministic Continuous Time

A continuous time model is the limit of a discrete time model with uniform increments Δt. Let $\Delta t \to 0$ and denote the increment as dt.

From the state equation we have.

$$Q(y(t), z(t)) = \frac{y(t + \Delta t) - y(t)}{\Delta t}, \tag{12.30}$$

and we can calculate.

$$\lim_{\Delta t \to 0} \frac{y(t + \Delta t) - y(t)}{\Delta t} \equiv \dot{y}. \tag{12.31}$$

The objective function we had $N = \sum_{t=0}^{T-1} \rho^t N(y_t, z_t) + \rho^T S(y_T)$ in continuous time becomes

$$N = \int_0^T e^{-\delta t} N(y(t), z(t))dt + e^{-\delta T} S(y(T)), \tag{12.32}$$

where $e^{-\delta t}$ is the continuous time discount factor.

Table 12.2 summarizes the nomenclature for this problem.

The problem with a fixed horizon T and a free state $y(T)$ where we choose values for the control variable $z(t)$ in continuous time becomes:

$$\max \ N = \int_0^T e^{-\delta t} N(y(t), z(t))dt + e^{-\delta T} S(y(T)) \tag{12.33a}$$

$$\text{Subject to } \dot{y} = Q(y(t), z(t)) \tag{12.33b}$$

$$y(0) = \bar{y}. \tag{12.33c}$$

We can formulate the Lagrangean for this problem as follows.

$$\mathcal{L} \equiv \int_0^T \left(e^{-\delta t} N(y(t), z(t)) + \Lambda(t)\left(Q(y(t), z(t) - \dot{y})\right)\right)dt + e^{-\delta T} S(y(T)). \tag{12.34}$$

In this Lagrangean, $\Lambda(t)$ is a discounted shadow price. We can solve the term $-\Lambda(t)\dot{y}$ using integration by parts:

$$\int_0^T -\Lambda(t)\dot{y}dt = -\left(\Lambda(T)y(T) - \Lambda(0)y(0)\right) + \int_{t=0}^T \dot{\Lambda}(t)ydt. \tag{12.35}$$

Then, (12.34) can be used in (12.35) to obtain:

$$\begin{aligned} \mathcal{L} \equiv & \int_0^T \left(e^{-\delta t} N(y(t), z(t)) + \Lambda(t)\left(Q(y(t), z(t))\right) + \dot{\Lambda}(t)y\right)dt \\ & -\left(\Lambda(T)y(T) - \Lambda(0)y(0)\right) + e^{-\delta T} S(y(T)). \end{aligned} \tag{12.36}$$

For notation ease we will omit the time arguments on the state $y(t)$ and control $z(t)$ variables.

Table 12.2 Nomenclature for the continuous problem

Variable	Description
$y(t)$	State variable in period t
$z(t)$	Control variable chosen in period t
$Q(y(t), z(t))\,\Delta t = y(t+\Delta t) - y(t)$	State equation, expressing the evolution of $y(t)$ over time
$N(y(t), z(t))$	Net benefit from $y(t)$ and $z(t)$ in period t
$S(y(t))$	Final Function or scrap value function of $y(t)$
δ	Instantaneous rate of discount

Definition 12.4 (Current-Value Hamiltonian, Continuous)

$$H(y,z,\mu) \equiv N(y,z) + \mu(t)Q(y,z), \tag{12.37}$$

where $\mu(t) = e^{\delta t}\Lambda(t)$.

The first order conditions (FOCs) for this problem can be expressed in terms of partial derivatives of the current-value Hamiltonian:

$$\frac{\partial \mathcal{L}}{\partial z} = 0 \Rightarrow \frac{\partial H}{\partial z} = N_z(\cdot) + \mu Q_z(\cdot) = 0, \tag{12.38a}$$

$$\frac{\partial \mathcal{L}}{\partial y} = 0 \Rightarrow -\frac{\partial H}{\partial y} = -\left(N_y(\cdot) + \mu Q_y(\cdot)\right) = \dot{\mu} - \delta\mu, \tag{12.38b}$$

$$\frac{\partial \mathcal{L}}{\partial \mu} = 0 \Rightarrow \frac{\partial H}{\partial \mu} = Q(y,z) = \dot{y}, \tag{12.38c}$$

$$\frac{\partial \mathcal{L}}{\partial y(T)} = 0 \Rightarrow -\Lambda(T) + e^{-\delta t}\frac{\partial S(y(T))}{\partial y(T)} = 0. \tag{12.38d}$$

Note that to obtain these first order conditions in terms of μ, we replaced $\Lambda(t) = \mu(t)e^{-\delta t}$ and $\dot{\Lambda}(t) = e^{-\delta t}(\dot{\mu}(t) - \delta\mu(t))$ Additionally, the boundary conditions are:

$$y(0) = \bar{y}, \qquad \frac{\partial S(y(T))}{\partial y(T)} = \mu(T). \tag{12.39}$$

The only change necessary for the infinite-horizon problem is to replace the terminal condition (12.38d) with a *transversality condition*:

$$\lim_{t\to\infty} e^{-\delta t}\mu(t)y(t) = 0 \tag{12.40}$$

12.1.3 Example: Optimal Replacement

Consider the decision to purchase and replace a capital good with deterministic depreciation. This is an abstraction of a user deciding to buy an ESS for use in a house, or a plug-in electric vehicle with a battery that can be used for both transportation and electricity usage, with decisions over a long time horizon. The constant cycling of chemical batteries reduces the usable energy capacity (see e.g. Shiau et al. (2009); Hu and Defourny (2017)), and hence the degradation assumed. This durable good has a quality level x_t and degrades over time at a rate γ per period. For the sake of simplicity, it is assumed that the number of states is finite. The quality of a new good is denoted by x_0. The utility of using this asset, $U(\cdot)$ depends on the quality level (e.g., less available capacity over time means more frequent charges, and therefore lower utility from the ESS). This good has a replacement cost, $C(\cdot)$ and an associated mapping from monetary cost to utility levels $\phi(\cdot)$. Other goods are ignored here. Let ρ denote the discount rate, and $B(x_t)=\{0, 1\}$ the replacement decision at time t. For this model, a threshold level R will indicate the minimum acceptable quality level (i.e., if $x_t<R$, the good needs to be replaced). The agent's problem in this case will be given by:

$$\begin{aligned} \max_{B(x_t)\in\{0,1\}} u &= \sum_{t=0}^{\infty} \rho^t \left[U\left(x_t - \phi\left(C(x_t)\right)\cdot B(x_t)\right)\right] \\ s.t.\ x_{t+1} &= \left(1-B(x_t)\right)\cdot(x_t-\gamma)+B(x_t)\cdot(x_0-\gamma) \\ & x_0 \text{ given.} \end{aligned} \tag{12.41}$$

The Bellman equation will be given by:

$$\begin{aligned} V(x_t) &= \max_{B(x_t)\in\{0,1\}} \left[U\left(x_t - \phi\left(C(x_t)\right)\cdot B(x_t)\right)+\rho\cdot V(x_{t+1})\right] \\ s.t.\ x_{t+1} &= \left(1-B(x_t)\right)\cdot(x_t-\gamma)+B(x_t)\cdot(x_0-\gamma) \\ & x_0 \text{ given.} \end{aligned} \tag{12.42}$$

The problem can then be rewritten as:

$$V(x_t) = \max_{B(x_t)\in\{0,1\}} \left[U\left(x_t - \phi(C(x_t))\cdot B(x_t)\right)+\rho\cdot V\left(\left(1-B(x_t)\right)\cdot(x_t-\gamma)+B(x_t)\cdot(x_0-\gamma)\right)\right]. \tag{12.43}$$

12.1.3.1 Optimality Conditions

This problem can be numerically solved by fixed point iteration over the possible values of the value function. The level of quality (x_t) is discretized, with 60 periods (e.g., months for 5 years). The cost of replacement is set as fixed (c \$/unit)

$$C(x_t) = cx_t, \tag{12.44}$$

and a per-period linear utility function ($a\times x_t$, a set to 2 as the marginal utility of quality):

$$U(x_t) = ax_t = 2x_t. \tag{12.45}$$

The discount rate (ρ) is set to 0.9, and the marginal utility of money is linear and numeraire ($m=1$). The degradation rate of the ESS (γ) is considered constant for all periods (γ_0).

For this case, Eq. (12.43) becomes

$$V(X_t) = \max_{B_t \in \{0,1\}} [ax_t - cmB_t + \rho V((1-B_t)(x_t - \gamma_0) + B_t(x_0 - \gamma_0))], \quad (12.46)$$

or replacing the numerical values

$$V(X_t) = \max_{B_t \in \{0,1\}} [2x_t - 50B_t + 0.9V((1-B_t)(x_t - 1) + B_t(x_0 - 1))]. \quad (12.47)$$

12.1.3.2 Results and Sensitivity

Given the likely changes to be observed in the ESS technologies Peterson et al. (2010), a sensitivity analysis of the value of the solution to Eq. (12.47) is performed. The two parameters analyzed are the cost of replacement, c, and the degradation rate, γ. Figure 12.1 shows the evolution of replacement time as a function of the (a) cost of replacement and (b) degradation over time, while holding the other parameters constant.

To see how the two studied parameters interact, Fig. 12.2 shows contour plots for changing values of the technology considered. The monetary unit values over which replacement cost vary are from 5 to 50. The variation on degradation are from 1 to 10.

The results show that the model accurately predicts the sensitivity of the mean ESS lifetime to advancements in technology: lower degradation rates and higher replacement costs of the storage sources lead to longer optimal replacement times. This stylized model then formalizes the investment decision for an energy storage, given that time spans are averaging over relatively long periods of time, e.g., months. Hour-to-hour decisions regarding usage will probably be more dependent on the possibility of arbitraging over inter-temporal price differences and the decision to offer ancillary services to the network than on the overall degradation of the ESS.

12.2 Stochastic Optimization

Consider a continuous time stochastic program. In this case the state equation (12.33b) is replaced by a stochastic process:

$$dy = Q(y,z)dt + \sigma(y)dz, \quad (12.48)$$

where $\sigma(y)$ denotes the standard deviation rate for the state variable y and dz is the increment of a *Weiner process*. The following derivations follow Pindyck (1991).

Definition 12.5 (Weiner Process (or Brownian Motion)) A Weiner process is a continuous-time Markov stochastic process. Due to the Markovian property, each

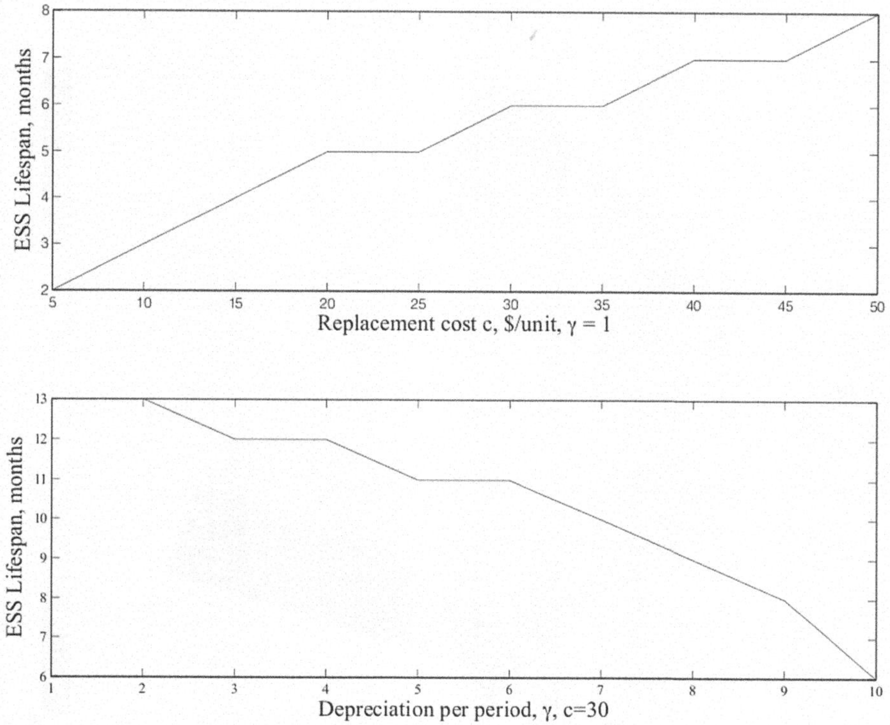

Fig. 12.1 Sensitivity to fixed technology parameters

increment is independent, regardless of the time interval length. All changes dz corresponding to a time interval dt satisfy:

1. dz is given by

$$dz = \varepsilon(t)\sqrt{dt}, \tag{12.49}$$

   where $\varepsilon(t)$ is normally distributed, $\varepsilon(t) \sim N(0, 1)$.
2. $\varepsilon(t)$ is serially uncorrelated

$$\mathbb{E}(\varepsilon(t)\varepsilon(\tau)) = 0, \quad \forall t \neq \tau \tag{12.50}$$

   Therefore, the values of dz for two different time intervals are independent.

An Itô process is generalization of the Wiener process:

Definition 12.6 The continuous time stochastic process $y(t)$ is an Itô process given by

$$dy = a(y,t)dt + b(y,t)dz, \tag{12.51}$$

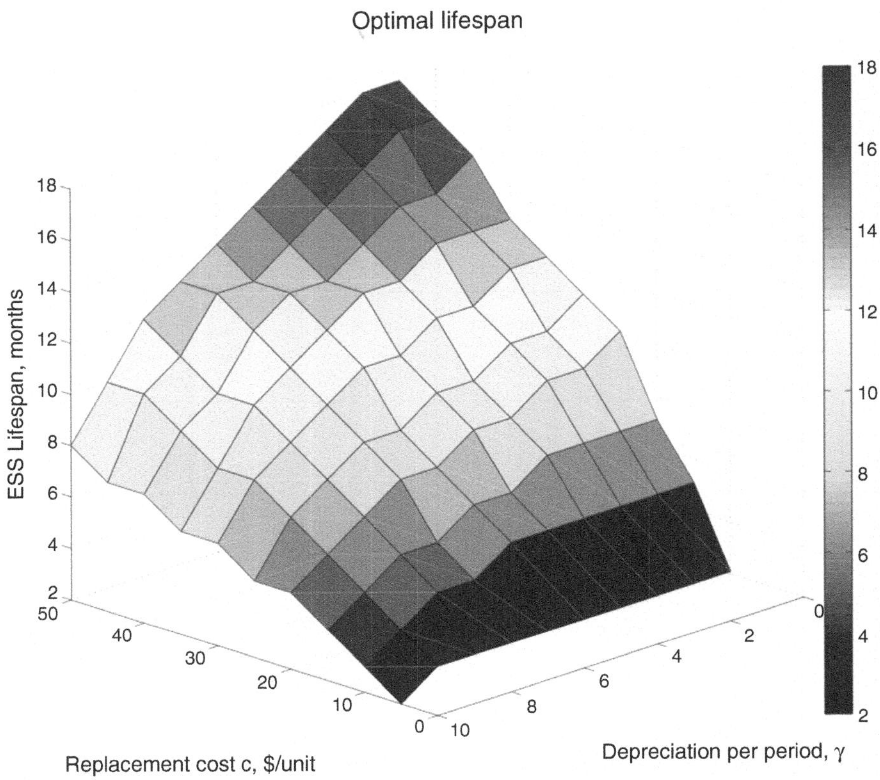

Fig. 12.2 Sensitivity to technology parameters

with dz as defined by (12.49). Due to the distribution of dz, $\mathbb{E}(dz) = 0$, and therefore the mean of the increments of this process is given by:

$$\mathbb{E}(dy) \quad = a(y,t)dt. \tag{12.52}$$

The variance of dy is given by

$$\begin{aligned}\mathbb{E}\left((dy - \mathbb{E}(dy))^2\right) &= \mathbb{E}\left((a(y,t)dt + b(y,t)dz - a(y,t)dt)^2\right) \\ &= \mathbb{E}\left((b(y,t)\varepsilon(t)\sqrt{dt})^2\right) = b^2(y,t)dt.\end{aligned} \tag{12.53}$$

Therefore, we refer to $a(y, t)dt$, the expected value of the Itô process, (12.52), as the expected drift rate. We refer to $b^2(y, t)dt$, the variance of the Itô process, (12.53), as the variance rate.

A particular case of the Itô process in (12.63) is a *Geometric Brownian Motion with drift*, where the expected drift rate, $a(y, t)dt$, and $b(y, t)$, are linear. Hence, (12.63) becomes

$$dy = \alpha z dt + \sigma y dz, \tag{12.54}$$

Consider the total differential of a twice-differentiable Itô process $F(y, t)$ following (12.63):

$$dF = F_y dy + F_t dt, \tag{12.55}$$

where F_y and F_t denote the derivatives of $F(\cdot)$ with respect to y, $\frac{\partial F(\cdot)}{\partial y}$, and t, $\frac{\partial F(\cdot)}{\partial t}$, respectively.

This differential can include higher order terms considering changes in y as a Taylor series:

$$dF = F_y dy + F_t dt + \frac{1}{2} F_{yy} (dy)^2 + \frac{1}{6} F_{yyy} (dy)^3. \tag{12.56}$$

As Pindyck (1991) shows, terms of y of order higher than two have a limit value of zero in the differential. This is due to the fact that additional terms include dt raised to a pawer greater than one. This allows the following result.

Definition 12.7 (Itô's Lemma) Let

$$dF = F_y dy + F_t dt + \frac{1}{2} F_{yy} (dy)^2. \tag{12.57}$$

We use this limit value of dF, (12.57), and the variance rate, (12.53) in the Itô process description, (12.63) to formulate:

$$\begin{aligned} dF &= F_y (a(y,t)dt + b(y,t)dz) + F_t dt + \frac{1}{2} F_{yy} (dy)^2, \\ &= F_y (a(y,t)dt + b(y,t)dz) + F_t dt + \frac{1}{2} F_{yy} b^2 (y,t) dt \\ &= (F_y a(y,t) + F_t + \frac{1}{2} F_{yy} b^2 (y,t)) dt + F_y b(y,t) dz. \end{aligned} \tag{12.58}$$

12.2.1 Stochastic Dynamic Programming

Consider the infinite horizon, stochastic equivalent of problem (12.33a), with a control variable $z(t)$:

$$\max \quad N = \mathbb{E}_0 \left(\int_0^\infty e^{-\delta t} N(y(t), z(t)) dt \right), \tag{12.59}$$

where y follows an Itô process (12.63):

$$dy = a(y,t)dt + b(y,t)dz. \tag{12.60}$$

Let $V(y)$ denote the net value assuming $z(t)$ is managed optimally:

$$V(y) \quad = \max_z \mathbb{E}_t\left(\int_t^\infty e^{-\delta\tau} N(y(\tau), z(\tau))d\tau\right). \tag{12.61}$$

The Bellman equation for this problem is given by

$$\delta V(y) \quad = \max_z \left(N(y(\tau), z(\tau)) + \frac{1}{dt}\mathbb{E}_t(dV) \right). \tag{12.62}$$

The Bellman equation (12.62) shows that the return on this asset has two components:

1. The income flow, $N(y(\tau), z(\tau))$
2. The expected rate of gain, $\frac{1}{dt}\mathbb{E}_t(dV)$

12.2.2 Example, Optimal ESS Management, Representative Agent

To study a stylized version of the optimal usage of an ESS with no network, a single agent problem with an ESS endowment is posited. This agent derives utility from consumption of the energy stored in the ESS. This can be thought of as an abstraction of an agent with an electric car that derives utility from the car usage, as well as from consumption of electricity for powering devices (e.g., appliances in the house).

The consumption good (electricity for transportation or for domestic usage) is homogeneous and denoted by c_t. This can be considered an instantaneous power drain from the battery (e.g., measured in kW). In this case, the period of consideration is averaged over a period of time (e.g., weeks), so that it can be approximated by a continuous-time process.

The energy capacity of the ESS can be modeled by a continuous time stochastic differential equation expressed as:

$$dB = a(B(t), c(t), t)dt + b(B(t), c(t), t)dz. \tag{12.63}$$

Equation (12.63) represents an Itô process in which $a(\cdot)$ is the drift rate of the process, $b^2(\cdot)$ is the variance rate and dz is the increment of a Wiener process. $B(t)$ is the energy capacity of the ESS (e.g., measured in kWh) and $c(t)$ is the consumption of electricity in period t. For notational simplification, these variables will be denoted by B and c. The agent's problem in this case will be given by:

$$\begin{aligned} \max_{c} W(c) &= E_0\left[\int_{t=0}^{\infty} U(c)e^{-\rho t}dt\right] \\ &s.t. \\ &dB = a(B,c,t)dt + b(B,c,t)dz \\ &\quad B(0) > 0 \text{ given.} \end{aligned} \tag{12.64}$$

where ρ denotes the discount rate. We assume the utility function is time invariant. The Hamilton-Jacobi-Bellman (HJB) equation and Itô's Lemma application for this problem is given in (12.65)

$$\begin{aligned} \rho V(B) &= \max_{c}\left[U(c) + \frac{1}{dt}E_t dV(B)\right], \Rightarrow \\ \rho V(B) &= \max_{c}\left[U(c) + a(B,c,t)\frac{dV}{dB} + \frac{1}{2}b^2(B,c,t)\frac{d^2V}{dB^2}\right]. \end{aligned} \tag{12.65}$$

Assuming that the technologies have discrete-size jumps with arrival times following a Poisson distribution, let q denote a Poisson process, with changes as follows.

$$dq = \begin{cases} 0 & \text{with probability } 1-\lambda dt \\ u & \text{with probability } \lambda\, dt \end{cases}$$

Then, combining both an Itô process with a jump process, the change equation can be expressed as:

$$dB = a(B(t),c(t),t)dt + b(B(t),c(t),t)dz + d(B(t),c(t),t)dq. \tag{12.66}$$

In Eq. (12.66), the term $d(\cdot)$ is a known function describing the change that may affect the state variable. In the case of a combined Itô and jump process, the HJB equation is given by:

$$\begin{aligned} \rho V(B) = \max_{c_t}\Big[&U(c) + a(B,c,t)\frac{dV}{dB} + \frac{1}{2}b^2(B,c,t)\frac{d^2V}{dB^2} \\ &+\varepsilon_u[\lambda\{V(B+d(B,c,t)u) - V(B)\}]\Big] \end{aligned} \tag{12.67}$$

12.2.2.1 Optimality Conditions

The First Order Conditions (FOC's) with respect to c_t for the Itô process problem are shown in Eq. (12.68).

$$\frac{dU(c)}{dc}+\frac{da(B,c,t)}{dc}\frac{dV}{dB}+b(B,c,t)\frac{db(B,c,t)}{dc}\frac{d^2V}{dB^2}=0. \tag{12.68}$$

In this case, the marginal utility will depend on the changes of the drift rate, and the associated variance of the process affecting the change in value of the storage resource through time. For the combined Itô and jump processes, the FOC's will depend on the relation describing the discrete movement. In situations as the one illustrated here, where the discrete movement does not depend on the control variable but is exogenous (e.g., innovation in batteries does not depend on the charging and discharging regimes adopted), the FOC's will have the form shown in (12.68).

12.2.2.2 Optimal Feedback Policy

An optimal feedback policy ($c=\phi(B)$) can be obtained according to the functions that describe $U(\cdot)$, $a(\cdot)$, $b(\cdot)$. The optimized HJB equation is shown in (12.69).

$$\rho V(B) = U(\phi(B))+a(B,\phi(B),t)\frac{dV}{dB}+\frac{1}{2}b^2(B_t,\phi(B),t)\frac{d^2V}{dB^2}. \tag{12.69}$$

In the combined Itô and jump processes modeling, and assuming that the jump process is independent of the control variable, the optimal feedback policy $c=\phi(B)$ is identical to that obtained in case of an Itô process with no jumps. After replacing the optimal feedback policy into Eq. (12.67), the optimized HJB obtained is shown in (12.70).

$$\begin{aligned}\rho V(B) &= U\left(\phi(B)\right)+a\left(B,\phi(B),t\right)\frac{dV}{dB}+\frac{1}{2}b^2\left(B_t,\phi(B),t\right)\frac{d^2V}{dB^2}\\ &+\varepsilon_u\left[\lambda\left\{V(B+d(B,c,t)u)-V(B)\right\}\right].\end{aligned} \tag{12.70}$$

12.2.2.3 Analytical Solution

Because this model is highly stylized, it allows for a closed form solution, depending on the selection of the functions describing the evolution of the process. Consider logarithmic decay of the capacity of an ESS $\left(a(B(t),c(t),t)=\left(rB\ln\left[\frac{k}{B}\right]-c_t\right), b(B(t),c(t),t)=\sigma B\right)$, and a logarithmic utility function, as described in (12.71a) and (12.71b) respectively.

$$dB = \left(rB \ln\left[\frac{k}{B}\right] - c \right) dt + \sigma B dz, \tag{12.71a}$$

$$U(c_t) = \alpha \ln c - \beta \frac{c}{B}. \tag{12.71b}$$

The utility function chosen allows for a concave and additive representation, while reflecting the decay observed in the capacity of the battery. In such a case, the optimal feedback policy obtained is $c = \phi(B) = \dfrac{\alpha B}{\dfrac{dV}{dB} B + \beta}$. Assuming a logarithmic value function $V(B) = M \ln B + N$, the optimal consumption obtained is given by Eq. (12.72)

$$c = \phi(B) = \frac{\alpha(\rho + r)B}{\alpha + \beta(\rho + r)}. \tag{12.72}$$

For a combined Itô and jump process with $d(B(t), c(t), t) = (1 + \gamma_i)B$ and parameters as in the Itô-only process, we obtain the following solution:

$$\begin{aligned} M &= \frac{\alpha}{\rho + r}, \\ N &= \alpha \left[\ln \alpha - \ln\left(\beta + \frac{\alpha}{\rho + r} \right) - 1 \right], \\ &+ \frac{\alpha}{\rho + r}\left(r \ln k - \frac{\sigma^2}{2} \right) \varepsilon_u \left[\lambda \frac{\alpha}{\rho + r} \ln\{1 + (1 + \gamma_i) u\} \right]. \end{aligned} \tag{12.73}$$

The main implication of equations (12.73), under the assumptions on the model is that usage of the energy in the storage resource for consumption follows a path similar to the degradation observed across time. This result follows the logic of pacing usage according to the technology available. Note that the optimal feedback policy in this case is described by Eq. (12.72), since the jump process is not affected by the control variable.

We use parameters consistent with those used in Sect. 12.1.3.1. Given the difference in the discounting, we choose $\rho = 0.05$, set the initial value $B(0) = 0.48$, $\alpha = 0.5$, $\beta = 1$ and perform 100 steps of a fixed point iteration. We use a normal random error as part of the process.

Figure 12.3 shows the time paths for the energy stored in the ESS and the consumption in each period. Figure 12.4 has a histogram for the Energy Stored with bin widths of 0.1.

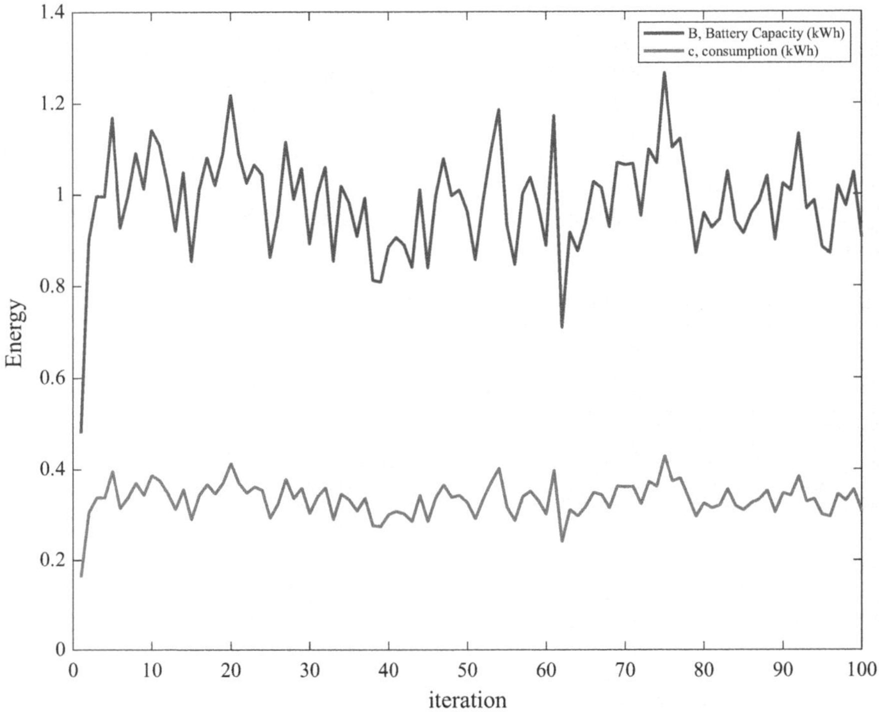

Fig. 12.3 Time trajectory, energy in the storage and consumption

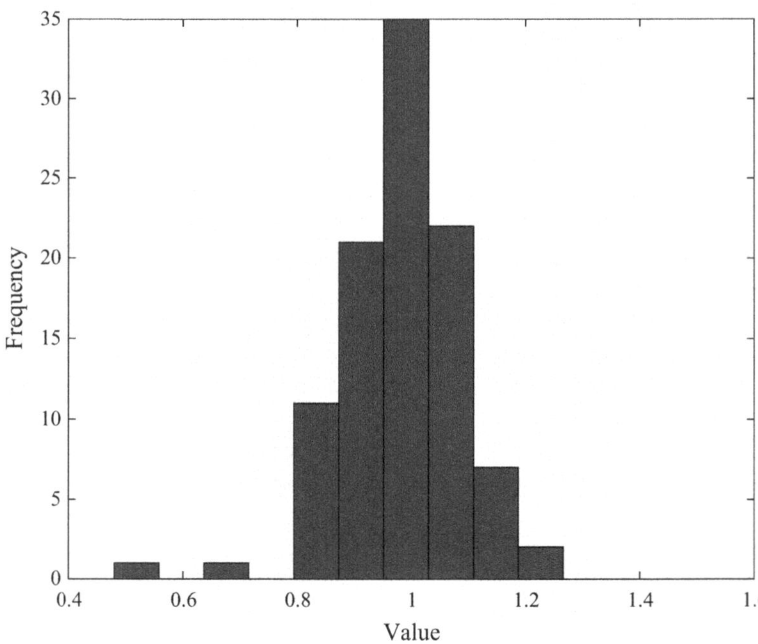

Fig. 12.4 Histogram of energy in the storage/battery

12.3 Stochastic Programming for Electric Power Systems

Stochastic programming is a method to model decisions that involve uncertainty. Generally, the parameters of a problem (e.g., the power available at any time, the energy demand realized) can be uncertain, and therefore are random variables. The objective of a stochastic program is to find a policy that is feasible for almost all the possible parameter realizations. It optimizes the expectation of the objective function of the decisions considering the random realizations of parameters. There are many sources with explanations of stochastic programs. In Birge and Louveaux (1997); Shapiro and Philpott (2007); Conejo et al. (2010) there are excellent explanations of these concepts. Here we summarize the main ideas for use in application to electric power systems. To solve a stochastic program, it is generally assumed that the probability distribution of the parameters included in the model can be somehow estimated or approximated. For example, even though we may not know for sure what the wind speed or solar irradiance will be at a given location and time in the future, we can estimate this based on historical data that we have collected. We can represent the set of all outcomes possible by Ω. There are possible subsets $\Omega_i \subset \Omega$ with collections of random events. For each event $\omega \in \Omega_i$, there is an associated probability ψ^{ω}.

12.3.1 *The Decision-Making Process*

Stochastic programs have data that is considered uncertain. These kind of programs can be applied to cases in which decisions are made only once, with a given representation of the uncertainty. The more common use of stochastic programs are as past of recourse programs. In recourse programs, the decision maker can take further actions, after uncertainty has been revealed. These recourse actions can adjust or compensate for the differences with the expected conditions, and mitigate possible negative effects. An implicit assumption is that we have a probability measure for the random variables. Let $\xi = \xi(\omega)$ denote the vector of realizations after the uncertainty is revealed. The most widely used form of recourse programs is the two-stage program. In such programs, the optimal policy includes a first stage decision and a set of second stage actions, in response to each possible random outcome. The underlying rationale for two-stage programs is that decisions should be made according to the data available at the time those specific decisions are made.

The decision sets in two-stage programs are then:

1. Decisions before the uncertainty is revealed, or *first stage decisions*. Let x denote the vector of first stage decisions
2. Decisions after the uncertainty is revealed, or *second stage decisions*. Let y, $y(\omega)$ or $y(\omega, x)$ represent the vector of second stage decisions

The definitions for first and second stages are therefore related to the revelation of the uncertainty. Each can have sequences of decisions and events.

The sequence of events and decisions is given by Eq. (12.74)

$$x \rightarrow \xi(\omega) \rightarrow y(\xi(\omega), x). \tag{12.74}$$

The general formulation of a two stage program is shown in (12.75).

$$\begin{aligned} \min_{x,y} \; & F(x) + \mathbb{E}_{\xi}[G(x, y(\xi(\omega)))] \\ \text{subject to} \; & h(\omega, x, y) \leq b(\omega). \end{aligned} \tag{12.75}$$

The linear formulation of this problem, due to Dantzig (1955), includes possible realizations of ω, affecting the recourse actions in response to the randomness affecting the decisions.

Due to the revelation of uncertainty and the information available at the time of making decisions, first stage decisions are sometimes called *here-and-now*. In such case, the decision-maker does not know the realization ξ of the uncertain data. Once this uncertainty is revealed and $\xi(\omega)$ is known, the decision-maker can take recourse actions, dependent on the outcome observed. For this reason, second stage decisions are sometimes called *wait-and-see* decisions.

Two common assumptions are imposed on two-stage programs:

1. the random events $\omega \in \Omega$ have *known* probability distributions. The objective function for the second stage decisions then involves calculating an expectation over possible realizations of the random events
2. The vector $\xi(\omega)$ has a finite number of possible realizations, with support Ω

This allows us to write a *Deterministic Equivalent Formulation*, with a finite number J of realizations or *scenarios*, $\xi^1, \ldots, \xi^J$, each one with an associated probability $\psi^1, \ldots, \psi^J$. We can thus write the second term in the objective function in (12.75):

$$\mathbb{E}_{\xi}[G(x, y(\omega))] = \sum_{j=1}^{J} \psi^j G\big(x, y\big(\xi(\omega)\big)\big). \tag{12.76}$$

For the sake of simplicity, we let the support set for the decision variables be a polyhedral set. Therefore we can rewrite the problem to obtain (12.77):

$$\begin{array}{llllr} \min_{x,y} & F(x) + \sum_{j=1}^{J} \psi^j G(x, y(\xi(\omega))) & & & \\ \text{subject to} & h^0(x) & & & \leq b \\ & h^x(\xi^1)x + h^y(\xi^1)y(\xi^1) & & & \leq b^c(\xi^1) \\ & h^x(\xi^2)x + & + h^y(\xi^2)y(\xi^2) & & \leq b^c(\xi^1) \\ & \vdots & \ddots & & \vdots \\ & h^x(\xi^J)x & & + h^y(\xi^J)y(\xi^J) & \leq b^c(\xi^J). \end{array} \tag{12.77}$$

In the formulation (12.77), there is a single decision vector x for the first stage variables, and there is a copy for each scenario of the second stage decision vector, satisfying the constraints in that scenario j.

Note that the stages are related to the revelation of the uncertainty. It is possible to have multiple periods in both the first stage and the second stage decisions.

12.3.2 Example: Unit Commitment with Economic Dispatch

Consider a simple form of the stochastic unit commitment problem with economic dispatch (see e.g., Constantinescu et al. (2011); Wood et al. (2013)). Table 12.3 defines the nomenclature for this problem.

We can formulate this Unit Commitment with Economic Dispatch problem as follows.

$$\min_x f(x) = f_{\text{uc}}(u,v,w) + f_p(p), \tag{12.78}$$

where

$$f_{\text{uc}}(u,v,w) = \sum_{t\in T}\sum_{i\in I}(C_P^{ti}(0)u^{ti} + C_\upsilon^{ti}\upsilon^{ti} + C_w^{ti}w^{ti}), \tag{12.79}$$

$$f_p(p) = \sum_{t\in T}\sum_{j\in J}\sum_{i\in I}\psi^j C_P^{ti}(p^{tij}). \tag{12.80}$$

Positive values of p^{tij} correspond to injections into the system (generation), whereas negative values correspond to withdrawals (demands).

In this formulation, the function $f_{\text{uc}}(\cdot)$ corresponds to the first stage evaluation, with integer variables determining the costs related to the fixed costs of running a unit $C_P^{ti}(0)$, startup and shutdown costs.

The function $f_p(p)$ corresponds to the second stage evaluation, with continuous variables determining the costs/benefits of the dispatches/withdrawals for each one of the generators/demands.

This minimization is subject to two categories of constraints, for all $t\in T$, all $j\in J$:

- *Standard Demand Balance and Limit Constraints*, including equality (e.g., power balance equations) and inequality (e.g., generator limits) constraints,

$$\sum_{i\in I} p^{tij} = 0, \tag{12.81}$$

$$u^{ti}P_{\min}^{tij} \le p^{tij} \le u^{ti}P_{\max}^{tij}. \tag{12.82}$$

- *Unit Commitment*, including startup and shutdown events, minimum up and down times and integrality constraints

Table 12.3 Nomenclature for the discrete problem

1. Variable and Parameters Indexing	
T	Set of time periods considered, n_t elements indexed by t
B	Set of buses in the system, n_b elements.
J	Set of scenarios in the system in period t, indexed by j
I	Set of all units available for dispatch, indexed by i
2. Optimization Variables	
p^{tij}	Active injection for unit i in scenario j at time t
u^{ti}	Binary commitment state for unit i in period t, 1 if unit is on-line, 0 otherwise
v^{ti}/w^{ti}	Binary startup and shutdown states for unit i in period t, 1 if unit has a startup/shutdown event in period t, 0 otherwise
3. Optimization Variables	
$C_P^{ti}(\cdot)$	Cost function for active injections for unit i at time t
C_v^{ti} / C_w^{ti}	Startup/shutdown costs for unit i at time t in $ per startup/shutdown
4. Constraint Functions and Parameters	
$P_{\min}^{tijk}, P_{\max}^{tijk}$	Limits on active injection for unit i in post-contingency state k of state j at time t
τ_i^+, τ_i^-	Minimum up and down times for unit i in number of periods
5. Other Parameters	
ψ^j	Probability of scenario j

$$u^{ti} - u^{(t-1)i} = v^{ti} - w^{ti}, \tag{12.83}$$

$$\sum_{y=t-\tau_i^+ +1}^{t} v^{yi} \le u^{ti}, \qquad \sum_{y=t-\tau_i^- +1}^{t} w^{yi} \le 1 - u^{ti}, \tag{12.84}$$

$$0 \le v^{ti} \le 1, \quad 0 \le w^{ti} \le 1, \quad u^{ti} \in \{0,1\}. \tag{12.85}$$

12.3.2.1 Scenario Construction

In stochastic programming, a central issue that can affect the quality of the solutions obtained is the construction of the scenarios to be used. Ideally, it would be desirable to have a limited number of scenarios, limiting the computational effort and helping to solve the problem in a reasonable time. Here it is useful to remember that it is unlikely that any of the scenarios will actually realize, and therefore the important decision here is the first stage variables determined (e.g., the commitment decisions in the example Sect. 12.3.2).

To construct scenarios we need to consider a couple of questions.

1. What is a good methodology to construct scenarios?
2. How many scenarios should be included?

We can use subject expert elicitation to build the scenarios. However, as the number of random parameters increases, so does the complexity to create the scenarios (see e.g., Jeyakumar et al. (2005)). Assume that uncertainty comes from a random vector $\omega \in \Omega \subset \mathbb{R}^n$. Further, assume that the components for ω, are orthogonal. We can built scenarios by asking experts to choose from options in a discrete space (e.g., $\{G, N, B\}$). The total number of scenarios in this case will grow exponentially, with $J=3^n$. Generally, as n increases, using expert elicitation is difficult due to this exponential complexity.

In case we know the distribution for the uncertainty (e.g., from past research and data available), we can devise an automated scenario generation method by sampling from this distribution. Yet, by having the probability density function function (pdf), we could use directly this information to calculate a policy function (e.g., using probabilistic constraints, Moarefdoost et al. (2016)). In most cases we may not have a closed form for the uncertainty generating process. Then we need to sample from the available data. This leads us to consider whether the moments of the sampling distribution in fact correspond to the moments of the underlying uncertainty generating process. The conditions for asymptotic convergence are relatively mild, so the main question for the researcher is how to select the scenarios generated automatically.

Monte Carlo techniques (see e.g., Niederreiter 1992) are a popular choice to represent this variability for large number of random variables, and its accuracy does not depend on the number of scenarios. Sampling from the joint probability mass function is an unbiased estimator for the mean, and the variance decreases with the sample size. Moreover, there are numerous examples of their application to energy problems (e.g., Billinton and Bai 2004; Cardell and Anderson 2014). However, these methods have slow convergence.

Let the function $G^*(x, y(\xi))$ denote the optimal value of the second stage problem. We can generate J replications of the random vector ξ, given by ξ^j, $j=1, \ldots, J$, each with the same probability distribution. Taking recursively independent and identically distributed (iid) draws ξ^j, we can approximate the expectation $\mathbb{E}[G^*(x, y(\xi))]$. In case all have the same probability of occurrence, this expectation will be given by:

$$\mathbb{E}[G^*(x, y(\xi))] = \frac{1}{N}\sum_{j=1}^{J} G^*(x, y(\xi^j)) \tag{12.86}$$

The selection of the number of scenarios relies on the statistical properties of the first two moments. In Glynn and Iglehart (1989), Infanger (1992) and references therein, it is shown that with a change in variables and using a probability mass function g, a heuristic can be derived to choose a g function such that priority is given to sampling in regions that provide more information about the uncertainty generating process. This *importance sampling* in turn improves the efficiency over Monte Carlo methods techniques. The last 40 years of research have produced a wealth of work in this area. Due to editor imposed space constraints, we cannot elaborate further. Hence this excerpt is provided for exhorting further research by the reader.

12.3.3 Markov Decision Processes

In some cases we may want to represent the uncertainty in the system in an alternative manner to scenarios. Reasons for this range from methodological, for example seeking possible methods that avoid the dimensionality curse associated with the creation of scenarios, to practical considerations, for example in the case of power systems we may want to dispatch the system in a secure manner that deals with worst-case scenarios.

A possible modeling framework is using a discrete Markov Decision Process. This framework allows us to study a sequential decision process with uncertainty. Table 12.4 defines the finite elements for this problem.

Due to the finite number of states and actions, we can define.

Definition 12.8 Let $i, j \in \mathbb{N}$ denote indices for the set of states, and let $k \in \mathbb{N}$ denote an index for the set of actions. Then.

$$\psi^{ijk} := \Psi(s^i, a^k, s^j). \tag{12.87}$$

$$b^{ijk} := \mathcal{B}(s^i, a^k, s^j). \tag{12.88}$$

Let t denote a time index. The elements $\mathcal{S}, \mathcal{A}, \Psi$ define a stochastic dynamic system given by.

$$s^{t+1} \sim \psi(\cdot \mid s^t, a^t). \tag{12.89}$$

Table 12.4 Elements for a Markov Decision Process

Element	Description
$\mathcal{S}$	The *state space*, set of states, n_s finite elements
$\mathcal{A}$	The *action space*, set of actions, n_a finite elements
$\mathcal{B}$	The *instantaneous reward function*, $\mathcal{B} : \mathcal{S} \times \mathcal{A} \times \mathcal{S} \mapsto \mathbb{R}$, given by $\mathcal{B}(s, a, s')$, providing the reward of being in state s, choosing action a, and moving to state s'
Ψ	The *probability transition function*, $\Psi : \mathcal{S} \times \mathcal{A} \times \mathcal{S} \mapsto [0,1]$, given by $\Psi(s, a, s')$, providing the probability of being in state s, choosing action a, and moving next to state s'

In this case, the conditional distribution of s^{t+1} given s^t and a^t is generated by Ψ following.

$$\mathbb{P}(s^{t+1} = s^j \mid s^t = s^i, a^t = a^k) = \psi^{ijk}. \tag{12.90}$$

The actions at time t are selected as a function of the current state, according to a decision rule.

$$A^{t\pi} : \mathcal{S} \mapsto \mathcal{A}, \tag{12.91}$$

where we choose the values $A^{t\pi} = a$ to optimize (e.g., minimize) a given objective function. In this case, a has the Markov property, with the conditional probability of future states depending only on the present state, and independent of past states.

The policy $\pi = \{A^{t\pi}\}_{n\geq 0}$ is a *Markov Policy*. The set of admissible policies is defined by.

$$\Pi = \{\pi = \{A^{t\pi}\}_{n\geq 0} : A^{t\pi}(s) \in \mathcal{A}(s) \quad \forall n, s\}. \tag{12.92}$$

Similar to the optimizations described in Sect. 12.1.1.1, the future rewards can be discounted using factor ρ and for a finite horizon problem we can have a scrap value function $S(s^i)$.

12.3.4 Further Reading

There is a rich literature in stochastic programs for electricity systems to extend the introduction presented in Sect. 12.3. The recent increase in renewable energy sources (RES) has prompted the research community to formulate models that recognize explicitly how increased uncertainty affects the work performed by System Operators. Some of this research deals with flexibility of the generation assets, and demand resources. Some representative works include (Bouffard et al., 2005; Wu et al., 2007; Bouffard and Galiana, 2008; Ortega-Vazquez and Kirschen, 2009; Tuohy et al., 2009; Parvania and Fotuhi-Firuzabad, 2010; Morales et al., 2011; Xie et al., 2011; Constantinescu et al., 2011; Lamadrid and Mount, 2012; Papavasiliou and Oren, 2013; Bakirtzis et al., 2014; Cheung et al., 2015; Warrington et al., 2016). Due to the copious amount of research, the aforementioned papers are just a starting point for the reader interested in this area. Moreover, there are useful survey articles on stochastic optimization for the optimal power flow problem and unit commitment problems (see e.g., Capitanescu et al., 2011; Zheng et al., 2015; Tahanan et al., 2015), in some cases with nonlinear constraints (see e.g., Capitanescu, 2016). There are also comprehensive books and articles (e.g., Conejo et al., 2010; Morales et al., 2014) covering this and other topics including robust optimization (see e.g., Ben-Tal

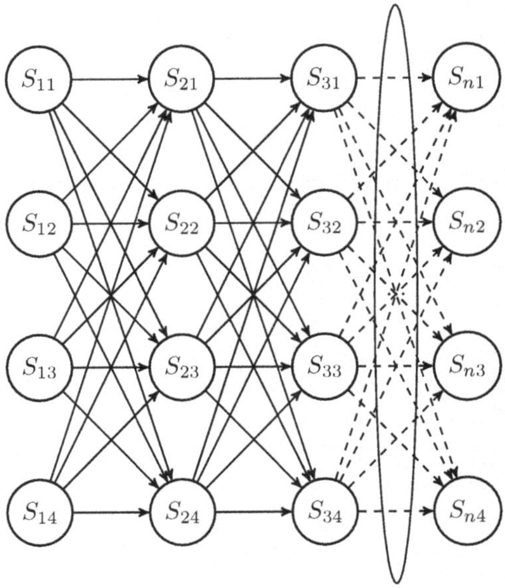

Fig. 12.5 Illustration of transitions over time with four states

et al., 2009; Bertsimas et al., 2013) and probabilistic constraints (see e.g., Bienstock et al., 2012; Moarefdoost et al., 2016).

Lamadrid et al., (2015) use a MDP model as the one presented in Sect. 12.3.3 to represent the transitions over time in the planning horizon, as an approximation of the multiple realizations that can occur inside an operating envelope. A simplification of this model uses a transition probability matrix, relating the possible states in period t with the states in period $t+1$, see Fig. 12.5. The model also enforces other inter-temporal constraints, allowing to cover for extreme variations.

Appendix 1: Discounted Value Hamiltonian

Consider the problem presented in (12.33a)–(12.33c) and the Lagrangean in (12.36). For notation ease we will omit the time arguments on the state $y(t)$ and control $z(t)$ variables.

Definition 12.9 (Discounted-Value Hamiltonian, Continuous)

$$H_d(y, z, \Lambda) \equiv e^{-\delta t} N(y, z) + \Lambda(t) Q(y, z). \tag{12.93}$$

We can redefine the Lagrangean in terms of the discounted-value Hamiltonian as follows.

$$\mathcal{L} \equiv \int_0^T \left(H_d(y,z,\Lambda) + \dot{\Lambda}(t)y\right)dt \\ -\left(\Lambda(T)y(T) - \Lambda(0)y(0)\right) + e^{-\delta T}S(y(T)). \tag{12.94}$$

Suppose we somehow can obtain the optimal paths for $y(t)$, $z(t)$ and $\Lambda(t)$ that maximize $\mathcal{L}$. Consider a *variation* around this optimal solution. In fact, we want to understand how changes in the control variable $z(t)$ affect the state variable $y(t)$, maintaining $\Lambda(t)$ constant, as well as the initial conditions, (12.33c). The change in the Lagrangean can be approximated by.

$$\Delta\mathcal{L} \equiv \int_0^T \left(\frac{\partial H_d(y,z,\Lambda)}{\partial y}\Delta y(t) + \frac{\partial H_d(y,z,\Lambda)}{\partial z}\Delta z(t) + \dot{\Lambda}(t)\Delta y(t)\right)dt \\ -\Lambda(T)\Delta y(T) + e^{-\delta T}\frac{\partial S(y(T))}{\partial y}\Delta y(T) < 0. \tag{12.95}$$

In this case $\Delta\mathcal{L}$ is nonpositive, as we assumed that $y(t)$, $z(t)$ are optimal and therefore any variation about the optimal paths would decrease $\mathcal{L}$. The following must hold.

$$\lim_{\Delta z \to 0} \Delta y = 0 \tag{12.96a}$$

$$\lim_{\Delta z \to 0} \Delta\mathcal{L} = 0. \tag{12.96b}$$

Hence, for all possible variations Δz, $T>t>0$, we have.

$$\frac{\partial H_d(y,z,\Lambda)}{\partial z} = 0 \tag{12.97a}$$

$$\frac{\partial H_d(y,z,\Lambda)}{\partial y} + \dot{\Lambda} = 0 \Rightarrow \dot{\Lambda} = -\frac{\partial H_d(y,z,\Lambda)}{\partial y} \tag{12.97b}$$

$$\frac{\partial H_d(y,z,\Lambda)}{\partial \Lambda} = Q(y(t),z(t)) = \dot{y} \tag{12.97c}$$

$$\Lambda(T) = e^{-\delta T}\frac{\partial S(y(T))}{\partial y} \tag{12.97d}$$

Recall Definition 12.4. Here $\mu(t)$ is the *current value multiplier, spot price*, or *shadow price*, defining what a consumer would pay in period t for an increment in

$y(t)$ delivered at instant t. The term $\Lambda(t)$ is a discounted multiplier ($\Lambda(t)=e^{-\delta t}\mu(t)$), denoting a future price. Note that.

$$\dot{\Lambda}(t) = e^{-\delta t}\dot{\mu}(t) - \delta e^{-\delta t}\mu(t) \tag{12.98}$$

Using (12.97b), we have that

$$\dot{\Lambda} = -\frac{\partial H_d(y,z,\Lambda)}{\partial y} = -e^{-\delta t}\frac{\partial H(y,z,\Lambda)}{\partial y} \tag{12.99}$$

And therefore

$$-\frac{\partial H(y,z,\Lambda)}{\partial y} = \dot{\mu}(t) - \delta\mu(t) \tag{12.100}$$

Compare this to (12.38b).

Appendix 2: Hotelling's Rule

The literature in energy and resource economics has a classical result dating back to Harold Hotelling 1931. In this article Hotelling posited that the extraction of a natural resource is economically efficient if the marginal profit increases at the discount rate. Consider a particular form of the problem presented in (12.33a)–(12.33c), where (12.33b) is given by.

$$\dot{y}(t) = -z(t) \tag{12.101}$$

The current-value Hamiltonian is given by $H(y, z, \mu) \equiv N(y, z) - \mu(t)z$. Using the first order necessary conditions for this problem, (12.38a)–(12.38d), we obtain.

$$\frac{\partial H}{\partial z} = N_z(\cdot) - \mu = 0, \tag{12.102a}$$

$$-\frac{\partial H}{\partial y} = -N_y(\cdot) = \dot{\mu} - \delta\mu, \tag{12.102b}$$

$$\frac{\partial H}{\partial \mu} = -z(t) = \dot{y}. \tag{12.102c}$$

We can use (12.102a) to calculate $\dot{\mu} = \dot{N}_z$ and replace this and (12.102a) into Eq. (12.102b) to obtain.

$$\dot{N}_z = \delta N_z - N_y. \tag{12.103}$$

Dividing (12.103) by N_z we obtain.

$$\frac{\dot{N}_z}{N_z} = \delta - \frac{N_y}{N_z}. \tag{12.104}$$

which is the general form of *Hotelling's Rule*. For production technologies without dependence on the state variable, this simplifies to.

$$\frac{\dot{N}_z}{N_z} = \delta. \tag{12.105}$$

References

Bakirtzis, E. A., Biskas, P. N., Labridis, D. P., & Bakirtzis, A. G. (2014). Multiple time resolution unit commitment for short-term operations scheduling under high renewable penetration. *IEEE Transactions on Power Systems, 29*(1), 149–159.

Ben-Tal, A., El Ghaoui, L., & Nemirovski, A. (2009). *Robust optimization*. Princeton series in applied mathematics. Princeton: Princeton University Press.

Bertsimas, D., Litvinov, E., Sun, X., Zhao, J., & Zheng, T. (2013). Adaptive robust optimization for the security constrained unit commitment problem. *IEEE Transactions on Power Systems, 28*(1), 52–63.

Bienstock, D., Chertkov, M., & Harnett, S. (2012). *Chance Constrained Optimal Power Flow: Risk-Aware Network Control Under Uncertainty*. Technical Report, Cornell University.

Billinton, R., & Bai, G. (2004). Generating capacity adequacy associated with wind energy. *IEEE Transactions on Energy Conversion, 19*(3), 641–646.

Birge, J., & Louveaux, F. (1997). *Introduction to stochastic programming*. Springer series in operations research series. London: Springer London, Limited.

Bouffard, F., & Galiana, F. (2008). Stochastic security for operations planning with significant wind power generation. *IEEE Transactions on Power Systems, 23*(2), 306–316.

Bouffard, F., Galiana, F., & Conejo, A. (2005). Market-clearing with stochastic security-part I: formulation. *IEEE Transactions on Power Systems, 20*(4), 1818–1826.

Capitanescu, F. (2016). Critical review of recent advances and further developments needed in AC optimal power flow. *Electric Power Systems Research, 136*, 57–68.

Capitanescu, F., Ramos, J. M., Panciatici, P., Kirschen, D., Marcolini, A. M., Platbrood, L., et al. (2011). State-of-the-art, challenges, and future trends in security constrained optimal power flow. Electric Power Systems Research, 81(8), 1731–1741.

Cardell, J., & Anderson, C. (2014). A flexible dispatch margin for wind integration. *IEEE Transactions on Power Systems, PP*(99), 1–10.

Cheung, K., Gade, D., Silva-Monroy, C., Ryan, S. M., Watson, J.-P., Wets, R. J., et al. (2015). Toward scalable stochastic unit commitment. *Energy Systems, 6*(3), 417–438. Copyright - Berlin/Heidelberg: Springer. Last updated 07 August 2015.

Conejo, A., Carrión, M., & Morales, J. (2010). *Decision making under uncertainty in electricity markets*. International series in operations research & management science. New York: Springer.

Conrad, J. (2010). *Resource economics*. Cambridge: Cambridge University Press.
Constantinescu, E., Zavala, V., Rocklin, M., Lee, S., & Anitescu, M. (2011). A computational framework for uncertainty quantification and stochastic optimization in unit commitment with wind power generation. *IEEE Transactions on Power Systems, 26*(1), 431–441.
Dantzig, G. B. (1955). Linear programming under uncertainty. *Management Science, 1*(3–4), 197–206.
Glynn, P. W., & Iglehart, D. L. (1989). Importance sampling for stochastic simulations. *Management Science, 35*(11), 1367–1392.
Hotelling, H. (1931). The economics of exhaustible resources. *Journal of Political Economy, 39*(2), 137–175.
Hu, Y., & Defourny, B. (2017). Optimal price-threshold control for battery operation with aging phenomenon: a quasiconvex optimization approach. *Annals of Operations Research*. https://doi.org/10.1007/S10479-017-2505-4
Infanger, G. (1992). Monte carlo (importance) sampling within a benders decomposition algorithm for stochastic linear programs. *Annals of Operations Research, 39*(1), 69–95.
Jeyakumar, V., Rubinov, A., Shapiro, A., & Nemirovski, A. (2005). On complexity of stochastic programming problems. In *Applied optimization* (pp. 111–146). New York: Springer US.
Lamadrid, A., Shawhan, D., Murillo-Sanchez, C., Zimmerman, R., Zhu, Y., Tylavsky, D., et al. (2015). Stochastically optimized, carbon-reducing dispatch of storage, generation, and loads. *IEEE Transactions on Power Systems, 30*(2), 1064–1075.
Lamadrid, A. J., & Mount, T. D. (2012). Ancillary services in systems with high penetrations of renewable energy sources, the case of ramping. *Energy Economics, 34*(6), 1959–1971.
Moarefdoost, M. M., Lamadrid, A. J., & Zuluaga, L. F. (2016). A robust model for the ramp-constrained economic dispatch problem with uncertain renewable energy. *Energy Economics, 56*, 310–325.
Morales, J. M., Conejo, A. J., Madsen, H., Pinson, P., & Zugno, M. (2014). *Integrating renewables in electricity markets*. International series in operations research & management science (Vol. 205). New York: Springer US.
Morales, J. M., Conejo, A. J., & Perez-Ruiz, J. (2011). Simulating the impact of wind production on locational marginal prices. *IEEE Transactions on Power Systems, 26*(2), 820–828.
Niederreiter, H. (1992). *Random number generation and quasi-monte carlo methods*. Society for industrial and applied mathematics. https://doi.org/10.1137/1.9781611970081
Ortega-Vazquez, M. A., & Kirschen, D. S. (2009). Estimating the spinning reserve requirements in systems with significant wind power generation penetration. *IEEE Transactions on Power Systems, 24*(1), 114–124.
Papavasiliou, A., & Oren, S. S. (2013). Multiarea stochastic unit commitment for high wind penetration in a transmission constrained network. *Operations Research, 61*(3), 578–592.
Parvania, M., & Fotuhi-Firuzabad, M. (2010). Demand response scheduling by stochastic SCUC. *IEEE Transactions on Smart Grid, 1*(1), 89–98.
Peterson, S. B., Apt, J., & Whitacre, J. (2010). Lithium-ion battery cell degradation resulting from realistic vehicle and vehicle-to-grid utilization. *Journal of Power Sources, 195*(8), 2385–2392.
Pindyck, R. S. (1991). Irreversibility, uncertainty, and investment. *Journal of Economic Literature, 39*(4), 1110–1148.
Shapiro, A., & Philpott, A. (2007). *A Tutorial on Stochastic Programming*. Technical Report, Georgia Institute of Technology.
Shiau, C.-S. N., Samaras, C., Hauffe, R., & Michalek, J. J. (2009). Impact of battery weight and charging patterns on the economic and environmental benefits of plug-in hybrid vehicles. *Energy Policy, 37*(7), 2653–2663.
Simon, C., & Blume, L. (1994). *Mathematics for economists*. New York: W. W. Norton, Incorporated.
Sioshansi, R., & Conejo, A. J. (2017). *Optimization in engineering: Models and algorithms*. Cham: Springer International Publishing.

Tahanan, M., van Ackooij, W., Frangioni, A., & Lacalandra, F. (2015). Large-scale Unit Commitment under uncertainty. *4OR, 13*(2), 115–171.

Tuohy, A., Meibom, P., Denny, E., & O'Malley, M. (2009). Unit commitment for systems with significant wind penetration. *IEEE Transactions on Power Systems, 24*(2), 592 –601.

Warrington, J., Hohl, C., Goulart, P., & Morari, M. (2016). Rolling unit commitment and dispatch with multi-stage recourse policies for heterogeneous devices. *IEEE Transactions on Power Systems, 31*(1), 187–197.

Weitzman, M. L. (2003). *Income, wealth, and the maximum principle*. Cambridge: Harvard University Press.

Wood, A. J., Wollenberg, B. F., & Sheble, G. B. (2013). *Power generation, operation and control*, 3rd ed. Hoboken: Wiley.

Wu, L., Shahidehpour, M., & Li, T. (2007). Stochastic security-constrained unit commitment. *IEEE Transactions on Power Systems, 22*(2), 800–811.

Xie, L., Carvalho, P., Ferreira, L., Liu, J., Krogh, B., Popli, N., et al. (2011). Wind integration in power systems: Operational challenges and possible solutions. *Proceedings of the IEEE, 99*(1), 214–232.

Zheng, Q. P., Wang, J., & Liu, A. L. (2015). Stochastic optimization for unit commitment; a review. *IEEE Transactions on Power Systems, 30*(4), 1913–1924.

Chapter 13
Reservoir Capacity Planning Using Stochastic Multiobjective Programming Integrated with MCMC Technique

Ho Wen Chen and Kieu Lan Phuong Nguyen

Abstract Determining the location and size of new reservoirs requires a risk-informed decision approach. The scope of risk management, which has been increasingly recognized today, should consider the resulting economic benefit and the local unique hydrological conditions while maintaining necessary water quality in the river. Thus, the incorporation of environmental and economic factors into the reservoir capacity planning process is essential. This chapter presents an optimization-based risk analysis framework to size a new reservoir in a river basin with focus on sustainable development. The framework is designed for a multidisciplinary assessment of reservoir sizing, simultaneously addressing natural patterns of stream flows, adequate water supply, and pristine water quality in the river watershed. A set of water quality parameters, that is, dissolved oxygen (DO) and biochemical oxygen demand (BOD), is used as a surrogate index to reflect water quality impacts. A dynamic stochastic optimization is applied to the problems in which the uncertainty is modeled using the Markov Chain Monte Carlo (MCMC) technique. A case study of Hou-Lung River Basin in Taiwan is used to illustrate the capability of the framework.

Keywords Reservoir capacity planning · Water quality · Water resources management · Stochastic programming · Markov Chain Monte Carlo technique

H. W. Chen (✉)
Department of Environmental Science and Engineering, Tung-Hai University, Taichung, Taiwan
e-mail: hwchen@thu.edu.tw

K. L. P. Nguyen
Department of Environmental Science and Engineering, Tung-Hai University, Taichung, Taiwan

Nguyen Tat Thanh University, Ho Chi Minh City, Vietnam

C. Chen et al. (eds.), *Pursuing Sustainability*, International Series in Operations Research & Management Science 301,
https://doi.org/10.1007/978-3-030-58023-0_13

13.1 Introduction

Adequate water supply is imperative and vital to sustain population growth, to maintain economy development, and to increase the standard of living in both developed and developing countries. Lack of access to clean water has been cited as one of impedances that affect public health in developing countries. With an increasing awareness of the consequences of changing environment, incorporating sustainability-related factors in water resource planning provides an important guidance that aids the traditional approaches in determining the size of a new reservoir.

Sizing a new reservoir is a daunting task for obvious reasons. The investment is long-run and irreversible. An undersized reservoir is unlikely to provide needed functions, while an oversized reservoir is doubtful to be economic desirable. Moreover, changes in stream flow due to short-term weather or long-term climate conditions coupled with urban development in adjacent areas and economy growth indicate that the problem is hardly deterministic. A number of approaches have been used to decide optimal size of a new reservoir, including critical period technique, probability matrix method, and simulation/optimization approaches. The critical period technique is the earliest method, which optimizes the size of a reservoir by considering an initial full reservoir that passes through a series of states, which are defined by historical inflows, outflows, and other factors. An example of the approach is the well-known mass curve method that estimates the required storage capacity of a reservoir (Alrayess et al. 2017). Others include the sequent peak method that determines the reservoir capacity required to meet a stipulated demand over a given record of inflows. The underlying assumption of this approach is that the daily water demand is fixed, and the inflow due to precipitation is known in advance. However, the approach is deterministic with complete information.

The second approach is the probability matrix method, which entails using statistics to analytically solve the storage-yield problems, which is defined as determination of the storage required to supply a given demand at a specified storage tolerance (Kottegoda 1980; McMahon and Mein 1986; Nagy et al. 2013). However, the approach needs numerical integration, which makes it computationally more demanding.

The third approach is system-based techniques or commonly referred to as systems analysis, using either simulation or optimization approaches. The approach provides an important set of tools that allows for characterizing the complex dynamics and trade-offs among different water uses considering various attributes related to water quantity and water quality. These attributes, whether physically, chemically, or biologically related, can favor, limit, or completely inhibit economic activity. Loucks et al. (1981) pioneer the approach and state that the technique is an indispensable tool for water resources management. Indeed, a significant amount of system-based research has addressed the related issues of reservoir capacity planning and design using mathematical programming models within the last decades. Examples include Revelle et al. (1969), Curry et al. (1973), Mutreja and Yevjevich (1978), Huock and Cohon (1978), and Helm et al. (1984). For instance, the paper by

(Buras 1985) develops a stochastic programming model to derive the optimal schedule of seasonal release under uncertainty in a reservoir planning study. Extending research efforts along this line cover marginal cost analysis (Miltz and White 1987), marginal benefit analysis (Oamek et al. 1990), nonconvexity analysis to determine the siting and sizing of the reservoirs (Turgeon and Brosseau 1987), an integrated simulation-optimization analysis for sizing potential reservoirs in a river basin (Barlishen et al. 1989), the emphasis on the generation of synthetic streamflows (Savic et al. 1989; Takeuchi and Sivaarthitkul 1995), reservoir reliability analysis (Plate 1989), the integration of expected water demand, storage loss due to sedimentation, and physical and hydrological characteristics of the watershed (Singh and Durgunoğlu 1990), and a hydro-meteorological forecasting technique applied to support a more efficient use of reservoirs (Takeuchi and Sivaarthitkul 1995). The later effort strives to size reservoirs in a multiple reservoir capacities design problem using a network flow programming model to optimize the flow in a multiperiod framework (Khaliquzzaman and Chander 1997). More detailed and elaborate analyses include the assessment of marginal negative environmental impact caused versus incremental benefit generated by sizing new reservoir (Takeuchi 1997), the global change impacts associated with the warm phase of El Nino, and positive anomalies with the cold phase (La Nina) (Mesa et al. 1996), and the use of optimal control theory for handling multireservoir systems for water supply (Mousavi and Ramamurthy 2000). Overall, the system-based approach is more versatile to study complicated problems raised from water resource planning.

Our focus in this chapter is on the sustainable reservoir sizing. Factoring sustainability, for example, environmental and biological integrity, into a reservoir sizing process turns out to be essential. Protection and enhancement of water quality in the downstream areas of many river basins has been identified as one of the most critical sustainable development issues facing in both developing and developed countries. For instance, some environmental management programs started assessing the intrinsic characteristics of a reservoir and the possible water quality impacts on downstream river reaches (Avogadro et al. 1997; Benoist et al. 1998). By using physics-based water quality simulation models, the impact due to siting and sizing a new reservoir may become understandable. However, appropriate tools are still needed to measure the health of rivers at scales sufficiently large for the outcomes to be useful for water resources management (Harris and Silveira 1999). Generally, the river health can be defined by three main physical and chemical attributes of a river (i.e., its energy sources, water quality, and flow regime) as well as their biota and habitats at all scales (Karr 1981, 1987, 1991). Indicators for assessing river health need to be ecologically based, efficient, rapid, and consistently applicable in different ecological regions (Harris and Silveira 1999).

Water resource planning is subject to an increasing degree of uncertainty, especially associated with warming climate. Changing climate implies that while the mean or average of a distribution is likely to be the same, there will be more extreme samples. Short-term meteorological processes, including evapotranspiration, precipitation, and temperature, are highly dependent on long-term climate change. On the supply side, the fact that warming climate would likely impact snow-water ratio and expedite snow melting processes implies that the pattern of stream flows in rivers in cold regions will likely to be changed. On the demand side, distribution of future populations, per capita water-usage rates, water consumption patterns, and priorities for water uses are likely to be impacted by climate change and other factors that are difficult to predict (Loucks and Van Beek 2017). All of these suggest that dealing with increasing level of uncertainty is a new norm in water resource planning.

There has been a torrent of studies that apply stochastic programming approaches to examine water resources under uncertainty (Li et al. 2015; Luo et al. 2007; Zeng et al. 2017). Uncertainties considered in these studies, for example, water demands, rainfall and available water resources, inflows, and stream flows, are typically estimated empirically (Dupačová et al. 1991). The stochastic processes underpinning the data formation of these uncertainties/parameters could be assumed to follow a Markov process (Loucks and Van Beek 2017). One of the most commonly used approaches is Markov Chain Monte Carlo (MCMC), which samples data from probability distributions using Markov chains. The approach can be used for Bayesian inference and numerical integration. Applications of MCMC in water resource management include a paper by (Chen et al. 2012) for mapping probabilistic characterizations of water leakage and the study by (Barandouzi et al. 2012) that assesses the uncertainty of watershed-scale water quality (Zheng and Han 2016).

This chapter aims at demonstrating a sustainable water resource planning framework based on simulation and optimization, considering hydrological, environmental, and biological factors in deciding optimal reservoir size. It advances the watershed management by combining environmental and biological criteria when meeting the planning objectives of reservoir sizing with respect to hydrological patterns. Moreover, the framework also allows for understanding the environmental feedbacks at diverse temporal and spatial scales in a coupled human and natural system: the reservoir and its linked river network. By integrating population, ecosystem, and socioeconomic aspects with watershed management, this framework provides a holistic approach in determining how development of a reservoir affects water resources use in a practical application. A case study in the Hou-Lung River Basin, Taiwan demonstrates the capability of the approach.

13.2 Methodology

13.2.1 Background

Taiwan is located in the west Pacific Rim of the Asian Continental Shelf with a small area of about 36,000 km^2 with a population of over 23 million people. In recent years, the relationship between land, water resources, and their human exploitation within river basins have been especially complex. The conservation of water resources is under increasing pressure due to tremendous demands from population growth, agricultural development, and industrialization. Facing a water resources shortage, the Taiwan government was planning to build an off-stream reservoir in the Hou-Lung River Basin located between the Hsin-Chu and Tai-Chung County in a mountainous area in central Taiwan in order to support regional economic development. Designing the surface off-stream reservoir, the Tien-Hua Lake, involves altering hydrologic natural regimes, managing active storage volume, and optimizing reservoir pumping-and-release policy.

The area is subject to extremely uneven stream flows between dry and wet seasons, complicating the development of a new reservoir and the conservation of water quality and ecosystem in the river basin. In particular, approximately 80% of available water resources in the area are concentrated between April and September while water during the rest of the year is relatively scarce. Local environmentalists also are concerned that the project might amend hydraulic conditions that could degrade water quality and affect ecological health. They therefore argue that any plan of diverting water should include a provision of environmental flows to preserve natural and biotic system.

The downstream segment of the Ho-Lung River system is well-known for its pollution loadings, predominantly from household wastewater, occasionally with industrial and farming discharges, together with runoffs and return flows from agricultural land. According to the river quality classification of Taiwan's Environmental Protection Administration, river water quality is divided into five categories, see Table 13.1. Along the Hou-Lung River, Taiwanese EPA classifies the water body into three polluting categories as shown in Table 13.2. The farther from the mouth of the river is the better the water quality: the first 9 km, 9–32 km, and 32 km—from the month of the river are classified as types 3, 2, and 1, respectively. The monitoring data shows that the current waste loading along the river system could result in a long-standing violation of water quality standards in the downstream segments.

Table 13.1 The water quality standard based on type of water body in Taiwan

Type of water body	Water quality standard						
	pH	Dissolved oxygen (DO) (mg/L)	Biochemical oxygen demand (BOD) (mg/L)	Suspended solids (SS) (mg/L)	Coliform group (CFU/100 ml)	Ammonia nitrogen (NH_4^+-N) (mg/L)	Total phosphorus (TP) (mg/L)
I	6.5–8.5	≥ 6.5	≤ 1	≤ 25	≤ 25	≤ 0.1	≤ 0.02
II	6.0–9.0	≥ 5.5	≤ 2	≤ 25	≤ 5000	≤ 0.3	≤ 0.05
III	6.0–9.0	≥ 4.5	≤ 4	≤ 40	≤10,000	≤ 0.3	–
IV	6.0–9.0	≥ 3.0	–	≤ 100	–	–	–
V	6.0–9.0	≥ 2.0	–	No floating substance and oil	–	–	–

Table 13.2 Information on monthly water rights in the Hou-Lung River Basin in cms

	Chuan-Long Canal (D_1)	Gui-Shan-Da-Po Canal (D_2)	Mang-Pu-Zhang-Li Canal (D_3)	Jia-Sheng-Wn-Zang-Li Canal (D_4)	Chang-Chun chemical industry (D_5)	Total
Jan.	1.980	0.900	0.062	0.140	0.242	3.694
Feb.	1.973	0.900	0.067	0.140	0.242	3.692
Mar.	1.961	0.982	0.020	0.166	0.242	3.741
Apr.	1.961	0.982	0.072	0.166	0.242	3.793
May.	1.961	0.900	0.063	0.141	0.242	3.677
Jun.	1.961	0.900	0.063	0.141	0.242	3.677
Jul.	1.961	0.900	0.073	0.166	0.242	4.342
Aug.	1.961	0.900	0.062	0.141	0.242	4.306
Sep.	1.961	0.900	0.062	0.141	0.242	4.306
Oct.	1.961	0.909	0.061	0.130	0.242	4.303
Nov.	1.423	0.466	0.035	0.125	0.242	2.661
Dec.	1.480	0.668	0.070	0.110	0.242	2.940
Total	22.544	10.307	0.71	1.707	2.904	45.132

Thus, the new reservoir should not exceed an essential capacity size (so as not to withdraw exceeding amount of water) in order to meet planning goals related to water supply, flood control, and hydroelectric generation while also allowing economically maintain the water quality and ecosystem integrity in the downstream. These highlight the challenges faced by the project and the strength of the framework, which considers both responses from the physical systems and conflicted objectives articulated by stakeholders using a system approach.

13.2.2 Framework

Figure 13.1 plots the schematics of the Hou-Lung river system. The system is comprised of eight waste discharge sites and five water intake sites. The non-tidal stream of the system is further divided into six river reaches based their hydraulic characteristics. Within each river reach, several segments with length of 0.5 km are defined for modeling purpose. Different scale of the system is represented by A, B, and C, respectively.

Figure 13.1 shows that Part A is the system in the macro level, with encompasses of Parts A, B, and C. Part B is the next level, which encloses Parts C. Each part is with a different scale, which we explain as follows. Part A gives the entire Hou-long River watershed, covering the Miao-li county in central Taiwan. Part B is a presentation of segments i from $i = 1$ to n in a river reach. Each segment within the Part B is a uniform reactor with homogeneous environmental characteristics, for example, deoxygenation rate, volumetric reaeration. This is the scale by which flow rate, water intake, and waste discharge are occurred and modeled in our analysis. This is also where various cooperation, confliction, and interaction concerning water

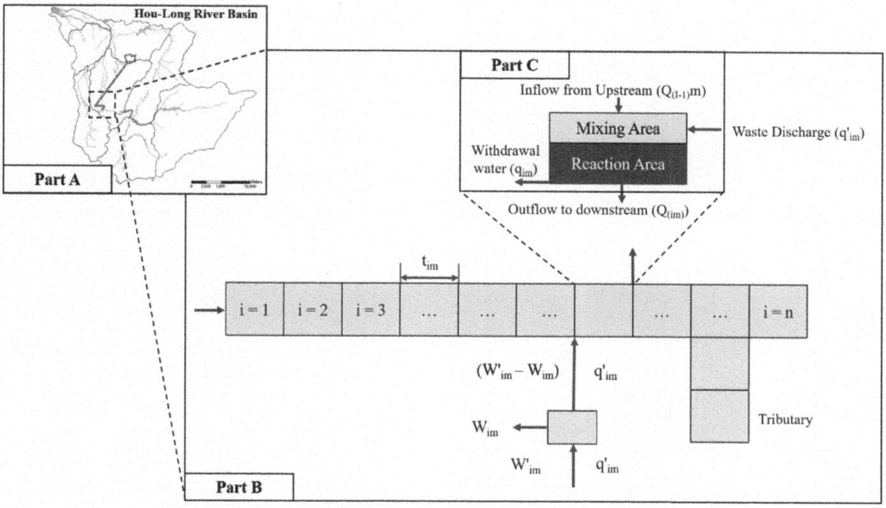

Fig. 13.1 A schematic representation of the Hou-Lung River

resources usage are simulated in both in temporal and spatial dimension. Each segment within Part B is defined in Part C, representing a smaller scale of the system that partitions into mixed and reaction layers. It is also where various physical or/and chemical reactions are taken place. With this defined conceptual framework, we can then apply simulation and optimization framework to identify the optimal management strategy while considering different sustainability-related factors.

13.2.3 Data Analysis

Figure 13.2 graphs the geographical location of the waste loadings, monitoring sites, and gauge stations in the Hou-Lung River system, where RK indicates the distance from the mouth of the river in km. The drain and water withdraw locations are marked in detail in Fig. 13.3. The waste loadings of each drain are in proportion to the size of the circle surrounding the square in Fig. 13.2. Figure 13.2 summarizes the information, including streamflow rate, hydrological pattern, drainage area, and stream flow duration, together with discharges and withdrawals.

13.2.3.1 Water Supply and Demand Analysis

We collect data of streamflow rate, hydrological pattern, drainage area, streamflow duration, and discharges and withdrawals from official open data in Taiwan. The monthly average streamflow rates in cms (cubic meter per second) during 1981 and 2005 are depicted in Fig. 13.3. Figure 13.3 clearly shows the natural variations of streamflow during seasons, suggesting the potential of pumping water from HL3 to the Tien-Hua

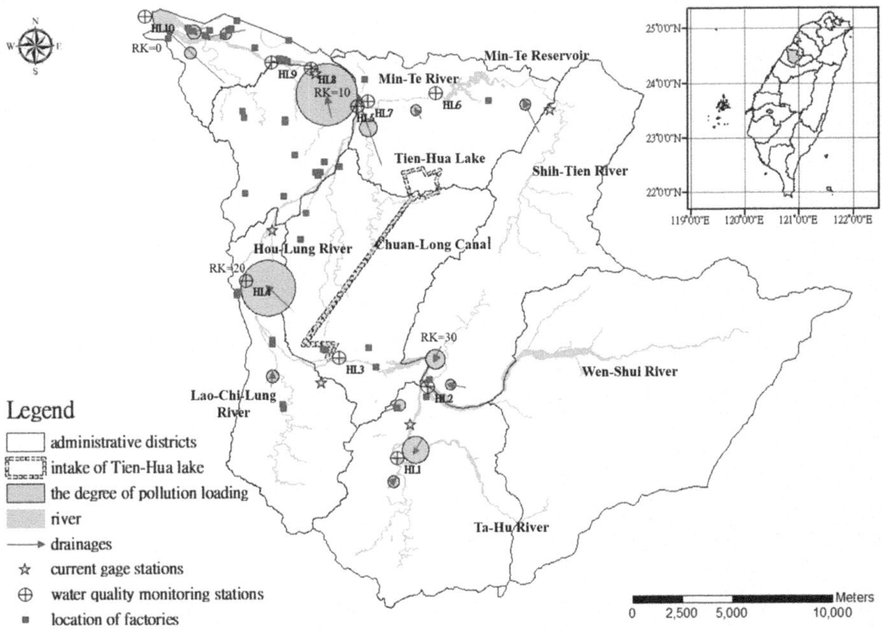

Fig. 13.2 Distribution of the waste loadings and monitoring stations in Hou-Lung River system

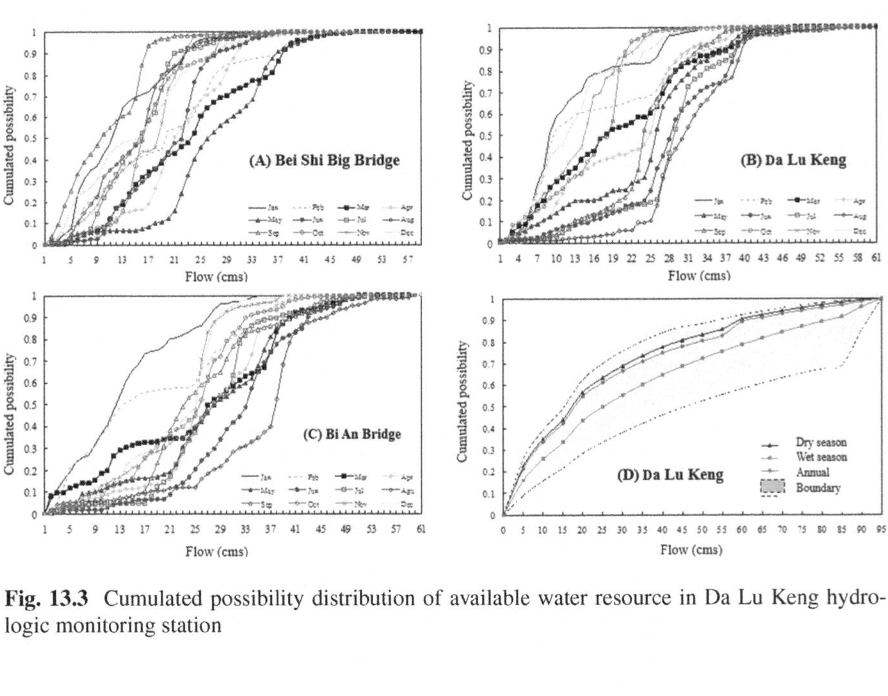

Fig. 13.3 Cumulated possibility distribution of available water resource in Da Lu Keng hydrologic monitoring station

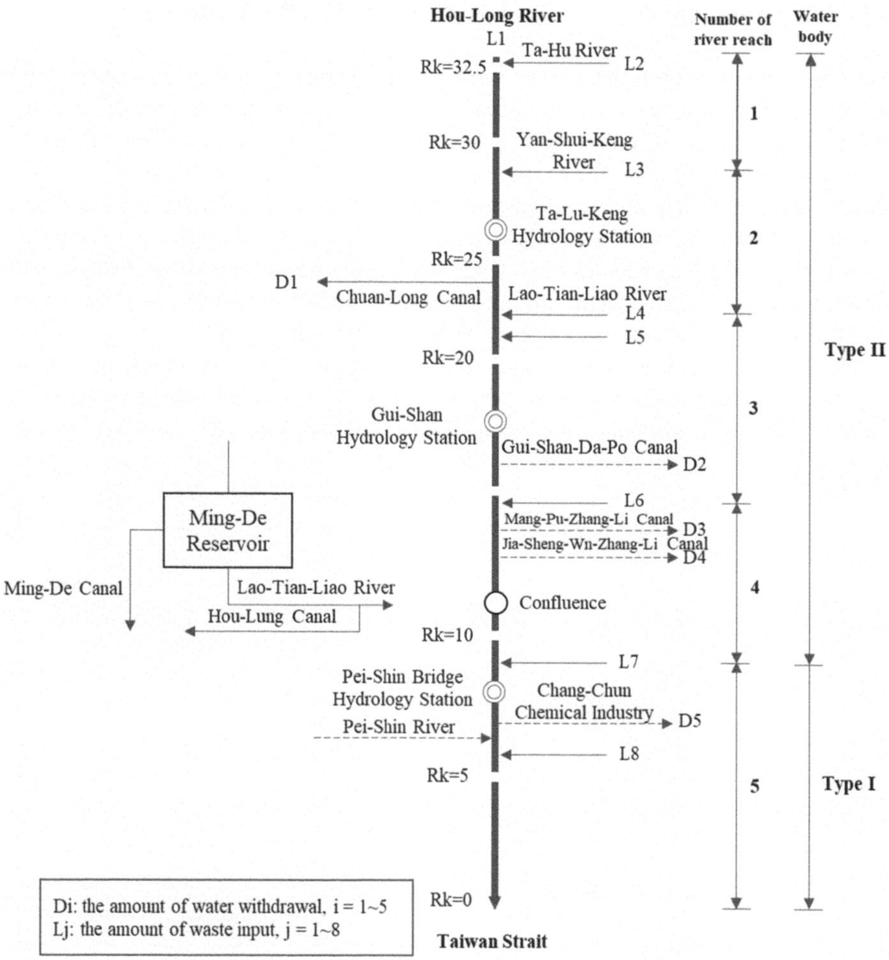

Fig. 13.4 The schematic diagram for modeling analysis in the Hou-Lung River Basin

during wet season. We identify five water right holders in the Hou-Lung river basin. Because of seasonal variation in hydrology, the amount of verified water resource use for each withdrawal gate is different and varied by months. We summarize the available water amount for each holder in Table 13.2. Their location from upstream to downstream of the river is also displayed in Fig. 13.4, where the locations marked by D, L, and I denote water inflow, water intake, and waste discharge sites, respectively.

13.2.3.2 Water Quality Monitoring and Water Quality Modeling

We first collected hydrology and water quality samples in order to understand changes in hydrological conditions and water quality over time. A total of ten sampling campaigns were conducted in ten locations along the Hou-Lung River to obtain water quality data, for example, BOD and DO. In particular sampling took place during both dry and wet seasons because of their uneven streamflow rates. Sampling during dry season with low streamflow gives a set of worst-case scenarios with high polluting concentrations, while sampling during high streamflow season presents an optimistic scenario, giving an upper bound of stream assimilative capacity. These sampling locations are marked by HL_1-HL_{10} in Fig. 13.2.

Based on riverbed conditions and river hydrological characteristics, this study divides Hou-Lung into six units (as shown in Fig. 13.5) and assumes that the river sections in each unit have the same deoxygenation rate (K_1) and volumetric

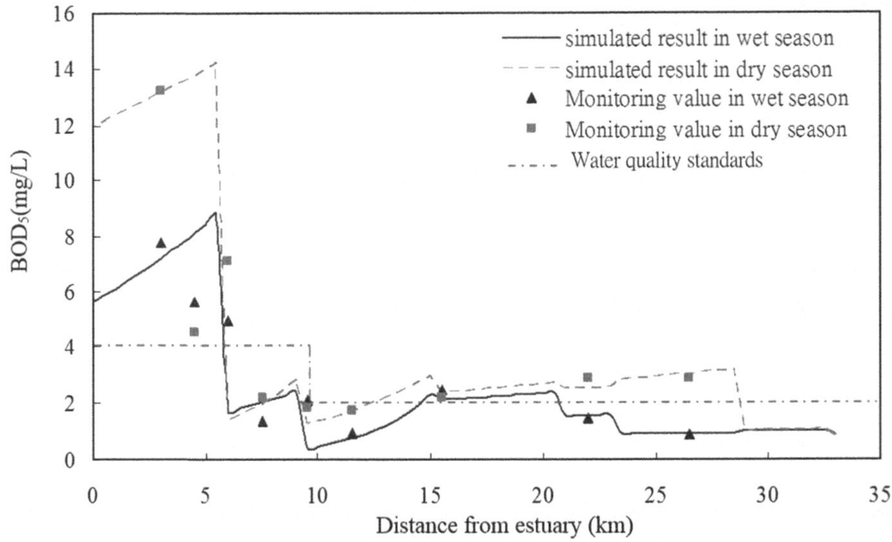

Fig. 13.5 Comparison of observations and the simulated results of the Streeter–Phelps model of BOD concentration in dry and wet seasons

Table 13.3 Parameter associated with water quality model

	Wet season		Dry season	
Number of river reach	k_1 (1/day)	k_2 (1/day)	k_1 (1/day)	k_2 (1/day)
1	0.002	0.7629	0.400	0.4629
2	0.001	0.4215	0.400	0.1215
3	0.001	1.2650	0.010	0.6650
4	0.350	1.1960	0.450	1.4960
5	0.200	0.6116	0.100	0.2616
6	0.100	0.8276	0.200	0.6876

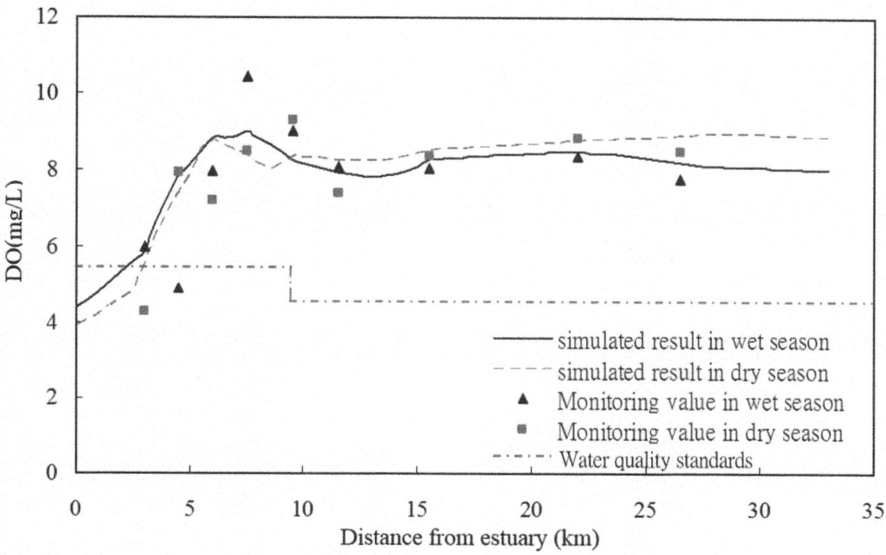

Fig. 13.6 Comparison of observations and the simulated results of the Streeter–Phelps model of DO concentration in dry and wet seasons

reaeration coefficient (K_2). The collected samples from ten sites along the river are then used to derive K_1 and K_2, respectively, for dry and wet seasons in Table 13.3. Overall, large deoxygenation rate (K_1) and low volumetric reaeration coefficient (K_2) are reported during the dry season as the river is subject to low DO and high loading from organic compounds, and a low TMDL is implemented. Figures 13.5 and 13.6 show the simulated results based on Streeter–Phelps model of the six river reaches for both wet and dry seasons. The observation is further verified by comparing them to the observations provided by the Taiwanese EPA. These figures indicate a clear violation of BOD standard in the river segment close to the mouth of the river during both dry and wet seasons. This suggests that withdrawal of water by the proposed reservoir might exacerbate its water quality.

13.2.3.3 Investigation of Cost and Benefit Databases

Total Cost

A total cost curve, which links costs to major design parameters, needs to be obtained. The total cost includes capital/fixed cost of initial project construction, and operation/variable costs. We combined cost information from the existing seventeen existing/planned reservoirs in Taiwan. The total cost of supplying or delivering water (NT$/m^3) in y-axis is then associated with the amount of guaranteed water stipulated under contracts (10^4 m^3) in x-axis as shown in Fig. 13.7.

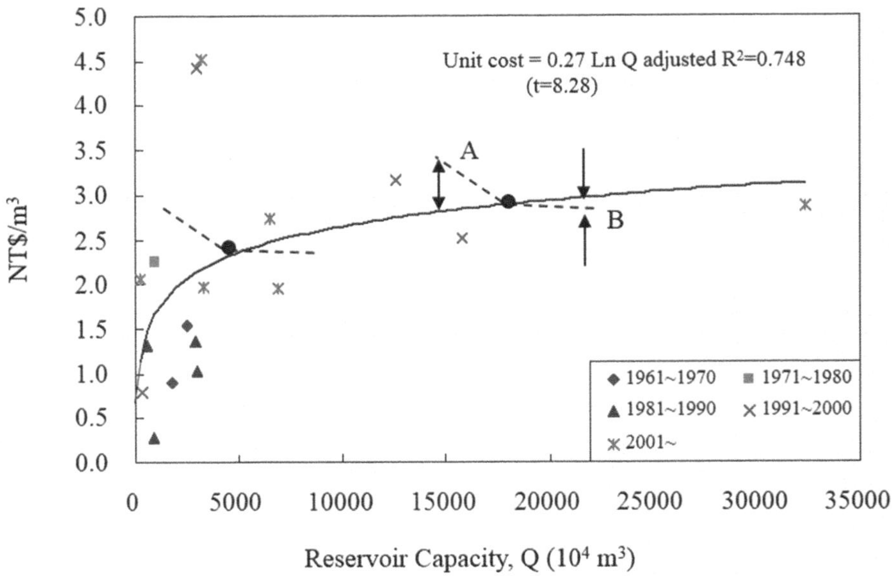

Fig. 13.7 Plot of marginal cost of existing reservoirs in Taiwan

Variable Costs

Deriving marginal (variable) cost is straightforward had there been an efficient water-right market in place. Because there is no active water right market in Taiwan, we rely on the notion of opportunity cost to estimate the marginal (variable) cost of water supply. In particular, as public water utilities in Taiwan are responsible for providing reliable water to the publics, the government needs to search for least-cost substituted sources when contracted water cannot be met during drought year. Surrounded by the Pacific Ocean, sea water desalinization is the most possible substitute when facing deficiency between contracted and available water. We therefore develop a cost function in Eq. (13.1) that contains three parts.

$$C_{1m} = \left\{ \begin{array}{cc} 0.27\mathrm{Ln}V + 24\left(q - q_m\right) & q_m < q \\ 0.27\mathrm{Ln}V & q_m = q \\ 0.27\mathrm{Ln}V - 4\left(q_m - q\right) & q_m > q \end{array} \right\} \forall m \tag{13.1}$$

where V denotes the capacity of reservoir, q is the planned capacity for monthly water supply from the Tien-Hua Lake, and q_m is the available water supply from the Tien-Hua lake at month m. In other words, when $q_m < q$, that is, shortage of water supply, additional cost per 10^4 m^3 of water is equal to 24 NT (slope A in Fig. 13.9). On the other hand, if $q_m > q$, the excessive water can then be sold elsewhere, thereby earning 4 NT$/10^4$m^3 (slope B in Fig. 13.9).

Wastewater Treatment Cost

In addition to capacity and water supply cost, this analysis also considers cost associated with treating wastewater. The costs of 48 domestic wastewater treatment plants and 29 industrial wastewater treatment plants were collected to derive wastewater treatment costs. Wastewater treatment plants are grouped into primary, secondary, and tertiary treatment (Chen and Chang 2002). This study selects the secondary treatment plants as the main treatment facilities because they are most general in Taiwan. The monthly wastewater treatment cost is displayed in Eq. (13.2), where W_{im} denotes the waste load deduction rate weight at i site in month m.

$$C_2 = \sum_m \sum_i 0.079 W_{im}^{\ 0.88} \tag{13.2}$$

13.2.4 *Optimization Analysis*

This chapter develops a chance-constrained multiobjective optimization model to determine the optimal size of a planning reservoir under dynamic hydrological conditions. The two objectives considered include maximization of the monthly water supply from reservoir in Eq. (13.3) and minimization of the total cost incurred from constructing and operating facilities in Eq. (13.4), where C_{1m} and C_2 are defined in Eqs. (13.1) and (13.2), respectively.

$$Maximize\ Z_1 = q \tag{13.3}$$

$$Minimize\ Z_2 = \sum_m C_{1m} + C_2 \tag{13.4}$$

in which q (10^4 m^3/month) is the pumping rate for monthly water supply from Hou-Lung River to the Tien-Hua Lake; C_{1m} (NT$) is the cost for construction of a reservoir for water supply in month m; C_2 (NT$) is the cost for water pollution control. We next discuss the constraints associated with the problem. These include water quality regulation constraints, water supply and demand constraints, reservoir operation and sizing constraints, and river health constraints.

13.2.4.1 Constraints for Water Resources Balance in Water Reaches

Streamflow balance constraints at confluence and withdraw or intake sites are included in Eqs. (13.5)–(13.8). In particular, Eq. (13.5), related the Part C in Fig. 13.2, is a mass-balancing condition, stating that $Q_{(i-1)m}$ plus q'_{im}, streamflow rate discharged into element i in mouth m, minus q_{im}, which is defined as the streamflow rate of withdrawing from element i in month m, is equal to Q_{im}, the outflow from element i to element $i + 1$ in mouth m. Equation (13.6) is a chance constraint that limits the variation of streamflow rate, where α_{im} is a risk measure, corresponding to the maximum in-stream flow rate. The larger the value the α_{im} is, the more risk-seeking a decision maker is. The value of α_{im} is determined only by $\alpha_{i(m-1)}$ through a state transition matrix in Fig. 13.10. The initial values of α_{im} are defined as 0.5, where a sensitivity analysis based on MCMC method is used to evaluate the impact of degree of the risk aversion on optimal solution. The α_{im} is limited to be between 0 and 1 in Eq. (13.7).

$$Q_{(i-1)m} + q'_{im} - q_{im} = Q_{im} \qquad \forall i,m \tag{13.5}$$

$$q'_{im} \geq \alpha_{im} q'_{im,\max} \qquad \forall i,m \tag{13.6}$$

$$1 \geq \alpha_{im} \geq 0 \qquad \forall i,m \tag{13.7}$$

$$Q_{im}, q'_{im}, q_{im}, q'_{im,\max} \geq 0 \qquad \forall i,m \tag{13.8}$$

13.2.4.2 Constraints for Sizing and Reservoir Operation

Constraints related to capacity sizing and reservoir operation are defined in Eqs. (13.9)–(13.14). Equation (13.9) states that mass balance in water flow, where q_{wm} is the water inflow collected from reservoir watershed in month m, and L_m is the evaporation loss in month m. Equation (13.10) limits the amount of water released in month m to be less than the capacity of the reservoir. Equation (13.11) requires that the ending storage at month m is less than the storage capacity of the reservoir, where V_m denotes the active storage in month m. Equation (13.12) defines the initial level of storage in the reservoir with a predetermined reliability ($\beta = 0.5$) while Eq. (13.13) confines β to be bounded between 0 and 1. The term V_0 gives the initial storage level of the reservoir. The value of β is subject to a MCMC-based sensitivity analysis.

$$V_m = V_{m-1} + \left(q_{im} + q_{wm} - q_m - L_m\right) \qquad \forall m \tag{13.9}$$

$$q_m + L_m \leq V_m \qquad \forall m \tag{13.10}$$

$$V_m \leq V \qquad \forall m \tag{13.11}$$

$$V_0 = \beta V \tag{13.12}$$

$$0 \leq \beta \leq 1 \tag{13.13}$$

$$V, V_m, V_0, q_{wm}, L_m \geq 0 \qquad \forall m \tag{13.14}$$

13.2.4.3 Constraints for Water Quality in River Reaches

Typically, water quality in a river is affected by both point sources, for example, industrial dischargers, wastewater treatment plants, and nonpoint sources, for example, farmland's run-off. A water body is also subject to a Total Maximum Daily Loading (TMDL) regulation that limits the maximal amount of a pollutant that can be discharged into a water body per day in order to meet water quality standards. Calculation of a TMDL includes waste load allocations from point and nonpoint sources and a safety margin to account for uncertainties (Piasecki 2000). Furthermore, an assimilative capacity in the downstream areas must also be accounted for when considering reservoir planning and pollution implication in order to avoid any impact on water quality due to upstream water pumping for the purpose of reservoir storage.

We use the Streeter–Phelps model to simulate physical and chemical reactions in the river body. Two Eqs. (13.15) and (13.16) are associated with BOD and DO, respectively.

$$C_{im,B} = C_{(i-1)m,B}e^{-k_{1,im}t_{im}} = \frac{C_{(i-1)m,B}q_{(i-1)m} + \left(W'_{im} - W_{im}\right)}{q_{(i-1)m} + q'_{im}}e^{-k_{1,i}t_{im}} \quad \forall i,m \qquad (13.15)$$

$$C_{im,D} = C_{im,DS} - \frac{k_{1,i}C_{im,B}}{k_{2,i} - k_{1,i}}\left(e^{-k_{1,i}t_{im}} - e^{-k_{2,i}t_{im}}\right) + \left(C_{im,DS} - C_{im,DO}\right)e^{-k_{2,i}t_{im}} \quad \forall i,m \ (13.16)$$

Their notations are summarized as follows. The term $C_{im,B}$ is BOD concentration in mg/L in compartment i at month m. $k_{1,i}$ denotes the BOD's deoxygenation rate in compartment i. t_{im} is the residence time in month of in-stream flow to pass compartment i; $C_{im,D}$ corresponds to the DO concentration mg/L in compartment i; $C_{im,DO}$ and $C_{im,DS}$ also in mg/L are the initial DO and saturated DO for compartment i. Coefficient $k_{2,i}$ in month^{-1} is the volumetric reaeration of DO in compartment i. Finally, $w'im$ is total waste loading in mg by weight in compartment i in month m.

Equations (13.15) and (13.16) are governing equations that determine the level of BOD and DO in the water body. They are affected by the amount of waste loadings and the assimilated capacity of the river. Equation (13.15) states that all the wastewater effluents generated in the natural drainage basins are treated for the compliance with the prescribed water quality standards in each river reach regardless of variable river flows conditions. The amount of waste loading deduction (W_{im}) in Eqs. (13.15) and (13.16) is the decision variables.

13.2.4.4 Constraints for Water Quality Regulation in River Reaches

Additional two constraints are imposed in Eqs. (13.17) and (13.18). Essentially, these two constraints ensure the compliance of water quality downstream, respectively, for BOD and DO once the reservoir project is built. The parameters $\hat{C}_{i,\mathrm{B}}$ and $\hat{C}_{i,\mathrm{D}}$ represent the water quality standards of BOD and DO in compartment i.[1]

$$C_{im,B} \leq \hat{C}_{i,B} \quad \forall i, m \tag{13.17}$$

$$C_{im,D} \geq \hat{C}_{i,D} \quad \forall i, m \tag{13.18}$$

The amount of upstream pumping into reservoir may affect the river's ability to accommodate unevenly seasonal streamflow (dry v.s. wet). This would also impact the extent to which the waste discharges at different locations might worsen the water quality in the river. Incorporation of physical and chemical principles in the optimization-based decision framework allows the impact of sizing and siting reservoir decisions to be explicitly studied.

13.2.4.5 Sensitivity Analysis

MCMC is a common tool to integrate posterior distributions when the marginal probability density function is not known. It is especially useful when a large amount of data is available. In a Markov Chain process, it is assumed that the probability distribution depends only on its marginal distribution and the transition probabilities form t to $t + 1$. Thus, Eq. (13.19) is used to estimate the probability in t given the marginal probably in $t - 1$ and the transition probability matrix are known.

$$p_t(x) = \sum_{x'} p_{t-1}(x') T_{t-1}(x', x) \tag{13.19}$$

$T_{t-1}(x', x)$ defines the transition probability that relates the likelihood of moving from state x' at time $t - 1$ to state x at time t. $p_t(x)$ is the probability of state x occurring at time t. Figure 13.8 displays the calibrated state transition probability matrix using historical streamflow data of the Hou-Lung River Basin. The model, defined by Eqs. (13.1)–(13.18) and Markov Chain Monte Carlo (MCMC), is developed using Visual in MS EXCEL®.

[1] In fact, the DO constraint in (13.18) might be redundant in Taiwan. This is mainly because its topological and geographic conditions that result in fast streamflow in a river system. As a result, compliance of BOD is typically a sufficient condition for BOD's compliance.

		Jan.									
		0.1	0.2	0.3	0.4	0.5	0.6	0.7	0.8	0.9	1.0
Feb.	0.1	10	9	12	3	2					
	0.2	1		5							
	0.3	1		3	3	4	1				
	0.4	1		2	1						
	0.5	1	1	2	1						
	0.6			2							1
	0.7	1			1						1
	0.8	4	1	2		3	2				1
	0.9	2	1	1				1	1	1	1
	1.0	2	1						1		1

		May									
		0.1	0.2	0.3	0.4	0.5	0.6	0.7	0.8	0.9	1.0
Jun.	0.1	2	4	4	1						
	0.2	2	1		1	1	2		1	1	1
	0.3	1	2		1		1	1	2		1
	0.4		2	1	2	1	1	1	1		1
	0.5	1	1	1	2			2		2	1
	0.6		1	1	2	1	1	1	1	1	1
	0.7			1	3	1		2	1	1	1
	0.8		2	1	1	1	1	1	1	1	1
	0.9		1		2	1	2	1	2	1	
	1.0		3			1	1	1	2	1	1

		Sep.									
		0.1	0.2	0.3	0.4	0.5	0.6	0.7	0.8	0.9	1.0
Oct.	0.1	2					1				
	0.2	2	5	2		1			1		
	0.3	4	2	3	3	2		1		1	
	0.4		2	2	1	4	1	1		1	
	0.5		1	2	1	1	2	4	3		
	0.6		2	1		2	2	2	1	1	
	0.7		1			1		2	3	1	1
	0.8		1	1	1	1			1	1	2
	0.9	2		1					2	2	1
	1.0		2							2	2

		Feb.									
		0.1	0.2	0.3	0.4	0.5	0.6	0.7	0.8	0.9	1.0
Mar.	0.1	11	4	7	2	1		2	7		
	0.2	5		3	2	1				1	
	0.3	1		1			1		1		
	0.4	1		1					1	1	
	0.5	2	1							1	
	0.6	3					1				
	0.7	4	1			1				1	2
	0.8	3							1	2	2
	0.9	3			1	1			1	2	1
	1.0	4						1	2		

		Jun.									
		0.1	0.2	0.3	0.4	0.5	0.6	0.7	0.8	0.9	1.0
Jul.	0.1	2	1	4	3	1	2	2	1		
	0.2	1	1	1	2		1	1	1	1	2
	0.3	1	1	1		1			1		1
	0.4	1	2		1	1	1	1	1		2
	0.5	2	2	1		2	1	2	1	1	2
	0.6	1		1	1	1		1		1	
	0.7	1			2	2			1	1	1
	0.8	1	1	1		1	1	1	1	2	2
	0.9	1	2	1	1	1	1		1	1	1
	1.0	1	1				1	1	2	2	

		Oct.									
		0.1	0.2	0.3	0.4	0.5	0.6	0.7	0.8	0.9	1.0
Nov.	0.1		1	1							
	0.2	2	9	8	2	2					
	0.3		1	4	6	3	1	1	1		
	0.4			1	2	5	1	1	1	2	
	0.5	1	1	1	1	2	2	1	2	3	2
	0.6					1					1
	0.7					1					1
	0.8					1	5	1	2	3	2
	0.9			1	1		1	4			
	1.0							1	1		

		Mar.									
		0.1	0.2	0.3	0.4	0.5	0.6	0.7	0.8	0.9	1.0
Apr.	0.1	5	2	2		1					
	0.2	5	1	1	1	2	1	1			
	0.3	3	3	1	1	1	1	1	1		
	0.4	4	4	1		1		2	2	1	1
	0.5	1	1						1		1
	0.6	3	1	1				1	2	2	
	0.7	3		1				1	1	1	
	0.8	2			1			1	2	6	1
	0.9	4	1			1				1	2
	1.0	3					1	2			1

		Jul.									
		0.1	0.2	0.3	0.4	0.5	0.6	0.7	0.8	0.9	1.0
Aug.	0.1	7	2		1	1		1	1	1	2
	0.2	1		1				1	1	1	
	0.3	2	1		1	2		1	2		1
	0.4	1	1	1	1	1		1	2	1	1
	0.5	2	2	2			1	1	1		
	0.6	1	1		2	1		1	1	1	
	0.7	1	2		2	2	1	2		1	
	0.8	1	1	1	2	2	1		1	2	
	0.9		1			1		1	1		4
	1.0		1		1	3	1	1		1	

		Nov.									
		0.1	0.2	0.3	0.4	0.5	0.6	0.7	0.8	0.9	1.0
Dec.	0.1	2	4	3							
	0.2		15	13	1	5	1	1			
	0.3		1		5	3			3		
	0.4		1		3	3			4		
	0.5		1		1	1					
	0.6		1	1		1					
	0.7									2	
	0.8				1	1			2	1	1
	0.9			1	1	2	1	1	3	2	
	1.0								2	2	

		Apr.									
		0.1	0.2	0.3	0.4	0.5	0.6	0.7	0.8	0.9	1.0
May	0.1	3	3								
	0.2	2	4	3	2		1	2	1	1	
	0.3	1	1	1	1	1	2	1	1		1
	0.4	1	2	2	2	1	3			2	2
	0.5		1	1	2		1		1	1	
	0.6			2	2		1	1		1	
	0.7	1		1	1	1		1	2	2	1
	0.8	1		1	2		1	1	2	1	1
	0.9	1			2			1	2		1
	1.0	1			1			1	3	1	

		Aug.									
		0.1	0.2	0.3	0.4	0.5	0.6	0.7	0.8	0.9	1.0
Sep.	0.1	2		2		2	1		1		
	0.2	6	1	1		1	1	2	2	1	
	0.3	3	1	1	2			1	1		1
	0.4			3	1	1				1	1
	0.5			3	2	3		2	1		1
	0.6			2	1	1	1	1		1	
	0.7	1	1	1	1	1	2			3	
	0.8	3	1	1			1	2		1	
	0.9				1	1	3	1	1	1	1
	1.0							2	1		1

		Dec.									
		0.1	0.2	0.3	0.4	0.5	0.6	0.7	0.8	0.9	1.0
Jan.	0.1		11	3	2		1	2	2	4	1
	0.2		8	2			1		2		
	0.3	7	10	3	7	1			2		
	0.4	1	5	2	1		1		1		1
	0.5	1	3	1						3	
	0.6		1							2	
	0.7				1					1	
	0.8			1						1	
	0.9										
	1.0		1		1						2

unit = %

Fig. 13.8 The transition probability matrixes of stream flow for Ta Lu Keng

13.3 Results

We summarize the results in this section. Our results suggest that (1) the maximum flow rate is equal to 10.2 (cms) at the pumping station associated with the diversion pipeline for water transfer from HL3 to the Tien-Hua Lake is, and (2) the optimal active storage capacity required for the Tien-Hua Lake is approximately equal to 48,013,040 m^3. This construction plan also is capable of supplying a total of 90 million tons of water per year for municipal, industrial, and agricultural consumption in this basin.

The results from the water quality simulation model show that most of river reaches in dry reason and some of river reaches that are close to estuary region in wet season might fail to comply with the regulation. As illustrated in Fig. 13.9, in absence of the proposed project, during both dry and wet seasons, the first 4-km segment of the river reach, the DO is below the standard while the rest of the river is in compliance with the standard. Because most rivers in Taiwan are characterized with fast descending topology, this suggests that DO is less likely to be a concern. However, Fig. 13.6 indicates that BOD, except in wet season, during which the middle segment of the river is in compliance with the standard, other segments are in violation of the BOD standard. It suggests the need to consider reducing waste loadings along these river reaches. Apparently, during dry season with an extremely low streamflow, additional managerial effort, for example, waste loading reductions, is needed to ensure compliance when the proposed reservoir withdraws water to replenish the Tien-Hua Lake.

Table 13.4 reports the monthly optimal waste removal rate as a percentage of waste loading at sub-drainage basin as well as pumping schedule in cms or cubic meter per second. Overall, the pumping is concentrated on wet season during March to September. We also observe that removal rate is higher during dry season when the river is with a less assimilative capacity to maintain water quality. For example, the removal rates in downstream areas are about 90% during dry season, they are 10% higher the wet season in average. Moreover, the removal rate tends to be larger for the river segments closer to the river mouth, reflecting that lower segments of the river are more polluting. The table also summarizes the pollution control cost. The results indicate a dip of the waste remove cost during wet (summertime) when river's natural assimilative capacity is at most. Additionally, as pumping to the Tien-Hua Lake is concentrated during the wet season of March to September, the pollution control cost is decreased during those wet months compared to the dry season when no pumping is recommended.

Figure 13.9a, b, respectively, report the results based on MCMC approach with 100 runs of simulations considering hydrological uncertainty during dry and wet seasons. We plot the simulated BOD and DO concentration in mg/L against the distance from the estuary along the river. Overall, the upstream segments of the river experience relatively less variation while the variation of BOD and DO in the downstream, that is, 20 km from the estuary, is more significant, especially during dry season. Compared to BOD, DO's variation is also relatively large.

We also report the variation of the optimized waste load reduction based on 100 simulations using MCMC approach in Fig. 13.10. In general, sites W_4–W_8 experience a larger variation compared to other sites. During the dry season, under various hydrological conditions, the waste removal rate of organic compounds needs to be between 60–95% in order to maintain water quality (Table 13.5). Thus, the reduction of organic pollutant becomes the key issue to ensue river health. These findings can be used to provide guidance to the government to develop an effective management plan of the Tien-Hua Lake.

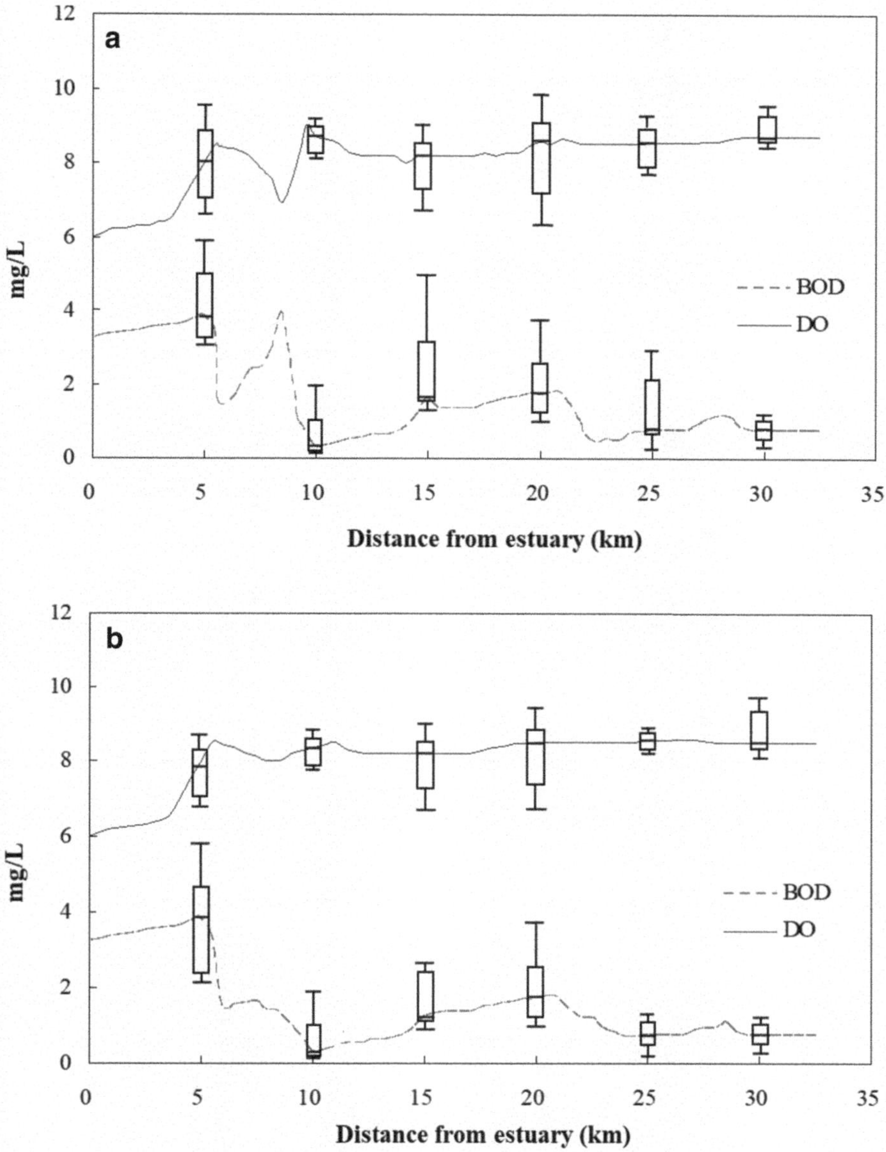

Fig. 13.9 (**a**) Uncertainty bars of BOD and DO in the Hou-Long River during dry season after the implementation of optimal management strategy. (**b**) Uncertainty bars of BOD and DO in the Hou-Long River during wet season after the implementation of optimal management strategy

Table 13.4 The optimal pollution removal strategy downstream and pumping strategy at water intake

	W_1		W_2		W_3		W_4		W_5		W_6		W_7		W_8		Pumping rate
	Weight[a]	%[b]	Weight[a]	%[b]	Weight[a]	%[b]	Weight[a]	%[b]	Weight[a]	%[b]	Weight[a]	%[b]	Weight[a]	%[b]	Weight[a]	%[b]	(cms)
Jan.	0.0	0%	0.0	0%	0.0	0%	0	0%	672.8	78%	493.3	90%	1539.3	64%	6537.8	88%	0.0
Feb.	0.0	0%	0.0	0%	0.0	0%	0	0%	690.1	80%	493.3	90%	1491.2	62%	6463.5	87%	0.0
March	0.0	0%	0.0	0%	0.0	0%	90.6	18%	793.6	92%	449.4	82%	986.1	41%	6463.5	87%	2.3
April	0.0	0%	0.0	0%	0.0	0%	135.9	27%	759.1	88%	394.6	72%	673.4	28%	6686.4	90%	3.4
May	0.0	0%	0.0	0%	0.0	0%	176.1	35%	776.3	90%	372.7	68%	529.1	22%	5794.8	78%	4.6
June	0.0	0%	0.0	0%	0.0	0%	211.4	42%	767.7	89%	334.3	61%	481.0	20%	6612.1	89%	10.2
July	0.0	0%	0.0	0%	0.0	0%	166.1	33%	733.2	85%	438.5	80%	1370.9	57%	6092.0	82%	4.2
Aug.	0.0	0%	0.0	0%	0.0	0%	161.0	32%	724.6	84%	301.4	55%	336.7	14%	6686.4	90%	6.5
Sep.	0.0	0%	0.0	0%	0.0	0%	110.7	22%	707.3	82%	389.1	71%	914.0	38%	6537.8	88%	5.8
Oct.	0.0	0%	0.0	0%	0.0	0%	100.6	20%	707.3	82%	383.7	70%	1154.5	48%	6612.1	89%	0.0
Nov.	0.0	0%	0.0	0%	0.0	0%	0	0%	681.4	79%	487.8	89%	1539.3	64%	6017.7	81%	0.0
Dec.	0.0	0%	0.0	0%	0.0	0%	0	0%	664.2	77%	482.3	88%	1587.4	66%	6834.9	92%	0.0

[a]Reduced amount of pollution loading (kg-BOD/month)
[b]Reduced percentage of pollution loading

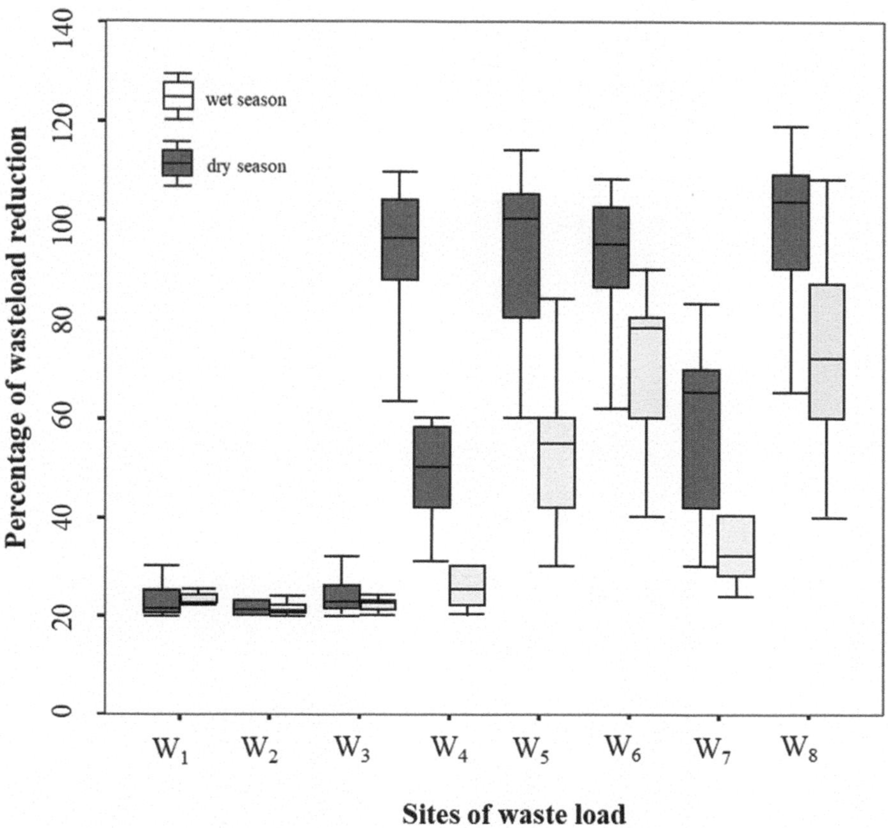

Fig. 13.10 The results of sensitive analysis for waste load reduction rate

13.4 Conclusions

Intelligent use, extraction, and reuse of water resources and the design of long-term management strategies require not only knowledge of the underlying hydrological systems and society's water demand but also appropriate quantitative-based approaches in order to study the trade-offs among different alternatives and objectives. This chapter illustrates using an optimization-based model embedded with hydrological systems to decide optimal sizing of a new reservoir. The analysis considers various aspects of water usages, including water quality, for example, BOD and DO, ecosystem service with minimal flows, and water quantity to meet the needs of municipal, industrial, and agricultural demand.

The framework is illustrated with a case study of optimal sizing of a new reservoir in the Hou-Lung River Basin in Taiwan. The framework successfully identifies alternatives for water resource management and to optimize the storage capacities

Table 13.5 The optimal pollution removal strategy downstream and pumping strategy at water intake sites

	W1 (kg-BOD/ day)		W2 (kg-BOD/ day)		W3 (kg-BOD/ day)		W4 (kg-BOD/ day)		W5 (kg-BOD/ day)		W6 (kg-BOD/ day)		W7 (kg-BOD/ day)		W8 (kg-BOD/ day)		Pumping rate (cms)	Pollution control cost (million NT$)
Jan.	0.0	0%	0.0	0%	0.0	0%	0.0	0%	698.7	81%	515.2	94%	1587.4	66%	7206.4	97%	0.91	152
Feb.	0.0	0%	0.0	0%	0.0	0%	0.0	0%	717.6	82%	515.2	94%	1587.4	66%	7206.4	97%	1.77	153
March	0.0	0%	0.0	0%	0.0	0%	140.9	28%	801.0	93%	463.8	85%	1135.2	47%	6952.7	94%	7.88	144
April	0.0	0%	0.0	0%	0.0	0%	359.8	37%	792.7	92%	415.5	76%	710.0	30%	6714.3	90%	17.94	138
May	0.0	0%	0.0	0%	0.0	0%	388.1	40%	807.6	94%	378.4	69%	683.2	28%	6908.6	93%	17.80	139
June	0.0	0%	0.0	0%	0.0	0%	501.5	43%	806.3	93%	350.6	64%	482.3	20%	6840.1	92%	29.30	138
July	0.0	0%	0.0	0%	0.0	0%	67.5	22%	811.8	94%	466.4	85%	1319.4	55%	7125.5	96%	17.70	149
Aug.	0.0	0%	0.0	0%	0.0	0%	554.9	42%	805.5	93%	333.0	61%	355.1	15%	6796.8	91%	25.73	135
Sep.	0.0	0%	0.0	0%	0.0	0%	275.9	38%	809.1	94%	409.9	75%	910.9	38%	6986.2	94%	12.98	144
Oct.	0.0	0%	0.0	0%	0.0	0%	135.0	27%	810.6	94%	441.9	81%	1141.9	48%	7065.0	95%	5.3	147
Nov.	0.0	0%	0.0	0%	0.0	0%	0.0	0%	702.0	81%	515.2	94%	1587.4	66%	7206.4	97%	1.36	154
Dec.	0.0	0%	0.0	0%	0.0	0%	0.0	0%	700.3	81%	515.2	94%	1587.4	66%	7206.4	97%	0.8	154

at the candidate site. The analysis applied an integrated simulation and optimization approach to generate the monthly operational pattern in terms of the pumping rate at the candidate site in the river system. The model also suggests the extent of waste reduction that is needed in order to maintain a healthy water system. The results provide useful information for decision-makers to understand the benefit and cost of the proposed projects.

The strength of the approach that allows for incorporating uncertainties associated with hydrological conditions in the optimized based model by coupling it with MCMC approach provides valuable information to water utilities to manage their water resources under changing climate. Our analysis implies the importance of considering uncertainties, for example, intra- and inter-year variation of rainfall, in a water system.

References

Alrayess, H., Zeybekoglu, U., & Ulke, A. (2017). Different design techniques in determining reservoir capacity. *European Water, 60*, 107–115.

Avogadro, E., Minciardi, R., & Paolucci, M. (1997). A decisional procedure for water resources planning taking into account water quality constraints. *European Journal of Operational Research, 102*(2), 320–334.

Barandouzi, M., Mahinthakumar, G., Ranjithan, R., & Brill, E. (2012). *Probabilistic mapping of water leakage characterizations using a Bayesian approach.* Paper presented at the World Environmental and Water Resources Congress 2012: Crossing Boundaries.

Barlishen, K. D., Simonovic, S. P., & Burn, D. H. (1989). A simulation–optimization algorithm for reservoir capacity calculation: the effect of inflow data set length. *Canadian Journal of Civil Engineering, 16*(4), 477–482.

Benoist, A., Brinkman, A., Van Diepenbeek, P., & Waalst, J. (1998). BEKWAAM, a model fit for reservoir design and management. *Water Science and Technology, 37*(2), 269–276.

Buras, N. (1985). An application of mathematical programming in planning surface water storage 1. *JAWRA Journal of the American Water Resources Association, 21*(6), 1013–1020.

Chen, H.-W., & Chang, N.-B. (2002). A comparative analysis of methods to represent uncertainty in estimating the cost of constructing wastewater treatment plants. *Journal of Environmental Management, 65*(4), 383–409.

Chen, H.-y., Teng, Y.-g., Wang, J.-s., Song, L.-t., & Zhou, Z.-y. (2012). Event source identification of water pollution based on Bayesian-MCMC. *Journal of Hunan University (Natural Sciences), 6*(015).

Curry, G. L., Helm, J. C., & Clark, R. A. (1973). Chance-constrained model of system of reservoirs. *Journal of the Hydraulics Division, 99*(12), 2353–2366.

Dupačová, J., Gaivoronski, A., Kos, Z., & Szantai, T. (1991). Stochastic programming in water management: A case study and a comparison of solution techniques. *European Journal of Operational Research, 52*(1), 28–44.

Harris, J., & Silveira, R. (1999). Large-scale assessments of river health using an Index of Biotic Integrity with low-diversity fish communities. *Freshwater Biology, 41*(2), 235–252.

Helm, J. C., Curry, G. L., & Hogg, G. L. (1984). Capacity expansion of linked reservoir systems. *Computers and Industrial Engineering, 8*(2), 119–128.

Huock, M. H., & Cohon, J. L. (1978). Sequential explicitly stochastic linear programing models: A proposed method for design and management of multipurpose reservoir systems. *Water Resources Research, 14*(2), 161–169.

Karr, J. R. (1981). Assessment of biotic integrity using fish communities. *Fisheries, 6*(6), 21–27.

Karr, J. R. (1987). Biological monitoring and environmental assessment: A conceptual framework. *Environmental Management, 11*(2), 249–256.

Karr, J. R. (1991). Biological integrity: A long-neglected aspect of water resource management. *Ecological Applications, 1*(1), 66–84.

Khaliquzzaman, & Chander, S. (1997). Network flow programming model for multireservoir sizing. *Journal of Water Resources Planning and Management, 123*(1), 15–25.

Kottegoda, N. T. (1980). *Stochastic water resources technology*. Berlin: Springer.

Li, W., Wang, B., Xie, Y., Huang, G., & Liu, L. (2015). An inexact mixed risk-aversion two-stage stochastic programming model for water resources management under uncertainty. *Environmental Science and Pollution Research, 22*(4), 2964–2975.

Loucks, D. P., Stedinger, J. R., & Haith, D. A. (1981). *Water resource systems planning and analysis*. Englewood Cliffs: Prentice-Hall.

Loucks, D. P., & Van Beek, E. (2017). *Water resource systems planning and management: An introduction to methods, models, and applications*. Cham: Springer.

Luo, B., Maqsood, I., & Huang, G. (2007). Planning water resources systems with interval stochastic dynamic programming. *Water Resources Management, 21*(6), 997–1014.

McMahon, T. A., & Mein, R. G. (1986). *River and reservoir yield*. Littleton, CO: Water Resources Publications.

Mesa, O., Poveda, G., Carvajal, L., & Salazar, J. (1996). *Reservoir reliability design under interannual climatic and hydrologic variability*. Paper presented at the Managing Water: Coping with Scarcity and Abundance.

Miltz, D., & White, D. C. (1987). Sedimentation and the economics of selecting an optimum reservoir size. *Water Resources Research, 23*(8), 1443–1448.

Mousavi, H., & Ramamurthy, A. (2000). Optimal design of multi-reservoir systems for water supply. *Advances in Water Resources, 23*(6), 613–624.

Mutreja, K., & Yevjevich, V. (1978). Sizing of reservoirs for periodic-stochastic input and periodic output. *Journal of Hydrology, 39*(3–4), 323–338.

Nagy, I. V., Asante-Duah, K., & Zsuffa, I. (2013). *Hydrological dimensioning and operation of reservoirs: Practical design concepts and principles* (Vol. 39). Dordrecht: Springer Science and Business Media.

Oamek, G., Schluntz, L., Bottorff, L., & Johns, E. (1990). *A methodology for analyzing alternative reservoir sizing, shortage, and operating criteria*. Paper presented at the Optimizing the Resources for Water Management.

Piasecki, M. (2000). TMDL: Development of an optimized WLA strategy. In *Building partnerships* (pp. 1–10).

Plate, E. J. (1989). Reliability concepts in reservoir DesignPaper presented at ICWRS-workshop on risk and uncertainty in hydrologic design Oslo, Norway, February-1989. *Hydrology Research, 20*(4–5), 231–248.

Revelle, C., Joeres, E., & Kirby, W. (1969). The linear decision rule in reservoir management and design: 1, Development of the stochastic model. *Water Resources Research, 5*(4), 767–777.

Savic, D. A., Burn, D. H., & Zrinji, Z. (1989). A comparison of streamflow generation models for reservoir capacity-yield analysis 1. *JAWRA Journal of the American Water Resources Association, 25*(5), 977–983.

Singh, K. P., & Durgunoğlu, A. (1990). Economic reservoir design and storage conservation by reduced sedimentation. *Journal of Water Resources Planning and Management, 116*(1), 85–98.

Takeuchi, K. (1997). Least marginal environmental impact rule for reservoir development. *Hydrological Sciences Journal, 42*(4), 583–597.

Takeuchi, K., & Sivaarthitkul, V. (1995). Assessment of effectiveness of the use of inflow forecasts to reservoir management. *IAHS Publications-Series of Proceedings and Reports-Intern Assoc Hydrological Sciences, 231*, 299–310.

Turgeon, A., & Brosseau, G. (1987). On the nonconvexity of the reservoir design problem. *Water Resources Research, 23*(8), 1689–1691.

Zeng, X., Zhang, S., Feng, J., Huang, G., Li, Y., Zhang, P., et al. (2017). A multi-reservoir based water-hydroenergy management model for identifying the risk horizon of regional resources-energy policy under uncertainties. *Energy Conversion and Management, 143*, 66–84.

Zheng, Y., & Han, F. (2016). Markov Chain Monte Carlo (MCMC) uncertainty analysis for watershed water quality modeling and management. *Stochastic Environmental Research and Risk Assessment, 30*(1), 293–308.

Chapter 14
High-Resolution Multiobjective Optimization of Sustainable Supply Chains for a Large-Scale Lignocellulosic Biofuel Industry

T. Edward Yu, Burton C. English, Jia Zhong, James A. Larson, Joshua S. Fu, Lixia He-Lambert, and Brad Wilson

Abstract This chapter introduces an integrative framework coupling GIS tools, biophysical model, enterprise budgeting tools, GHG emission models, and optimization mechanism in determining a sustainable switchgrass biofuel supply chain network. The integrative framework is applied to a case study of replacing 30% of gasoline used in transportation in Tennessee. Using high-resolution spatial data in a multiobjective mixed-integer programming model, we find that land use choice makes substantial impacts on the deployment of the supply chains under different objectives. When considering private cost alone, hay and pasture land concentrated in the east and central Tennessee will be the major source for switchgrass production. If targeting GHG emission minimization solely, more than 500 thousand hectares of the state's cropland is converted to switchgrass for biofuel production. Moreover, the trade-off between cost and GHG emissions in the supply chains shows that the marginal rate of substitution between total cost and GHG emissions on the frontier curve increases at an accelerating rate. Our findings illustrate the importance of land resource management on the sustainability of a dedicated energy crop supply chain.

T. E. Yu (✉) · B. C. English · J. A. Larson · B. Wilson
Department of Agricultural and Resource Economics, University of Tennessee, Knoxville, TN, USA
e-mail: tyu1@utk.edu

J. Zhong
Department of Agricultural and Consumer Economics, University of Illinois , Urbana-Champaign, IL, USA

J. S. Fu
Department of Industrial and Environmental Engineering, University of Tennessee, Knoxville, TN, USA

L. He-Lambert
Department of Agricultural Economics, Oklahoma State University, Stillwater, OK, USA

C. Chen et al. (eds.), *Pursuing Sustainability*, International Series in Operations Research & Management Science 301,
https://doi.org/10.1007/978-3-030-58023-0_14

14.1 Introduction

The desire for reducing the usage of fossil fuels and resultant greenhouse gas (GHG) emissions has driven the development of a biofuel sector in the USA. The Renewable Fuel Standard 2 (RFS 2) in the Energy Independence and Security Act of 2007 mandated 136 billion liters of biofuels to be produced annually in the USA by 2022, with at least 79 billion liters of advanced biofuels (US Congress 2007). Advanced biofuels generated from the lignocellulosic biomass (LCB), such as agricultural and forest residues, short-rotation woody crops, and herbaceous grasses, has been advocated by the US Environmental Protection Agency (EPA) as a strategy to reduce GHG emissions (US EPA 2014). Switchgrass, a native perennial species to North America, is suggested as a promising LCB feedstock for biofuel production, particularly in the Southeastern USA given the high biomass yield and relative low rent-to-value agricultural land within the region (Wright and Turhollow 2010; English et al. 2006). Biofuels produced from switchgrass could reduce GHG emissions compared with gasoline, also alleviate the issues of soil erosion, excess irrigation, and direct linkage to food supply associated with the first-generation feedstock, for example, corn (Kort et al. 1998; Monti et al. 2009, 2012).

Although LCB feedstock such as switchgrass has potential advantages over the first-generation feedstock for biofuel production, the supply chains of switchgrass encounter more technical challenges than corn. Unlike corn, currently, there is no established market for switchgrass so large scale production of switchgrass for biofuel production is not available. Also, the collection, storage, and transportation technologies for switchgrass are not as mature as that for field crops due to its low bulk density. These issues associated with switchgrass could result in higher cost of biofuel production and influence the economic sustainability of the biofuel supply chain network. In addition, fluctuations in GHG emissions due to cropping system change and the production and transportation of bulky switchgrass could affect the environmental sustainability of the biofuel supply chain (Yu et al. 2014). Thus, determining the area of switchgrass production, the location of biorefineries, and the routes connecting farms, biorefineries, and end-users to meet transportation fuel demand regarding the criteria of economic and environmental sustainability is crucial to the development of the biofuel industry.

A large number of studies have focused on analyzing and optimizing LCB biofuel supply chains in the literature driven by the rising interest (Ghaderi et al. 2016; Ba et al. 2016). The majority of related research has applied mathematical programming tools to determine the optimal supply chain network and facility locations that meet a given objective. Growing attention has also given to incorporating multiple criteria, such as private cost, environmental outcome, or ecological welfare, in the biofuel supply chains and the potential trade-off between these objectives in optimization (Yu et al. 2014; Cambero and Sowlati 2016; Zhong et al. 2016; Petridis et al. 2018; Field et al. 2018). One advantage of considering both economic and environmental objectives in the supply chains is to estimate the imputed cost of improving environmental quality associated with LCB-based biofuels (Zhong et al. 2016). The imputed cost could serve as an indicator of the ecosystem value obtained from the LCB supply chains under varying objectives, therefore providing relevant information to decision makers while designing a sustainable LCB biofuel industry.

To adequately assess multiple sustainable criteria of LCB biofuel supply chains, high-resolution spatial data generated from geospatial technologies, such as geographic information system (GIS) or remote-sensing techniques, serve an important role. High-resolution spatial data provide essential information on resource endowment and attributes (e.g., land availability, soil quality) and physical features (e.g., transport network, power lines) for LCB supply chain optimization. Also, using the spatial data in biophysical models can measure the potential effect on the ecosystem, for instance, soil carbon stock, from changes in the cropping system and land management, which is highly relevant to LCB feedstock development. Combining GIS tools and multiobjective optimization mechanism outperforms the traditional spatial multiobjective model as the latter approach overlooks the details in natural and physical resource location and attributes, hence potentially misleading decision makers with partial information of the resources and ultimately distressing the sustainability of the biofuel supply chains. Thus, the integrative framework provides a logical framework in designing a spatially explicit sustainable LCB biofuel supply chain network (Nguyen et al. 2019).

The enhanced precision regarding resource management in the supply chains from the integrative framework does not come without cost. Coupling high-resolution spatial data from GIS tools with a multiobjective optimization model implies a substantial demand for spatially detailed data and computation power in the analysis (Ekşioğlu et al. 2009; He-Lambert et al. 2018). Thus, most existing studies utilizing the integration approach with high-resolution spatial data typically focus on a particular site or area (Field et al. 2018; Nguyen et al. 2019), which may not offer a more comprehensive view of a large-scale biofuel supply chain network.

This present chapter provides an overview of combining GIS-based information with a multiobjective optimization framework in determining land resource utilization for feedstock and biorefineries in a large-scale sustainable biofuel supply chain network. This integrative framework, including high-resolution spatial data, biophysical model, GHG emission model, optimization mechanism, and other tools is applied to a case study of replacing 30% of transportation fuel use with biofuels in Tennessee. The feedstock considered in the study is switchgrass, and cellulosic ethanol is the biofuel product. The location of feedstock production area and biorefineries with varying sizes, and the imputed cost of reducing GHG emissions will be determined through the multiobjective optimization process.

The remaining chapter is organized into three sections. Section 14.2 describes the supply chain boundary and assumptions, the integrative analytical framework, and the multiobjective model. Section 14.3 presents the outputs under different objectives and the derived trade-off relationship between economic and environmental criteria. Finally, Sect. 14.4 summarizes the key features of this study and discusses probable extensions for future research.

14.2 Method

14.2.1 Biofuel Supply Chain Configurations and Assumptions

The system boundary for cost and GHG emissions assessments in the switchgrass biofuel supply chains originates from the farm field through biorefineries to the gate of blending facilities (Fig. 14.1). Once the perennial switchgrass is established in the field, farmers manage the maintenance, harvest, and storage of feedstock annually. Harvested switchgrass in the square-bale form is delivered to the contracted biorefineries through semi-trucks during harvest season (November – February), while the remaining square-baled switchgrass is put to storage and delivered during off harvest. Biorefineries grind the baled switchgrass at the site and convert the feedstock to biofuels that are finally hauled to blending facilities. The system boundary ends at the gate of blending sites because isolating the cost and GHG emissions for the blending operations between fossil fuels and biofuels is not practical. Switchgrass is established at year 0 and year 11, and biorefineries investment is one-time expenses, whereas other activities are annual operations.

The location and production area of feedstock, also the location and capacity of biorefineries, is simultaneously determined through optimization. The top three metropolitans in Tennessee, Memphis, Nashville, and Knoxville, is assigned as the location of blending facilities. The state is decomposed into a 13 km^2 hexagon resolution to capture the geographical variations within the study area: there are a total of 8753 hexagons in Tennessee. Biorefineries are assumed to be located in industrial parks within the state and a total of 298 potential industrial parks with enough capacity are identified in the study area. Combining all those nodes together for large scale supply chain network (8753 potential origin nodes, 298 intermediate

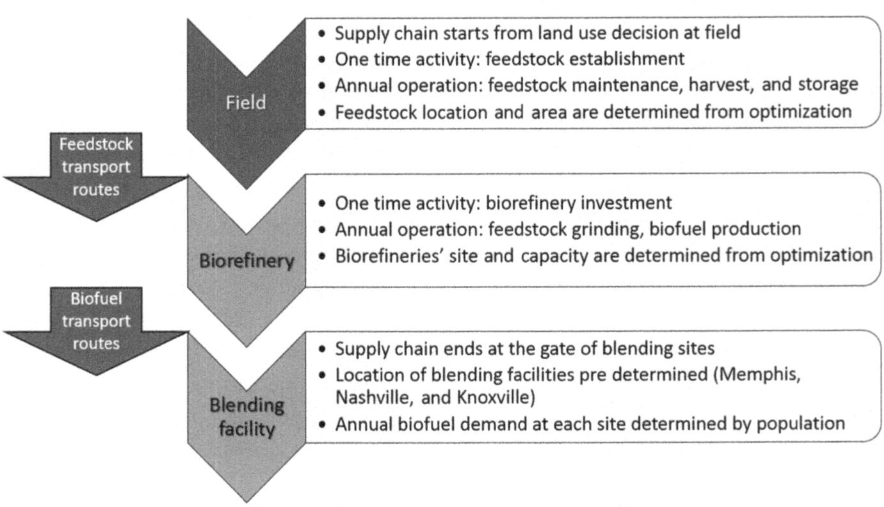

Fig. 14.1 Switchgrass biofuel supply chains boundary in the study

nodes, and 3 destination nodes) can generate nearly four million arcs, which is not feasible to be solved. Thus, the distance between farms and biorefineries is constrained up to 80.5 km, while biorefineries cannot be located beyond 242 km from blending facilities.

Assuming 30% of annual gasoline consumption in transportation (12.64 billion liters equivalents in 2017) is replaced by biofuel, the annual demand for biofuel is then 3.78 billion liters. Demand at each blending site is then determined based on the share of the population among the three metropolitans (Memphis—32.9%, Nashville—39.0%, Knoxville—28.1%). The assigned capacity of biorefineries includes 189 million liters per year (MLY) or 378 MLY, and biofuel is assumed to be evenly produced across 12 months with a conversation rate of 288 liters per Mg from switchgrass.

14.2.2 Integrative Modeling Framework

Figure 14.2 presents the integrative framework incorporating the GIS tool, biophysical model, enterprise budget tool, GHG emission simulation model, and optimization model to determine the multiobjective large-scale biofuel supply chains in this study. The high-resolution data generated from the GIS tool are used as inputs in a biophysical model, DayCent (Parton et al. 2001), to simulate the effects of cropping system change (i.e., from original crops to switchgrass) and related land management practices on soil carbon cycling and stock (arrow A). The simulated ecosystem outputs are then employed as inputs of GHG emissions associated with land-use changes in the multiobjective optimization model (arrow B). Similarly, the spatial data of resource availability and physical characteristics are also used as inputs in the optimization model for location decisions (arrow C). Moreover, the budget for capital investment and annual operations at the farms and biorefineries generated from an enterprise budget model (Larson et al. 2010) serves as an additional input

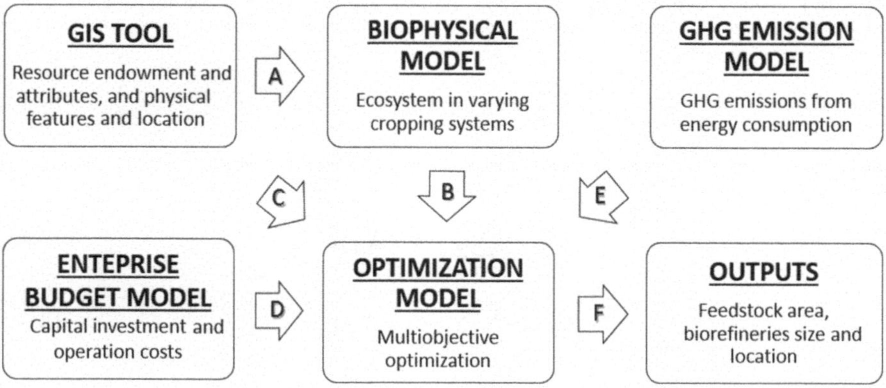

Fig. 14.2 Modeling framework of the large-scale switchgrass biofuel supply chains

for economic decisions in the optimization model (arrow D). Finally, two GHG emissions models, the GREET (Argonne National Laboratory 2013) and MOVES (US EPA 2013) simulate the GHG from energy use for operations at the farms, in the biorefineries, and on the roads, and export the estimation to the optimization models (arrow E). Outputs from the GIS tool, biophysical model, enterprise budget tool, and GHG emission simulation model are all generated at the spatial resolution (13 km^2). Given the outputs from those different types of models and tools, the multiobjective model is then solved to determine the optimal location and size of biorefineries, and the associated location and production area of feedstock in corresponding to the multiple objectives (arrow F).

14.2.3 Model Structure

Two objectives are considered in the development of large-scale switchgrass biofuel supply chains in this study: total private cost minimization and total GHG emission minimization. Optimizing the total private cost (or the total GHG emissions) over the whole supply chain network implies that multiple players in the system (i.e., farmers and biorefineries) reach the decisions with the common interest to achieve the overall goal. This assumption is practically relevant to the case study as the State agencies eager to establish a biomass-based energy sector and actively coordinate the collaboration between farmers and the biofuel industry (Tiller 2011).

The total private cost (TC) of switchgrass biofuel supply chains is composed of ten components, including switchgrass establishment (C_{est}^{swi}), opportunity cost of land use for switchgrass (C_{opp}^{swi}), switchgrass maintenance and management (C_{mnt}^{swi}), switchgrass harvesting (C_{hrv}^{swi}), switchgrass storage (C_{stg}^{swi}), switchgrass transportation (C_{tran}^{swi}), switchgrass grinding (C_{grd}^{swi}), biorefinery investment (C_{inv}^{fac}), biofuel conversion (C_{conv}^{bio}), and biofuel transportation (C_{tran}^{bio}). Equation (14.1) presents the objective of minimizing private cost solely, while Eqs. (14.2)–(14.11) further define each cost component in the objective function. The definitions of indices, parameters, and variables in the equations from hereon are provided in Table 14.1.

$$\min.\mathrm{TC} = C_{est}^{swi} + C_{opp}^{swi} + C_{mnt}^{swi} + C_{hrv}^{swi} + C_{stg}^{swi} + C_{tran}^{swi} + C_{grd}^{swi} + C_{inv}^{fac} + C_{conv}^{bio} + C_{tran}^{bio} \tag{14.1}$$

where

$$C_{est}^{swi} = \sum_{i \in I} \sum_{h \in H} \left(\alpha \times X_{ih} \right), \tag{14.2}$$

$$C_{opp}^{swi} = \sum_{i \in I} \sum_{h \in H} \left(\beta_{ih} \times X_{ih} \right), \tag{14.3}$$

Table 14.1 Definitions of indices, parameters, and variables

Category	Unit	Definition
Indices		
$i \in I$		Location of switchgrass production field
$j \in J$		Location of biorefinery facility
$b \in B$		Location of blending facility
$g \in G$		Capacity of biorefinery facility
$m \in M$		Season of the year
$M_{on} \in M$		Harvest season of the year
$M_{off} \in M$		Off-harvest season of the year
$h \in H$		Crop type (pasture, corn, soybean, wheat, cotton)
s		Switchgrass
Parameters		
y	Mg/ha	Commodity yield
α	\$/ha	Amortized establishment cost of switchgrass
β	\$/ha	Opportunity cost of cultivating switchgrass
ξ	\$/ha	Switchgrass maintenance and management cost
μ	\$/plant	Amortized investment cost of conversion facility
ω	\$/Mg	Switchgrass harvest cost
γ	\$/Mg	Switchgrass storage cost
θ	\$/Mg	Switchgrass transportation cost
ψ	\$/Mg	Switchgrass grinding cost
ρ	\$/L	Biorefinery operation cost
δ	\$/L	Biofuel transportation cost
A	Ha	Available cropland
λ_1	%	Dry matter loss during storage
λ_2	%	Dry matter loss during transportation
σ	L/Mg	Switchgrass-biofuel conversion rate
D	L/season	Seasonal demand for biofuel
ϵ_1	CO_2e Mg/ha	GHG emission factor of energy use for maintenance
ϵ_2	CO_2e Mg/Mg	GHG emission factor of energy use for harvesting
ϵ_3	CO_2e Mg/Mg	GHG emission factor of energy use for storage
ϵ_4	CO_2e Mg/route	GHG emission factor of fuel use for hauling switchgrass
ϵ_5	CO_2e Mg/Mg	GHG emission factor of energy use for grinding
ϵ_6	CO_2e Mg/L	GHG emission factor of energy use for biofuel conversion
ϵ_7	CO_2e Mg/route	GHG emission factor of fuel use for hauling biofuel
ε	CO_2e Mg/L	GHG emission factor for biorefinery construction
Variables		
Z		Binary variable: 1 for biorefinery selection, 0 otherwise
X	Ha	Switchgrass area harvested during harvest season
XNS	Mg	Switchgrass used during harvest season
XS	Mg	Switchgrass stored at the harvest site after harvest

(continued)

Table 14.1 (continued)

Category	Unit	Definition
XQ	Mg	Switchgrass delivered to the biorefinery
XO	L	Biofuel transported from biorefinery to blending facility

$$C_{mnt}^{swi} = \sum_{i \in I} \sum_{h \in H} \left(\xi_{ih} \times X_{ih} \right), \tag{14.4}$$

$$C_{hrv}^{swi} = \sum_{i \in I} \sum_{h \in H} y_{is} \times X_{ih} \times \omega, \tag{14.5}$$

$$C_{stg}^{swi} = \sum_{i \in I} XS_i \times \gamma, \tag{14.6}$$

$$C_{tran}^{swi} = \sum_{m \in M} \sum_{i \in I} XQ_{mi} \times \theta, \tag{14.7}$$

$$C_{grid}^{swi} = \sum_{m \in M} \sum_{i \in I} XQ_{mi} \times \psi, \tag{14.8}$$

$$C_{inv}^{fac} = \sum_{j \in J} \sum_{g \in G} \left(\mu_g \times Z_{jg} \right), \tag{14.9}$$

$$C_{conv}^{bio} = \sum_{m \in M} \sum_{j \in J} \sum_{b \in B} XO_{mjb} \times \rho, \tag{14.10}$$

$$C_{tran}^{bio} = \sum_{m \in M} \sum_{j \in J} \sum_{b \in B} XO_{mjb} \times \delta, \tag{14.11}$$

In Eqs. (14.2)–(14.4), X_{ih} is a decision variable that represents the area of a given type of agricultural land, *h*, at a hexagon *i* converted for switchgrass production. Switchgrass establishment cost (α) in Eq. (14.2) is amortized with an interest rate of 3%. The opportunity cost of land use (β) is calculated as the profit from existing agricultural activities on a hexagon or its land rent before land conversion to switchgrass production. Decision variable *XS* is the amount of harvested switchgrass that is placed to storage for the off-season use, while *XQ* refers to the amount of switchgrass delivered to the biorefineries every season. The unit cost for switchgrass harvesting (ω), storage (γ), transportation (θ), and grinding (ψ) includes equipment, labor, fuel, and materials. The investment cost for biorefineries (μ) is also amortized with a 3% interest rate and incorporates the scale economies for the larger biorefinery (378 MLY) with a scaling factor of 0.7 from the investment cost of the 189-MLY biorefinery. The binary variable *Z* equals one when the particular biorefinery is selected, and zero if not. Similar to the operation cost for switchgrass, biofuel conversion cost (ρ) and transportation cost (δ) also consider the expenses of equipment, labor, fuel, and materials.

Equations (14.12)–(14.20) present related constraints for the cost minimization problem. Equation (14.12) sets the upper bound of switchgrass area in each hexagon (i.e., the available agricultural land, *A*). Equation (14.13) constrains switchgrass production in a hexagon to be either directly shipped to the biorefineries (*XNS*) or to be stored (*XS*). Equations (14.14)–(14.17) present the mass balance constraints: harvested switchgrass at each hexagon (*i*) should be moved to the biorefineries (*j*) throughout the year and converted to biofuels for blending facilities (*b*) to meet the demand each season (*D*). Equation (14.18) limits at most one biorefinery at each candidate site. Equation (14.19) defines the binary variable for biorefinery location, while Eq. (14.20) is the nonnegativity condition for continuous decision variables.

$$X_{ih} \leq A_{ih} \forall i, h, \tag{14.12}$$

$$\sum_{h \in H} y_{is} \times X_{ih} = XNS_i + XS_i \forall i, \tag{14.13}$$

$$XNS_i = \sum_{m \in M_{on}} \sum_{j \in J} \frac{XQ_{mij}}{(1 - \lambda_1)} \quad \forall i, \tag{14.14}$$

$$XS_i = \sum_{m \in M_{off}} \sum_{j \in J} \frac{XQ_{mij}}{(1 - \lambda_2) \times (1 - \lambda_1)} \quad \forall i, \tag{14.15}$$

$$\sum_{i \in I} \sum_{j \in J} XQ_{mij} = \sigma \sum_{j \in J} \sum_{b \in B} XO_{mjb} \forall m, \tag{14.16}$$

$$\sum_{j \in J} \sum_{b \in B} XO_{mjb} = D_m \quad \forall m, \tag{14.17}$$

$$\sum_{g \in G} Z_{jg} \leq 1 \forall j, \tag{14.18}$$

$$Z_{jg} \in \{0,1\} \quad \forall \quad j, g, \tag{14.19}$$

$$X, XNS, XS, XQ, XO \geq 0. \tag{14.20}$$

Similarly, total GHG emissions (TE) include soil carbon stock related to land use change, GHG emissions from energy consumption for switchgrass maintenance, harvest, storage, and biofuel conversion, and fuel use for feedstock and biofuel transportation. GHG emissions associated with biorefineries construction are also considered in the estimation of TE. Equation (14.21) presents the TE minimization objective as follows:

$$\min.\text{TE} = E_{luc}^{swi} + E_{mnt}^{swi} + E_{hrv}^{swi} + E_{stg}^{swi} + E_{tran}^{swi} + E_{grd}^{swi} + E_{inv}^{fac} + E_{conv}^{bio} + E_{tran}^{bio} \tag{14.21}$$

where

$$E_{luc}^{swi} = \sum_{h \in H} \left(\sum_{i \in I} \left(\Delta CO_{2ih} + \Delta CH_{4ih} + \Delta N_2O_{ih} \right) \times X_{ih} \right) \tag{14.22}$$

$$E_{mnt}^{swi} = \sum_{i \in I} \sum_{h \in H} \left(X_{ih} \times \epsilon_1 \right), \tag{14.23}$$

$$E_{hrv}^{swi} = \sum_{i \in I} \sum_{h \in H} y_{is} \times X_{ih} \times \epsilon_2, \tag{14.24}$$

$$E_{stg}^{swi} = \sum_{i \in I} XS_i \times \epsilon_3, \tag{14.25}$$

$$E_{tran}^{swi} = \sum_{m \in M} \sum_{i \in I} (XQ_{mi} / \left(wt \times \left(1 - \lambda_2 \right) \right) \times \epsilon_4, \tag{14.26}$$

$$E_{grid}^{swi} = \sum_{m \in M} \sum_{i \in I} XQ_{mi} \times \epsilon_5, \tag{14.27}$$

$$E_{inv}^{fac} = \sum_{j \in J} \sum_{g \in G} \left(\varepsilon \times g \times Z_{jg} \right), \tag{14.28}$$

$$E_{conv}^{bio} = \sum_{m \in M} \sum_{j \in J} \sum_{b \in B} XO_{mjb} \times \epsilon_6, \tag{14.29}$$

$$E_{tran}^{bio} = \sum_{m \in M} \sum_{j \in J} \sum_{b \in B} (XO_{mjb} / \left(wt \times \left(1 - \lambda_2 \right) \right) \times \epsilon_7, \tag{14.30}$$

Equation (14.22) estimates GHG emissions from land use change by applying the emission factors for CO_2, CH_4, and N_2O to the agricultural lands converted to switchgrass. Equations (14.23)–(14.25) assess GHG emissions from energy consumption for switchgrass maintenance, harvest, storage, and grinding using operation emissions factors from the GREET database. Transportation GHG emissions associated with switchgrass and biofuels are determined based on the number of truckloads and associated emission factors generated from the MOVES model. The GHG emissions resulting from constructing biorefineries by capacity are estimated in Eq. (14.28). Similar to the cost minimization scenario, constraints in Eqs. (14.12)–(14.20) are imposed when minimizing TE. The sources of costs and GHG emissions parameters in the analysis are available in Table 14.2.

An improved augmented ε-constraint method developed by Mavrotas and Florios (2013) is employed to solve the multiobjective optimization problem and derive the trade-off between TC minimization and TE minimization. The augmented ε-constraint method minimizes one objective while setting the other objective as a constraint to generate efficient solutions. Assuming the decision makers' main objective is to minimize the total cost of the supply chains and use GHG emissions as constraints, the model is formulated as follows:

Table 14.2 Data sources

Category	Data source
Land use change	
Land rents	USDA NASS (USDA 2013–2015a)
Crop yields	USDA, NRCS SSURGO (USDA 2012)
Crop price and area	USDA NASS (USDA 2013–2015b)
Crop production budget Switchgrass yield	POLYSIS (Ugarte and Ray 2000), USDA ERS (USDA 2015) Jager et al. (2010)
Switchgrass production and harvest budget	Larson et al. (2010), University of Tennessee (2015)
Land use GHG emissions	Daycent (Parton et al. 2001)
Weather data	DayMet (Oak Ridge National Laboratory 2015)
Land management—crops	Bowling et al. (2006)
Land management—switchgrass	Muir et al. (2001)
Feedstock production	
Switchgrass establishment budget	American Agricultural Economics Association (2000)
Switchgrass maintenance budget	American Society of Agricultural and Biological Engineers (2006)
Operation GHG emissions	GREET (Argonne National Laboratory 2013)
Feedstock harvest and storage	
Fuel and labor budget	University of Tennessee (2015)
Operation GHG emissions	GREET (Argonne National Laboratory 2013)
Covers and pallets budget	University of Tennessee (2015)
Feedstock grinding	
Operation GHG emissions	GREET (Argonne National Laboratory 2013)
Feedstock and biofuel transport	
Trailer, fuel and labor budget	University of Tennessee (2015)
Transport GHG emissions	MOVES (US EPA 2013)

$$\begin{aligned} &\min\left(\mathrm{TC} - \varepsilon \times \frac{s}{r}\right), \\ &\text{s.t.} \\ &\mathrm{TE} + s = e_p, \end{aligned} \tag{14.31}$$

where ε is a small number (in this study ε was set to 10^{-3}), s is a nonnegative slack variable that prevents weakly efficient solutions, r is the range of the GHG emission objective between the minimum GHG emissions level (e_l) and the upper-bound GHG emissions level (e_u), and e_p refers to the level at the pth grid point in the range of TE [$e_p = e_u - \left(\frac{e_u - e_l}{m}\right) \times p,\ p = 0,1,\ldots,m$]. In this study, we have six grid points in TE and five equivalent intervals. Changing the level of e_p with the grid points determines the trade-off between the two objectives and derives the imputed cost of reducing GHG emissions. The upper bound of e (e_n) is an ex-post estimate after minimizing TC subject to biofuel demand and mass-balance without adding any constraints on GHG emissions. The lower bound of e (e_0) is obtained from minimizing TE subject to biofuel demand and mass-balance without considering costs. The total cost under the TE minimization scenario can be then calculated.

14.3 Empirical Results

We code the multiobjective model in GAMS 24.2 and solve it with CPLEX using a high-performance workstation (Dell Precision T5810 with 128GB RAM and 3.70 GHz CPU). The cost minimization scenario, consisting of more than three million nonzero variables and 296 binary variables, is solved using 2.23 h CPU time with a relative gap of 1.5%; while solving the GHG emission minimization scenario with the same number of decision variables consumes 10.03 h CPU time with a relative gap of 1%. Solving the multiobjective model that includes four grid points between the cost minimization and GHG emission minimization levels by the augmented ε-constraint method consumes 49.54 h CPU time with a relative gap of 3.6%.

Figure 14.3 presents the levels of annual cost and GHG emissions in the switchgrass biofuel supply chains at varying grid points from multiobjective optimization. Under the cost minimization scenario (TC_{min}), the annual cost of the supply chains that meet the demand of 3.78 billion liters of biofuels is nearly \$3.63 billion (at 2015 constant dollar). The associated GHG emissions generated from the supply chains are about 2.28 billion CO_2e Mg, which is the upper bound of the GHG emissions in the supply chains. The level of GHG emissions at the GHG emission minimization scenario (TE_{min}) is less than 2.00 billion CO_2e Mg with a higher total cost that

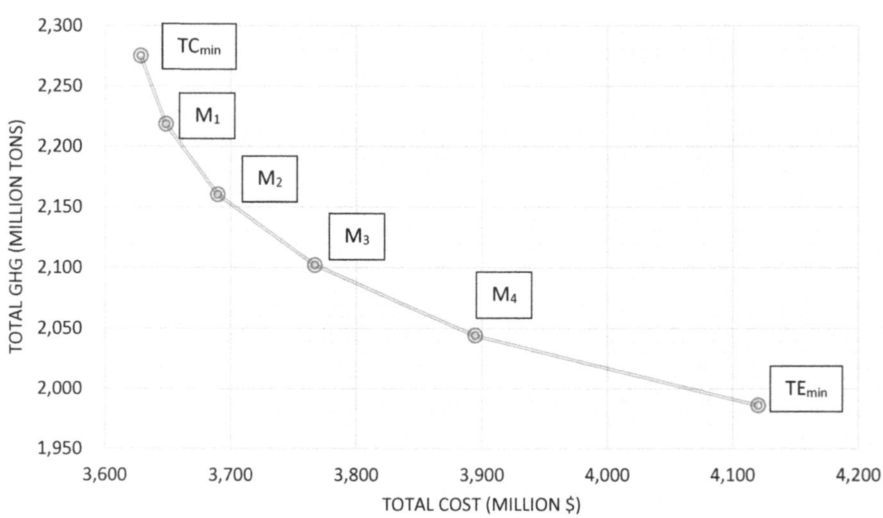

Fig. 14.3 The trade-off curve between cost and GHG emissions of switchgrass supply chain in Tennessee

exceeds \$4.12 billion for the switchgrass biofuel supply chains. The trade-off between the total cost and overall GHG emissions is observed, implying that we cannot reduce GHG emissions without increasing total cost in the supply chains. Also, the reduction in GHG emissions between each adjacent grid point is identical (58 million CO_2e Mg); however, the total cost rises at an accelerating pace. Reducing 58 million CO_2e Mg from the cost minimization point (TC_{min}) to the adjacent point (M_1) leads to an increase of \$20 million in total cost, implying a marginal cost of 35 cents per CO_2e Mg. When further lowering the amount of GHG emissions from grid point M_1 to M_2, the total cost in the supply chains increases by \$41 million (more than double than the previous interval). Finally, the total cost increases by \$226 million from grid point M_4 to the GHG emission minimization point (TE_{min}), suggesting the marginal cost of GHG emissions reduction soars to \$3.89 per CO_2e Mg.

One main factor related to the accelerating marginal cost of GHG emissions reduction is the land use choice for switchgrass production. Figure 14.4 summarizes the type of land converted to switchgrass production. Under the total cost minimization scenario (TC_{min}), the majority (>81%) of agricultural land used for switchgrass production is from hay and pasture lands because the opportunity cost is much lower than that for cropland. However, the share of cropland for switchgrass production starts increasing when departing from the cost minimization scenario to the adjacent grid points and eventually dominates under the GHG emission minimization scenario (TE_{min}). The shift of land use among those scenarios is due to the difference in soil carbon stock from cropping system: converting cropland to switchgrass increases soil carbon stock because of switchgrass' deeper roots, whereas replacing hay and pasture with switchgrass releases more soil carbon (Cherubini and Jungmeier 2009; Monti et al. 2012).

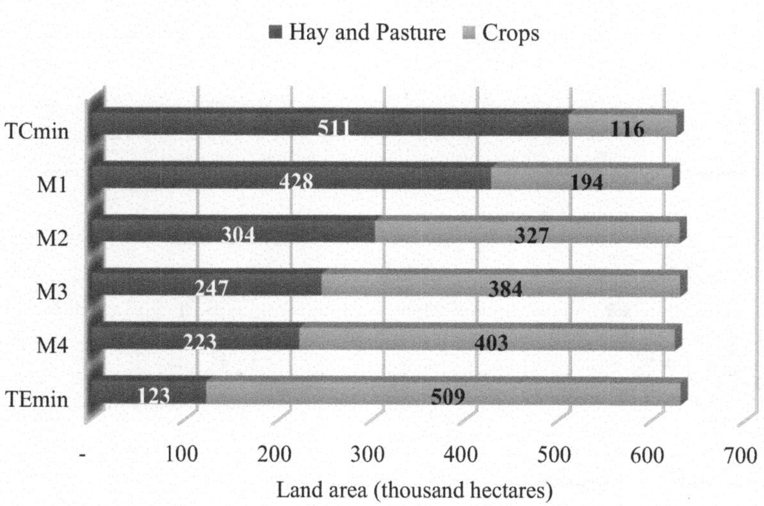

Fig. 14.4 Area of crop type converted for switchgrass production to meet 3.78 billion liters of biofuels in Tennessee. (**a**) Cost minimization scenario; (**b**) GHG emission minimization scenario

The choice of land use for switchgrass production has substantial impacts on the deployment of the biofuel supply chain system in Tennessee. Figure 14.5 shows the location of biorefineries and switchgrass in the supply chains under the (a) cost minimization scenario and (b) GHG minimization scenario. As pasture and hay land is primarily located on the east side of the state, the model selects this region as the major switchgrass production area in the cost minimization scenario. A total of ten biorefineries are placed within the state: biorefineries A–C serve biofuel demand in Knoxville, biorefineries C–G for Nashville, and biorefineries G–J for Memphis. Note that biorefinery C delivers biofuels to blenders in both Knoxville and Nashville, with Knoxville as the major destination. Similarly, biorefinery G serves blending facilities in both Nashville and Memphis while the major share of biofuels is brought to Nashville. All 10 biorefineries have a capacity of 378 MLY due to the scale economies in the investment cost.

Under the GHG emission minimization scenario, the majority of switchgrass is produced in west Tennessee which is the state's central crop production region. A total of 16 biorefineries are established with 12 units running at a smaller capacity (189 MLY) plus four operating larger capacity (378 MLY). Locating smaller-size biorefineries near the scatter cropland in the east and central Tennessee reduces GHG emissions from feedstock transportation although the distance between biorefineries and blending facilities increases. Biorefineries 1–6 meet the blending demand in Knoxville, biorefineries 7–12 for Nashville, and biorefineries 12–16 for Memphis. Also, including a large number of smaller biorefineries in the supply

(a) Cost minimization scenario

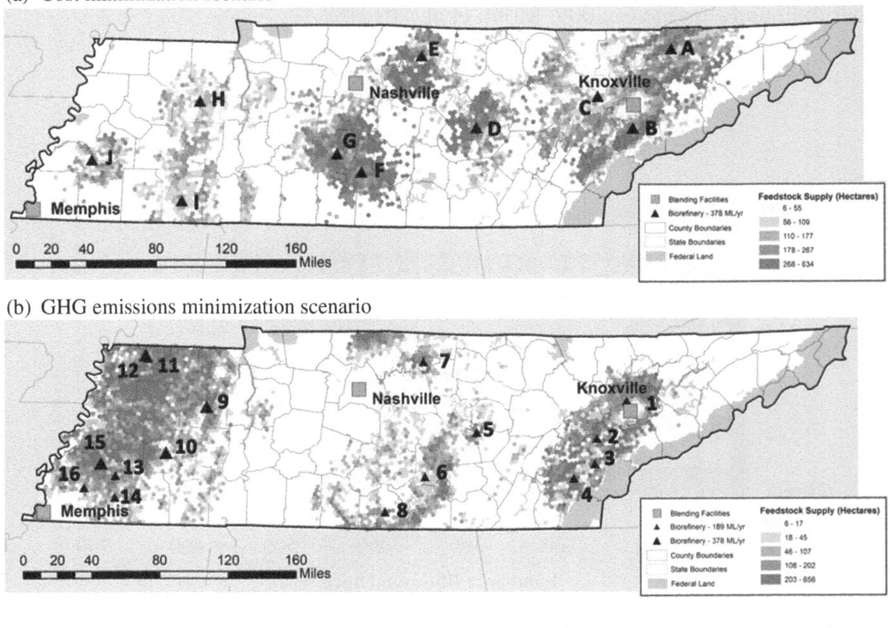

Fig. 14.5 Location of biorefineries and switchgrass production area meeting 3.78 billion liters of biofuel in Tennessee under (**a**) cost minimization and (**b**) GHG emission minimization scenarios

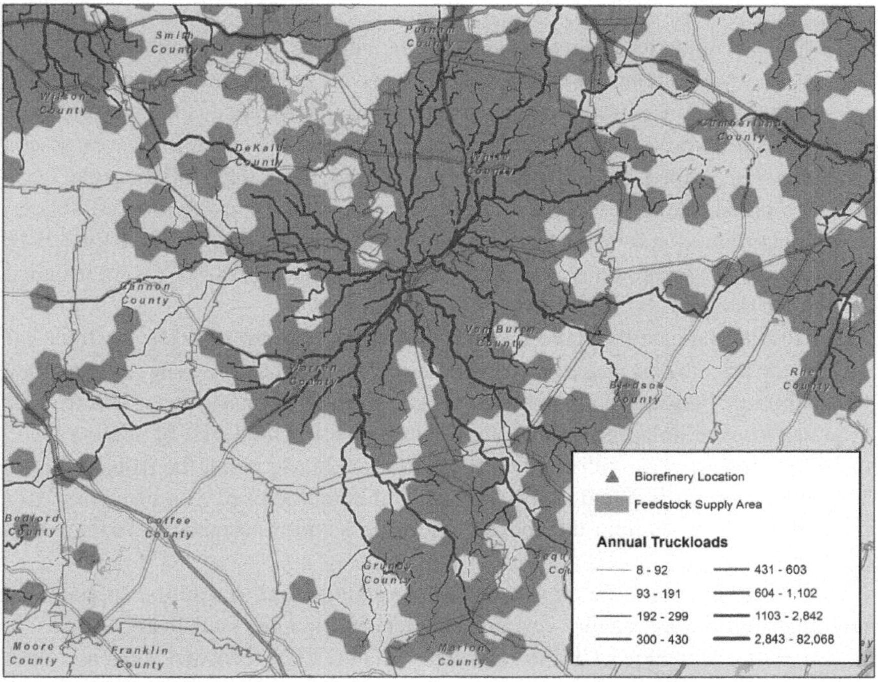

Fig. 14.6 Annual truck flows of switchgrass to the biorefinery D in Fig. 14.5a under the cost minimization scenario

chains increases the investment cost compared to the cost minimization scenario that employs larger-size biorefineries with scale economies, hence resulting in an accelerating marginal cost to abate GHG emissions.

Identifying the location for biorefineries and feedstock production areas for biofuel production implies the demand for the local road network. Figure 14.6 demonstrates the additional traffic from feedstock transportation on the selected routes linking fields to a given biorefinery under the cost minimization scenario (site D in Fig. 14.5a). The busiest road linking the biorefinery gate could encounter more than 82,000 truckloads annually. Related information could help economic developers and regional agencies identify potential weak links and manage the quality and safety of the road network.

14.4 Conclusions

This study presents an integrative framework that combines GIS tools, biophysical models, optimization mechanisms, and other tools in determining a large-scale sustainable biofuel supply chain network. The integrative approach is applied to a case study of replacing 30% of gasoline used for transportation in Tennessee. Our results suggest that land resource choice heavily influences the design of the supply chains under different objectives, and consequently the imputed cost of abating GHG emissions in the supply chains. When considering private cost minimization alone, hay and pasture land concentrated in the east and central Tennessee is the major source for switchgrass production. A total of 10 larger biorefineries (378 MLY) are placed across the state in the supply chains. On the contrary, more than 500 thousand hectares of cropland primarily located in the western region are converted to switchgrass for biofuel production if GHG emission minimization is the sole target. A total of 16 biorefineries, 12 smaller ones (189 MLY) and 4 large ones (378 MLY), are installed to serve the demand.

Our results indicate that land use and biorefinery capacity lead to the trade-off between the private cost and GHG emissions in the supply chains. The marginal rate of substitution between annual private cost and GHG emissions on the frontier curve increases at an accelerating rate: the marginal imputed cost of abating GHG emissions increases from 35 cents to \$3.89 per CO2e Mg along the frontier curve. The information on imputed cost is crucial as it could be used as a benchmark for potential subsidies to help the industry develop a more sustainable biofuel supply chain.

This study illustrates the importance to incorporate high-resolution spatial data, including soil type and quality, resource availability, and physical characteristics, in the design of sustainable LCB biofuel supply chains. LCB feedstock currently does not have an established market, thus considering land resources at a high-resolution spatial level will provide more adequate information to identify the potential production areas and assist the management of natural resources. The integrative approach enhances the traditional aspatial multiobjective optimization approach by connecting the multiobjective decision support system with detailed spatial information and provide more comprehensive support.

Enhanced precision will result in higher demand for computational power and capacity, thus developing a more efficient way to handle large-scale high-resolution optimization model warrants more attention in future research. Besides, a possible extension of the present study is to incorporate uncertainties in the availability and quality of natural resources if the demand for computation power and capacity could be addressed.

References

American Agricultural Economics Association. (2000). *Commodity cost and returns handbook*. Ames, IA: American Agricultural Economics Association.

American Society of Agricultural and Biological Engineers. (2006). *Agricultural machinery standards*. St. Joseph, MI: American Society of Agricultural and Biological Engineers.

Argonne National Laboratory. (2013). *The Greenhouse gases, regulated emissions, and energy use in transportation model (GREET)*. Retrieved from http://www.transportation.anl.gov/publications/index.html

Ba, B. H., Prins, C., & Prodhon, C. (2016). Models for optimization and performance evaluation of biomass supply chains: An operations research perspective. *Renewable Energy, 87*, 977–989.

Bowling RG, McKinley TL, Rawls EL, Tennessee Forage Budgets, UT extension, (2006) 23 Available at: http://economics.ag.utk.edu/budgets/PB1658ForageBudgets07.pdf.

Cambero, C., & Sowlati, T. (2016). Incorporating social benefits in multi-objective optimization of forest-based bioenergy and biofuel supply chains. *Applied Energy, 178*, 721–735.

Cherubini, F., & Jungmeier, G. (2009). LCA of a biorefinery concept producing bioethanol, bioenergy, and chemicals from switchgrass. *The International Journal of Life Cycle Assessment, 15*, 53–66.

De La Torre Ugarte DG, Ray DE, Biomass and bioenergy applications of the POLYSYS modeling framework Paper prepared for presentation at the Modeling Tools for Biomass and Bioenergy Conference, 8–10 April 1997 Knoxville, TN, *Biomass and Bioenergy, 18* (2000) 291–308.

Ekşioğlu, S. D., Acharya, A., Leightley, L. E., & Arora, S. (2009). Analyzing the design and management of biomass-to-biorefinery supply chain. *Computers and Industiral Engineering, 57*, 1342–1352.

English, B. C., De La Torre Ugarte, D., Walsh, M. E., Hellwinckel, C. M., & Menard, R. J. (2006). Economic competitiveness of bioenergy production and effects on agriculture of the southern region. *Journal of Agricultiral and Applied Economics*, 389–402.

Field, J. L., Evans, S. G., Marx, E., Easter, M., Adler, P. R., Dinh, T., Willson, B., & Paustian, K. (2018). High-resolution techno–ecological modelling of a bioenergy landscape to identify climate mitigation opportunities in cellulosic ethanol production. *Nature Energy, 3*, 211–219.

Ghaderi, H., Pishvaee, M. S., & Moini, A. (2016). Biomass supply chain network design: An optimization-oriented review and analysis. *Industrial Crops and Products, 94*, 972–1000.

He-Lambert, L., English, B. C., Lambert, D. M., Shylo, O., Larson, J. A., Yu, T. E., & Wilson, B. (2018). Determining a geographic high resolution supply chain network for a large scale biofuel industry. *Applied Energy, 218*, 266–281.

Jager HI, Baskaran LM, Brandt CC, Davis EB, Gunderson CA, Wullschleger SD, Empirical geographic modeling of switchgrass yields in the United States, *GCB Bioenergy, 2* (2010) 248–257.

Kort, J., Collins, M., & Ditsch, D. (1998). A review of soil erosion potential associated with biomass crops. *Biomass and Bioenergy, 14*, 351–359.

Larson, J. A., Yu, T. E., English, B. C., Mooney, D. F., & Wang, C. (2010). Cost evaluation of alternative switchgrass producing, harvesting, storing, and transporting systems and their logistics in the Southeastern USA. *Agricultural Finance Review, 70*, 184–200.

Mavrotas, G., & Florios, K. (2013). An improved version of the augmented ε-constraint method (AUGMECON2) for finding the exact pareto set in multi-objective integer programming problems. *Applied Mathematics and Computation, 219*, 9652–9669.

Monti, A., Barbanti, L., Zatta, A., & Zegada-Lizarazu, W. (2012). The contribution of switchgrass in reducing GHG emissions. *GCB Bioenergy, 4*, 420–434.

Monti, A., Fazio, S., & Venturi, G. (2009). Cradle-to-farm gate life cycle assessment in perennial energy crops. *European Journal of Agronomy, 31*, 77–84.

Muir JP, Sanderson MA, Ocumpaugh WR, Jones RM, Reed RL, Biomass Production of 'Alamo' Switchgrass in Response to Nitrogen, Phosphorus, and Row Spacing, *Agronomy Journal, 93* (2001) 896.

Nguyen, T. H., Granger, J., Pandya, D., & Paustian, K. (2019). High-resolution multi-objective optimization of feedstock landscape design for hybrid first and second generation biorefineries. *Applied Energy, 238*, 1484–1496.

Oak Ridge National Laboratory. (2015). *DAYMET database*. https://daymet.ornl.gov/getdata

Parton, W. J., Holland, E. A., Del Grosso, S. J., Hartman, M. D., Martin, R. E., Mosier, A. R., Ojima, D. S., & Schimel, D. S. (2001). Generalized model for NO_x and N_2O emissions from soils. *Journal of Geophysics Resource, 106*, 17403–17420.

Petridis K, Grigoroudis E, and Arabatzis E. A goal programming model for a sustainable biomass supply chain network, *International Journal of Energy Sector Management, 12*(1) (2018) 79–102.

Tiller, K. (2011). Tennessee case study in full-scale development and integration of switchgrass feedstocks with biorefinery operations: Genera Energy and UT biofuels initiative. *Industrial Biotechnology, 7*, 357–364.
U.S. Congress. (2007). *Energy Independence and Security Act of 2007*. Reterived from http://frwebgate.access.gpo.gov/cgi-bin/getdoc.cgi?dbname=110_cong_bills&docid=f:h6enr.txt.pdf
U.S. Department of Agriculture ERS. (2015). *Commodity costs and returns*. Retrieved from https://www.ers.usda.gov/data-products/commodity-costs-and-returns/
U.S. Department of Agriculture NASS. (2013–15a). *Cash rents by county*. Retrieved from https://www.nass.usda.gov/Surveys/Guide_to_NASS_Surveys/Cash_Rents_by_County/
U.S. Department of Agriculture NASS. (2013–15b). *Crop values annual summary*. Retrieved from http://usda.mannlib.cornell.edu/MannUsda/viewDocumentInfo.do?documentID=1050
U.S. Department of Agriculture NRCS. (2012). *SSURGO structural metadata*. Retrieved from https://www.nrcs.usda.gov/wps/portal/nrcs/detail/soils/survey/geo/?cid=nrcs142p2_053631
U.S. Environmental Protection Agency. (2014). *Carbon pollution emission guidelines for existing stationary sources: Electric utility generating units*. http://yosemite.epa.gov/opei/rulegate.nsf/byRIN/2060-AR33
U.S. Environmental Protection Agency. (2013). *Motor vehicle emission simulator (MOVES) model*. https://www.epa.gov/moves/latest-version-motor-vehicle-emission-simulator-moves
University of Tennessee. (2015). *Field crop budgets*.http://economics.ag.utk.edu/budgets.html
Wright, L., & Turhollow, A. (2010). Switchgrass selection as a "model" bioenergy crop: A history of the process. *Biomass and Bioenergy, 34*, 851–868.
Yu TE, Wang Z, English BC, Larson JA, Designing a dedicated energy crop supply system in Tennessee: a multiobjective optimization analysis, *Journal of Agricultural and Applied Economics, 46* (2014) 357–373.
Zhong, J., Yu, T. E., Larson, J. A., English, B. C., Fu, J. S., & Calcagno, J. (2016). Analysis of environmental and economic tradeoffs in switchgrass supply chains for biofuel production. *Energy, 107*, 791–803.

Chapter 15
Mathematical Models for Evolving Natural Gas Markets

Felipe Feijoo, Sriram Sankaranarayanan, Charalampos Avraam, and Sauleh A. Siddiqui

Abstract In the past years, natural gas has expanded its significance as an energy source mainly due to its low carbon emissions and low competitive prices as result of new technologies. Furthermore, the regional and global natural gas markets have also been significantly influenced by domestic and international socioeconomic conditions and politics. This chapter discusses the main drivers for the evolution of natural gas markets and the modeling approaches that have been taken to understand such evolutions at regional and global levels. Thereafter, we focus on two well-known natural gas models, a global model (World Gas Model) and a regional model (North American Natural Gas Model). We provide the mathematical formulations and discuss the modeling paradigms that are behind both models. In order to demonstrate the usefulness of these models, we present a numerical case study using the North American Natural Gas Model (NANGAM). We conclude with an outlook of future research and examples of historical energy system transformations due to the appearance of natural gas as a competitive energy source.

F. Feijoo (✉)
School of Industrial Engineering, Pontificia Universidad Católica de Valparaiso, Valparaiso, Chile
e-mail: felipe.feijoo@pucv.cl

S. Sankaranarayanan
Department of Mathematics and Industrial Engineering, Polytechnique Montreal, Montreal, Canada

C. Avraam
Department of Civil Engineering, The Johns Hopkins University, Baltimore, MD, USA

S. A. Siddiqui
Department of Environmental Science, American University, Washington, DC, USA
e-mail: sauleh@american.edu

C. Chen et al. (eds.), *Pursuing Sustainability*, International Series in Operations Research & Management Science 301,
https://doi.org/10.1007/978-3-030-58023-0_15

15.1 Introduction

The global demand for natural gas has rapidly increased over the past decades. However, local supply in many regions has been constantly outpaced by demand, which has created the need for a global market. Natural gas has gained great attention due to its environmental credentials, potential to improve local air quality, higher efficiency in power generation compared to oil and coal, lower greenhouse gas (GHG) emissions relative to other fossil fuels, suitability for partnering with renewables as a back-up fuel, and lower prices compared to other fossil fuels (Colombo et al. 2016; Mignone et al. 2017). Hence, the expansion of natural gas in the global energy mix has been justified by multinational environmental and climate policies, where natural gas has been identified as the "bridge fuel" toward a fully decarbonized energy sector (McJeon et al. 2014; Levi 2013; Davis and Shearer 2014; Jacoby et al. 2012). These advantages have led to progressively consider natural gas a crucial component of the world's current and future energy supply and a factor in geopolitical competition (Colombo et al. 2016).

The natural gas markets have gone through significant changes in the last decades. These changes were mainly driven by technological development, which has resulted in evolved patterns for production, altering traditional energy independence of countries. Technological development helped to overcome obstacles related to infrastructure, particularly, in technologies and processes necessary to extract, process, store, and transport, and distribution to consumers. A clear example of the evolving gas market is the shale gas "boom," which allowed economically feasible production of unconventional shale gas. Global unconventional natural gas resources are believed to be at least as large as conventional ones, with great potential not only in the US market, but also in China, India, Australia and Indonesia. Indeed, in the US natural gas surpassed coal to become the leading source of electricity generation in 2016, the most important sector consuming natural gas. In addition, US pipeline and liquefied natural gas (LNG) exports have increased significantly over the last 5 years and are expected to continue to increase through the mid-century. Furthermore, LNG is projected to dominate US natural gas exports, increasing total US liquefaction capacity by roughly ten times (US Energy Information Administration 2019) between 2016 and 2019. In this context, the USA is expected to become a net exporter of total energy by 2020. Another example of increased importance of natural gas is the Chinese energy market (Colombo et al. 2016; Lin and Wang 2012; Xu et al. 2017; Zhang et al. 2016). Natural gas production in China has grown rapidly in the last decade, increasing by 500% between 2000 and 2016. However, a higher increase in the demand has been observed over the same period (Feijoo et al. 2016, 2018). Further details of the Chinese gas market, as well as other examples of international markets, will be provided in Sect. 15.3.

15.1.1 Understanding the Evolution of Natural Gas Markets

The study of the evolution of the natural gas markets requires accounting for and understanding the main drivers. Projections made both for global and regional (geographical) markets, such as those for the cases of the USA and China, are subject to significant variability due to technological breakthroughs that are hard to quantify and predict. Historically, energy markets have seen a variety of new technologies with the potential of replacing existing practices. For instance, hydraulic fracturing for extracting natural gas from shale formations is the most recent example that has had a dramatic impact. Other drivers that have shown significant contribution to the evolving markets include the character of technology deployment in demand sectors such as the electric power sector, demographics, economic growth patterns, energy and environmental policy landscapes, and energy security (Mignone et al. 2017; Feijoo et al. 2018; Cole et al. 2016).

Understanding the impact of all the above factors and their interactions poses a challenge for energy policy makers. Existing models created to study natural gas markets vary in levels of technological, sectorial, and geographical details and hence conclusion may vary significantly. Quantification of these variabilities in natural gas models have been performed for instance by the Energy Modeling Forum (EMF) 31, where global and regional models where used to study the transitions of the North American natural gas market (Bernstein et al. 2016; Yeh et al. 2016; Huntington 2016). Studies have also looked into infrastructure access. For instance, the Chinese natural gas market, policies and its infrastructure deployment has been widely studied (Lin and Wang 2012; Xu et al. 2017; Zhang et al. 2016; Paltsev and Zhang 2015). These studies conclude that import prices are important to determine the infrastructure development and interregional flows within China. However, high import costs compared to the low natural gas capped prices set by the government are likely to exert great pressure on China's price reforms. Hence, pipeline capacity scarcity must be properly managed by the Chinese government. The European gas infrastructure has also been studied in detail (Dieckhöner et al. 2013; Egging et al. 2008; Holz and von Hirschhausen 2013; Egging and Holz 2016; Holz et al. 2016). Studies show that the European natural gas market presents high integration. However, network congestion and a need for new pipeline capacity in Germany, Denmark and eastern Europe have been identified. Authors have also found that Europe will depend on exports from Africa and Caspian region, leading to added import pipeline capacity (Dieckhöner et al. 2013; Egging et al. 2008; Holz and von Hirschhausen 2013; Egging and Holz 2016). However, in deep-decarbonization scenarios, Europe's import infrastructure and intra-European transit capacity currently in place or under construction are largely sufficient to accommodate the import needs of the decarbonization scenarios, despite the reduction of domestic production and the increase of import dependency (Holz et al. 2016).

Several studies have also focused on the US and North American natural gas sector. The US Department of Energy (DOE) analyzed the US gas infrastructure under different demand scenarios from the power sector (U.S. Department of Energy

2015). Increased demand for natural gas in the power sector will lead to pipeline capacity additions. Authors in (Feijoo et al. 2016; Sankaranarayanan et al. 2018) studied the effect of increased Mexican natural gas demand from the power sector. Results show higher US pipeline exports to Mexico, which are possible under a shift of flows within the USA and pipeline capacity expansions in both the USA and Mexico. It has also been shown that lack of US pipeline capacity has resulted in network congestion and increased transportation costs. The increased prices could be managed by increased storage or additional pipeline capacity (Oliver et al. 2014; Brown and Yücel 2008; Oliver 2015). In a global context, authors in (Egging et al. 2010) show that the share of LNG and pipeline flows changes over time and region. The European region will require new pipeline import capacity due to proximity to major gas suppliers while LNG will play a major role in the Asian market.

15.1.2 Modeling Approaches for Natural Gas Markets: Examples of Global and Regional Studies

The importance of the interactions between decision makers with conflicting objectives along the supply chain led modelers to develop game-theoretic models. Nonetheless, econometric models are still used for forecasting purposes, focusing either at global shocks or at short term effects. In the USA, the Short Term Energy Outlook is produced using the IHS Markit macroeconomic model. Similarly, forecasts of global natural gas trends have been conducted using stochastic models like PROMETHEUS (Fragkos et al. 2016). Depending on their purpose, game-theoretic and optimization-based models often varied in terms of their geographical, technological, and temporal resolution, assumptions on capacity expansion and benchmarking data, to name a few. One of the first natural gas models with focus in North America is the Gas Trade Model (GTM) (Beltramo et al. 2018). The GTM was developed in the late 1980s and comprised of regions covering the USA, Canada, and Mexico. A large scale linear complementarity model for North America was presented in (Gabriel et al. 2005). The linear complementarity framework allowed to analyze equilibrium among several market participant, including producers, storage and peak gas operators, third party marketers and four end-use sectors. Yet a major drawback of the formulation is that it did not consider endogenous capacity expansion decisions. The World Gas Model (WGM) described in (Egging et al. 2010), is indeed developed based on the work presented in (Gabriel et al. 2005), considers six regions in the USA, adds Mexico as a single region to the model, and allows for endogenous capacity investment in the upstream and midstream sectors. Another global model is the Rice World Gas Trade Model (RWGTM) (Cole et al. 2016; Hartley and Medlock 2006). The RWGTM is a dynamic spatial partial equilibrium model with detail geographical resolution, including over 290 global demand regions, 140 supply basins connected in the midstream sector by both pipeline and LNG deliveries (Cole et al. 2016). The GaMMES (Gas Market Modeling with Energy Substitution) model is a generalized Nash Cournot model with a focus on the European market. The model considers endogenous decisions for capacity expansion and long-term contracts and it has been previously used to study the northwestern European natural gas trade. Other models that also particularly model the European market include GASMOD (Holz et al. 2008) and GASTALE (Lise et al. 2008; de Joode and Özdemir 2010).

At the same time, the fact that the natural gas is a natural resource and is an input for other sectors, including power sector and demand sectors, implies that its correct functioning is critical for the performance of broader energy system. Therefore, natural gas modules exist in many multimarket or macroeconomic models. Computable General Equilibrium models often represent the natural gas sector just like all the other sectors, namely, as the response of a single representative producer who substitute between labor, capital and other inputs to decide the least costly production plan. This is particularly the case for both global and regional models. Among global models, we distinguish the Global Trade Analysis Project (GTAP) (Aguiar et al. 2016) and the World Induced Technical Change Hybrid (WITCH) (Bosetti et al. 2006). Many European studies on climate change have been conducted using GEM-E3 (Capros et al. 2013), whereas many studies related to the USA have been conducted using the Emissions Prediction and Policy Analysis (EPPA) model. Other models include multimarket models or Integrated Assessment Models, such the Global Change Assessment Model (GCAM) (Feijoo et al. 2018; Kim et al. 2006). These types of models provide a broader socioeconomic context for natural gas supply and demand and consider the interactions with other sectors of the economy in an unified framework. However, they often do not consider sufficient details of the technological production process and are built based on simplified assumptions regarding the pipeline (midstream) system. The importance of the first feature is highlighted in Golombek et al. (1995), while the impact of pipeline prices is studied in Bernard et al. (2002). The authors in (Feijoo et al. 2018) attempted to close this gap and presented a US focused study where the GCAM model is linked to a natural gas infrastructure model. Finally, even decades old energy models that integrate an operational submodel of natural gas assume that infrastructure capacity expansion is exogenous. That is the case for example with the Price-Induced Market Equilibrium System (PRIMES) (Capros et al. 2018) and Natural Gas Market Module (NGMM) of the National Energy Modeling System (NEMS) (US Energy Information Administration 2018a).

15.2 Components of Natural Gas Supply Chain: Mathematical Formulation and Solutions Approaches

There exists a significant amount of modeling approaches for natural gas markets, as described in Sect. 15.1. Generally, natural gas market models consider three levels of the supply chain. These include the upstream sector where natural gas is

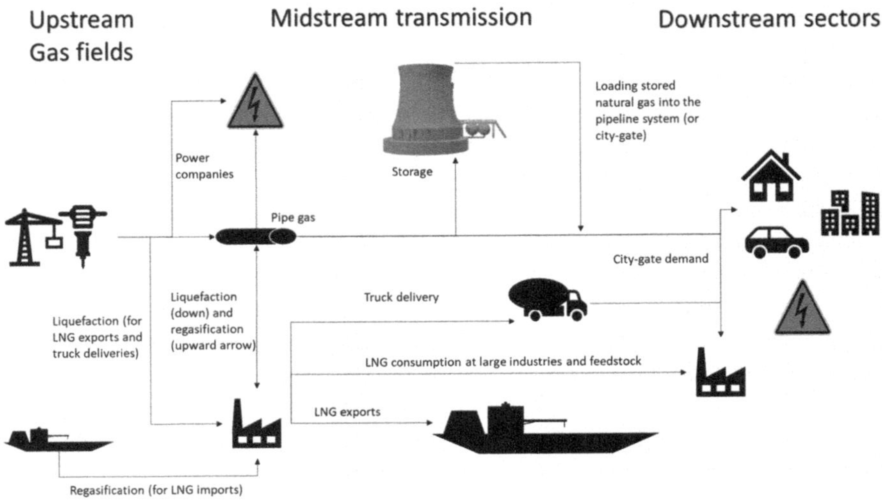

Fig. 15.1 Representation of the natural gas supply chain

produced; the midstream sector where natural gas is either transformed into secondary energy forms (e.g., electricity), stored or sent via different delivery modes to other nodes; and finally, the downstream sector where natural gas is sold to the different demand sectors (e.g., residential, commercial, and industrial). We next present general mathematical formulations to the three stages of the supply chain of natural gas markets and then discussed the solution approaches. Particularly, the two formulations (NANGAM and WGM) are discussed as generalization of models (Fig. 15.1).

15.2.1 Upstream Sector

In natural gas market models, the upstream sector is generally composed of natural gas suppliers. The suppliers produce various types of natural gas (for instance shale gas, coalbed-methane gas, or conventional gas) and use infrastructure services (midstream sector) purchased from service providers to transport and/or transform them into other types of energy (e.g., LNG or secondary energy like electricity or heat), and/or to store across seasons. Suppliers profit from selling the natural gas resource to the demand sectors (other models, like the WGM (Egging et al. 2010) model a "trader" player who settles the equilibrium quantities between suppliers and the demand sectors) while facing a cost for production and for having access to the midstream services. Some models (such as (Feijoo et al. 2016; Huppmann and Egging 2014)) also include endogenous decisions for capacity investment, incurring in an investment cost if suppliers find profitable. The upstream sector may be modeled via Nash-Cournot players, who could exert some source market power or behave as price-takers (perfectly competitive case). The formulation below (objective function and constraints), based on the NANGAM built from the MultiMod framework (Huppmann and Egging 2014), models the upstream sector described above (see Tables 15.1 and 15.2 for sets, variable and parameter definition).

$$\max_{\substack{q^P, q^A, q^C, z^P \\ q^{-o}, q^{+o}, q^D}} \sum_{y,h,n,e} df_y dur_h \begin{pmatrix} \sum_{d \in D} \left[cour^s_{ysnd} \pi^D_{yhnde}(\cdot) + \left(1 - cour^s_{ysnd}\right) p^D_{yhsnde} \right] \\ q^D_{yhsnde} - \text{cost}(\cdot)^P_{yhsne} - \\ \sum_{a \in A} p^A_{yhae} q^A_{yhae} - \sum_{o \in O} \left(p^{-o}_{yhno} q^{-o}_{yhno} + p^{+o}_{yhno} q^{+o}_{yhno} \right) \\ - \sum_{g \in G} p^g_{yng} emiss^P_{ysneg} q^P_{ysneg} - inv^P_{ysne} z^P_{ysne} \end{pmatrix} \tag{15.1}$$

Table 15.1 General sets and parameters used by NANGAM-MultiMod

Sets	
$y \in Y$	Years
$h \in H$	Hours/days/seasons/representative periods (time slices)
$v \in V$	Loading cycles of storage (grouped time slices for injection and extraction)
$s \in S$	Suppliers
$n, k \in N$	Nodes (spatial disaggregation)
$d \in D$	Final demand sectors
$a \in A$	Transportation arcs
$o \in O$	Storage operators/technologies
$e, f \in E$	Energy carriers/fuels
$r \in R$	Regions
$g \in G$	Emission types (greenhouse gases)
	Mappings and subsets
$n, k \in N_r$	Node-to-region mapping
$r \in R_n$	Region-to-node mapping (any node can be part of several regions)
$a \in A^+_{ne}$	Subset of arcs ending at node *n* transporting fuel *e*
$a \in A^-_{ne}$	Subset of arcs starting at node *n* transporting fuel *e*
$e \in E^A_a$	Fuel(s) transported via arc *a*
$n^{A+}(a)$	End node of arc *a* (singleton)
$n^{A-}(a)$	Start node of arc *a* (singleton)
$e^O(o)$	Fuel stored by technology *o* (singleton)
$o \in O^E_e$	Subset of technologies storing fuel *e*
$h \in H^V_{vo}$	Mapping between loading cycle and hour/day/season/time slice
$v^H(h, o)$	Loading cycle of hour/day/season/time slice (singleton)
dur_h	Relative duration of hour/day/season *h* (with $\sum_h dur_h = 1$)

Table 15.2 Set of parameters and variables for the suppliers' problem in NANGAM-MultiMod

Parameters for the supplier problem	
df^{S}_{ys}	Discount factor of supplier s
$cour^{S}_{ysnd}$	Cournot market power parameter of supplier s at node n regarding sector d
$\cos t^{P}_{yhsne}(\bullet)$	Production cost function faced by supplier s at node n for fuel e
lin^{P}_{ysne}	Linear term of the production cost function ($lin^{P} \geq 0$)
qud^{P}_{ysne}	Quadratic term of the production cost function ($qud^{P} \geq 0$)
gol^{P}_{ysne}	Logarithmic (Golombek) term of the production cost function ($gol^{P} \geq 0$)
cap^{P}_{ysne}	Gross initial production capacity
avl^{P}_{yhsne}	Availability factor of production capacity
exp^{P}_{ysne}	Production capacity expansion limit
inv^{P}_{ysne}	Production capacity expansion (per-unit) costs
$dep^{P}_{yy'sne}$	Production capacity expansion depreciation factor
hor^{P}_{sne}	Production horizon (reserves)
$loss^{P}_{sne}$	Loss rate during production of fuel e at node n
ems^{P}_{ysneg}	Emission of type g during production of fuel e at node n by supplier s
Variables for the supplier problem	
q^{P}_{yhsne}	Quantity produced of fuel e by supplier s at node n
q^{A}_{yhsae}	Quantity transported through arc a
q^{O-}_{yhsno}	Quantity injected into storage o
q^{O+}_{yhsno}	Quantity extracted from storage o
q^{D}_{yhsnde}	Quantity sold to final demand sector d
z^{P}_{ysne}	Expansion of production capacity
α^{P}_{yhsne}	Dual for production capacity constraint
α^{O}_{yysno}	Dual for injection/extraction constraint
γ^{P}_{sne}	Dual for production horizon constraint
ζ^{P}_{ysne}	Dual for production capacity expansion limit
ϕ_{yhsne}	Dual for mass-balance constraint

where $\text{cost}(\cdot)^{P}_{yhsne}$ represent the supply cost function. There are several choices for the form of this cost function depending on where the gas is being extracted (onshore or offshore) as well as the type of gas (conventional or unconventional). In general, this function should be increasing in its argument (producing more gas costs more money). A nonlinear production cost curve (Golombek cost function) has been introduced for instance by the Rice World Gas Model—GASTALE (Lise et al. 2008; Boots et al. 2004), for which the marginal cost function is shown below

$$\text{Marginal Cost} = \alpha + \beta q + \gamma \ln\left(1 - \frac{q}{\text{capacity}}\right) \tag{15.2}$$

where α is the minimum per unit cost term, β the per unit linearly increasing cost term, and γ represents a term that induces high marginal costs when production is close to full capacity (Egging et al. 2010). As discussed in previously, the suppliers may either behave as competitive players (i.e., price-taking behavior) or as a Cournot (oligopolistic) player and exert some degree of market power. In the competitive case, the parameter $cour^{S}_{ysnd}$ is set to 0, and the suppliers faces the corresponding market equilibrium price (p^{D}_{yhsnde}). Otherwise, if suppliers behave as pure Cournot players ($cour^{S}_{ysnd} = 1$), suppliers are aware of the inverse demand function (π^{D}_{yhnde}) and hence the impact of their sales on the market price (Huppmann and Egging 2014). The negative components following the revenue terms represent the costs incurred by the suppliers for transporting natural gas via pipelines (p^{A}_{yhae}), for extracting and injecting natural gas from/to natural gas storage ($p^{-o}_{yhno}, p^{+o}_{yhno}$), cost associated to emissions (p^{g}_{yng}), and investments cost (inv^{P}_{ysne}). The prices charged by the service operators (pipeline and storage operators) are derived from market clearing constraints (to be defined later). The physical and engineering constraints that suppliers face are described next.

$$q^{P}_{yhsne} \le avl^{P}_{yhsne}\left(cap^{P}_{ysne} + \sum_{y'<y} dep^{P}_{y'ysne} z^{P}_{y'sne}\right) \tag{15.3}$$

$$\sum_{h\in H} dur_h q^{+o}_{yhsno} = \sum_{h\in H} dur_h \kappa q^{-o}_{yhsno} \tag{15.4}$$

$$\left(1 - loss^{P}_{sne}\right) q^{P}_{yhsne} - \sum_{a\in A+}\left(1 - loss^{A}_{a}\right) q^{A}_{yhsa} - \sum_{a\in A-} q^{A}_{yhsa} + \sum_{o\in O}\left(q^{+o}_{yhsno} - q^{-o}_{yhsno}\right) = 0 \tag{15.5}$$

$$z^{P}_{yno} \le \exp^{P}_{yno} \tag{15.6}$$

$$\sum_{y\in Y, h\in H} dur_h q^{P}_{yhsne} \le Hor^{P}_{sne} \tag{15.7}$$

Equation (15.3) limits at every model-period the amount of natural gas resource to be produced to the initial capacity plus the investments made up to the corresponding period, given the appropriate depreciation of historical capacity. Constraint (15.4) ensures the balance between injection and extraction of natural gas storage (where κ represents the corresponding losses). Equation (15.5) corresponds to the mass-nodal balance for every period. The upper limits on capacity investments and resource supply are given by constraints (15.6) and (15.7), respectively.

It is worth mentioning that the choice of the mathematical formulation highly depends on the natural gas market that is to be modeled and on the assumptions of the modelers. The formulation presented here differs, for instance, from the WGM, which models the suppliers as the producers of the natural gas resources, who then sell the primary resource to traders, who are then responsible to store and transport the gas to the demand regions (hence negotiate with pipeline and storage operators and with the final demand markets). This representation then considers the objective function of suppliers to be the classical profit maximization formulation (revenue minus the cost of production) subject to limits of resource extraction or capacity production. The WGM also considers traders to behave as Cournot players (opposite to NANGAM where suppliers have this ability). Following the notations in the WGM (Egging et al. 2010), the suppliers problem is then formulated as follows,

$$\max_{\mathrm{SALES}^P_{pdm}} \sum_{m\in M} \gamma_m \left\{ \sum_{d\in D} \mathrm{days}_d \left[\pi^P_{n(p)dm} \mathrm{SALES}^P_{pdm} - \mathrm{c}^P_{pm}\left(\mathrm{SALES}^P_{pdm}\right)\right]\right\} \tag{15.8}$$

$$s.t. \quad \mathrm{SALES}^P_{pdm} \le \mathrm{CAP}^{PR}_{pm} \quad \forall d,m \tag{15.9}$$

$$\sum_{d\in D} \mathrm{days}_d \mathrm{SALES}^P_{pdm} \le \mathrm{CAP}^P_p \quad \forall m \tag{15.10}$$

$$\mathrm{SALES}^P_{pdm} \ge 0 \quad \forall d,m \tag{15.11}$$

where SALES^P_{pdm} corresponds to the actual primary resource production by the suppliers who also incurred in a cost due to production. The WGM, like NANGAM, also considers a Golombek-type cost function (Golombek et al. 1995). The revenue obtained by the suppliers comes from the market-clearing equilibrium price $\pi^P_{n(p)dm}$, the dual variable of the market clearing constraint that links the supplier and trader problem. As mentioned above, the role of the traders in the WGM relies on buying gas from one or more producers and sell gas to one or more final consumption markets, facing the costs of transportation and congestions of natural gas pipelines. The type of formulation where producers or suppliers and traders are modeled independently has, for instance, also been considered to model the European natural gas market (Egging et al. 2008).

15.2.2 Midstream Sector

Several studies have focused on the midstream sector of natural gas markets. Studies include, for instance, a focus on North America (Feijoo et al. 2016, 2018; Sankaranarayanan et al. 2018; Siddiqui and Gabriel 2016), Europe (Egging et al. 2008; Egging and Holz 2016; Holz et al. 2016), and China (Xu et al. 2017; Zhang et al. 2016; Rioux et al. 2019). The midstream sector involves the service providers, such as the transportation (e.g., pipeline or truck), storage, and liquefaction. Pipeline operators maximize profit over a single or a group of owned pipelines. They incurred in a contract with either natural gas suppliers or traders (as discussed in Sect. 15.2.1) to transport natural gas across regions. To maximize profit, pipeline operators are responsible for assigning available capacity to suppliers/traders. Although pipeline network operations, regulations, and third-party access are different across countries or markets (see Rioux et al. 2019 for a study of price controls and third-party access to infrastructure on China's natural gas supply chain) a simplified approach that allows to model the contracts is to permit pipeline operators to allocate capacity based on a willingness to pay basis (Egging et al. 2010). The willingness to pay is modeled via a market clearing constraint linking the suppliers' optimization problem to the pipeline operator problem. The dual variable to this market clearing constraint represent the equilibrium price that suppliers must pay to the pipeline owners if they want to have access to available capacity. The pipeline owners then maximize their profit from transporting natural gas considering an operational cost or tariff. This modeling approach (following (Feijoo et al. 2016; Huppmann and Egging 2014)) is represented through Eqs. (15.12)–(15.15). A variant that can be considered [as shown in objective function (15.12)] is to model endogenous capacity investments in pipeline capacity and to penalize or tax emissions (e.g., methane leakage).

$$\max_{f^A, z^A} \sum_{y,h} df^A_{ya} dur_h \left(\left[\left(p^A_{yha} - trf^A_{ya} \right) f^A_{yha} \right] - \sum_{g \in G} p^g_{yng} emiss^A_{yag} f^A_{yha} - inv^A_{ya} z^A_{ya} \right) \quad (15.12)$$

$$s.t. \quad f^A_{yha} \leq cap^A_{ya} + \sum_{y' < y} dep^A_{y'ya} z^A_{ya} \quad (15.13)$$

$$z^A_{ya} \leq \exp^A_{ya} \quad (15.14)$$

$$\sum_s q^A_{yhsa} = f^A_{yha} \qquad \left(p^A_{yha} \right) \quad (15.15)$$

The formulation maximizes the profit of pipeline owners subject to capacity constraints (15.13) and limits to capacity expansion (15.14). The market clearing constraints are shown in Eq. (15.15), where the variable in parenthesis p^A_{yha} represents the equilibrium price (which appears in both suppliers and pipeline operator's optimization problems). Table 15.3 shows the parameter and variable definitions for the above problem.

Table 15.3 Set of parameters and variables for the arc operators' problem in NANGAM-MultiMod

Parameters for the Arc operator	
df^{A}_{ya}	Discount factor of arc operator a
trf^{A}_{yae}	Tariff for using arc a to transport fuel e
cap^{A}_{ya}	Gross initial capacity of arc a
exp^{A}_{ya}	Arc capacity expansion limit
inv^{A}_{ya}	Arc capacity expansion (per-unit) costs
$dep^{A}_{yy'a}$	Arc capacity expansion depreciation
wgt^{A}_{ae}	Weighting factor for distinct fuels in arc capacity
$loss^{A}_{ae}$	Loss rate during transportation through arc a of fuel e
ems^{A}_{yaeg}	Emission of type g during transportation through arc a of fuel e
Variables for the arc operator	
f^{A}_{yhae}	Quantity transported by the arc operator
z^{A}_{ya}	Expansion of arc capacity
τ^{A}_{yha}	Dual for arc capacity constraint
ζ^{A}_{ya}	Dual to arc capacity expansion limit
p^{A}_{yhae}	Market-clearing price of arc capacity

One assumption of the formulation is that constraints of the pressure that natural gas generates on pipelines are not considered. This assumption allows to have linear constraints which are amenable to the algorithms for solving these problems (to be discussed in Sect. 15.2.4). This assumption is unrealistic for offshore transportation networks which often consist of very long pipelines without compression, for which it is crucial to describe the pressure drops in the pipeline systems (Midthun 2007). Pressure constraints can be modeled using the *Weymouth* equation as follows,

$$W_{ij}\left(p_{ij}^{in}, p_{ij}^{out}\right) = K_{ij}^{W}\sqrt{\left(p_{ij}^{in}\right)^{2} - \left(p_{ij}^{out}\right)^{2}} \tag{15.16}$$

where $W_{ij}\left(p_{ij}^{in}, p_{ij}^{out}\right)$ is the flow on a pipeline between nodes $i - j$ because of the pressure difference between $p_{ij}^{in}, p_{ij}^{out}$. The constant K_{ij}^{W} represents the *Weymouth* factor for the pipeline. This constant depends (among others) on the pipelines length and its diameter and is used to relate the correct theoretical flow to the characteristics of the specific pipeline (Midthun 2007; Hellemo et al. 2012). Considering the *Weymouth* constraints brings nonlinearities into the constraints of the model. Different approaches have been proposed to deal with this, including the linearization through the Taylor series expansion (Midthun 2007). Note that given the structure of constraint (15.14), the mathematical programs including the *Weymouth* equation can be formulated (relaxed) as a second order cone programming (SOCP) model. Solving these problems with SOCP has become popular with the advent of quick interior-point method based solvers and their inclusion into popular optimization suites (Alizadeh and Goldfarb 2003; Andersen et al. 2003).

Another important component of the midstream sector is the natural gas storage. Storage operators allow suppliers to transfer natural gas between different seasons/days/hours (e.g., low and high demand seasons and peak demand days or hours) considered by the models. Transfer of natural gas is needed to account for daily balancing and price arbitrage, and seasonal balancing, and as a strategic backup supply to meet peak-demand as a result of cold winter days or hot summer days (air conditioning).

Here, we discuss two distinct representation of natural gas storage operators that are used by NANGAM and the WGM. The formulations differ based on the strategic decisions and hence the role that storage operators take in natural gas markets. In NANGAM (Eqs. (15.17)–(15.25) describe the mathematical formulation) (see Table 15.4 for parameters and variable definition), storage operators received

Table 15.4 Set of parameters and variables for the storage operators' problem in NANGAM-MultiMod

Parameters for the storage technology operator	
df^{O}_{yno}	Discount factor of operator of storage technology o at node n
trf^{O-}_{yno}	Tariff for injecting into storage technology o
cap^{O}_{yno}	Gross initial capacity for fuel stored in technology o over one loading cycle
exp^{O}_{yno}	Yearly storage capacity expansion limit
inv^{O}_{yno}	Yearly storage capacity expansion (per-unit) costs
$dep^{O}_{yy'no}$	Yearly storage capacity expansion depreciation
cap^{O-}_{yno}	Initial capacity for fuel injection into storage
exp^{O-}_{yno}	Storage injection capacity expansion limit
inv^{O-}_{yno}	Storage injection capacity expansion (per-unit) costs
$dep^{O-}_{yy'no}$	Storage injection capacity expansion depreciation
cap^{O+}_{yno}	Initial capacity for fuel extraction rate from storage technology o
exp^{O+}_{yno}	Storage extraction capacity expansion limit
inv^{O+}_{yno}	Storage extraction capacity expansion (per-unit) costs

(continued)

Table 15.4 (continued)

$dep^{O+}_{yy'no}$	Storage extraction capacity expansion depreciation
$loss^{O-}_{o}$	Loss rate of storage technology o (accounted at injection)
ems^{O-}_{yog}	Emission of type g of storage technology o (accounted at injection)
Variables for the storage technology operator	
f^{O-}_{yhno}	Quantity injected into storage
f^{O+}_{yhno}	Quantity extracted from storage
z^{O}_{yno}	Expansion of yearly storage capacity
z^{O-}_{yno}	Expansion of injection capacity
z^{O+}_{yno}	Expansion of extraction capacity
ζ^{O}_{yno}	Dual to yearly storage capacity expansion limit
ζ^{O-}_{yno}	Dual to injection capacity expansion limit
ζ^{O+}_{yno}	Dual to extraction capacity expansion limit
p^{O-}_{yhno}	Market-clearing price for injection into storage
p^{O+}_{yhno}	Market-clearing price for extraction from storage
τ^{O}_{yvno}	Dual for capacity constraint of storage technology in loading cycle v
κ^{O-}_{yhno}	Dual for injection capacity constraint of storage technology
κ^{O+}_{yhno}	Dual for extraction capacity constraint of storage technology

payments from suppliers in order for them to get access (injection) to underground storage facilities and then (in a later model period) to extract the natural gas. Storage operators face a cost (or tariff) for extracting natural gas whereas the suppliers face the cost of injection (which transfer to payments to storage operators). This dynamic is represented in the objective function (15.17) by the terms $\left(p^{-O}_{yhno} - trf^{-O}_{yno}\right) f^{-O}_{yhno} + p^{+O}_{yhno} f^{+O}_{yhno}$, where the first term (in parenthesis) represents the extraction payment and cost, and the second term representing the injection payment from suppliers to the storage operator. Note that the term $p^{-O}_{yhno} f^{-O}_{yhno} + p^{+O}_{yhno} f^{+O}_{yhno}$ appears explicitly in the objective function of suppliers (as costs that supplier face), where p^{-O}_{yhno} and p^{+O}_{yhno} are the equilibrium prices or dual variables of the market clearing constraints (15.24) and (15.25). The rest of the components of the objective function (15.17) correspond to the emissions cost during the injection process, investment in storage capacity, injection capacity and extraction capacity, respectively. Constraints (15.18) through (15.20) limit the storage, injection and extraction capacities which depend on historical and current capacity investment decisions.

Constraints (15.21) through (15.23) put upper bounds on the capacity investment decision variables.

$$\max_{\substack{f^{-o},f^{+o} \\ z^{O},z^{-O},z^{+O}}} \sum_{y,h} df^{O}_{yno} dur_h \begin{pmatrix} \left[\left(p^{-O}_{yhno} - trf^{-O}_{yno}\right) f^{-O}_{yhno}\right] + p^{+O}_{yhno} f^{+O}_{yhno} - \sum_{g \in G} p^{g}_{yng} emiss^{-O}_{yog} f^{-O}_{yhno} - \\ inv^{O}_{yno} z^{O}_{yno} - inv^{-O}_{yno} z^{-O}_{yno} - inv^{+O}_{yno} z^{+O}_{yno} \end{pmatrix} \quad (15.17)$$

$$\text{s.t.} \quad \sum_{h \in H^{V}_{vo}} dur_h f^{-O}_{yhno} \leq cap^{O}_{yno} + \sum_{y'<y} dep^{O}_{y'yno} z^{O}_{y'no} \quad (15.18)$$

$$f^{-O}_{yhno} \leq cap^{-O}_{yno} + \sum_{y'<y} dep^{-O}_{y'yno} z^{-O}_{y'no} \quad (15.19)$$

$$f^{+O}_{yhno} \leq cap^{+O}_{yno} + \sum_{y'<y} dep^{+O}_{y'yno} z^{+O}_{y'no} \quad (15.20)$$

$$z^{O}_{yno} \leq \exp^{O}_{yno} \quad (15.21)$$

$$z^{-O}_{yno} \leq \exp^{-O}_{yno} \quad (15.22)$$

$$z^{+O}_{yno} \leq \exp^{+O}_{yno} \quad (15.23)$$

$$\sum_{s} q^{-O}_{yhsno} = f^{-O}_{yhno} \qquad \left(p^{-O}_{yhno}\right) \quad (15.24)$$

$$\sum_{s} q^{+O}_{yhsno} = f^{+O}_{yhno} \qquad \left(p^{+O}_{yhno}\right) \quad (15.25)$$

A different modeling paradigm is used in the WGM. In this case, the strategic actions of storage operators are associated to the maximization of profit by selling stored gas to the marketers (Eq. (15.26), component $\pi^{W}_{n(s)dm}\text{SALES}^{S\rightarrow M}_{sdm}$) at the equilibrium price $\pi^{W}_{n(s)dm}$ (as opposed to NANGAM, where suppliers pay to storage operators to access stored natural gas, which will be then delivered to demand markets). Storage operators incur two types of purchase cost; one with Traders ($\pi^{T}_{n(s)1m}\text{PURCH}^{S\leftarrow T}_{sm}$) and another with regasifier agents ($\pi^{R}_{n(s)1m}\text{PURCH}^{S\leftarrow R}_{sm}$). Hence the modeling paradigm in the WGM assumes that storage operators behave as intermediate agents who buy natural gas from traders (hence, can be considered by traders as a different marketer) and regasifiers in low-demand seasons to profit from selling this gas to marketers when required in high demand periods. Note that the WGM also considers an injection cost ($c^{S}_{sm}\left(\text{PURCH}^{S\leftarrow T}_{sm} + \text{PURCH}^{S\leftarrow R}_{sm}\right)$) that is faced by the storage operators. Also, similar to NANGAM, investment in injection, extraction and storage capacity are considered as strategic decisions. The complete objective function used by the WGM is shown in Eq. (15.24). The constraints used in the WGM are not shown here. However, these are similar in nature to the constraints used in NANGAM shown in Eqs. (15.18)–(15.25).

$$\max_{\substack{\text{SALES}_{sm}^{S\to M}\\ \text{PURCH}_{sm}^{S\leftarrow R}\\ \text{PURCH}_{sm}^{S\leftarrow T}\\ \Delta_{sm}^{S,INJ},\Delta_{sm}^{S,EXT},\Delta_{sm}^{S,WG}}} \sum_{m\in M}\gamma_m \left\{ \begin{array}{l} \sum_{d=2,3} \text{days}_d \pi_{n(s)dm}^{W} \text{SALES}_{sdm}^{S\to M} \\ -\text{days}_1 \left[\begin{array}{l} \pi_{n(s)1m}^{T} \text{PURCH}_{sm}^{S\leftarrow T} \\ \quad + \pi_{n(s)1m}^{R} \text{PURCH}_{sm}^{S\leftarrow R} + \\ c_{sm}^{S} \left(\begin{array}{l} \text{PURCH}_{sm}^{S\leftarrow T} \\ \quad + \text{PURCH}_{sm}^{S\leftarrow R} \end{array} \right) \end{array} \right] \\ -\left[b_{sm}^{S,INJ}\Delta_{sm}^{S,INJ} + b_{sm}^{S,EXT}\Delta_{sm}^{S,EXT} + b_{sm}^{S,WG}\Delta_{sm}^{S,WG} \right] \end{array} \right\} \tag{15.26}$$

There are other midstream agents that are important to consider when global market models and secondary energy sectors (such as the power sector) are considered. For global markets, liquefiers and regasifier agents are important since most of the natural gas global trade is in liquified form (liquified natural gas, LNG). LNG provides the advantage that it uses a smaller volume than natural gas in its gaseous form. Hence, LNG is used to facilitate the transport of natural gas over long distances. LNG import/export facilities must then account for regasification (if LNG is transformed into its gas form and sent delivered pipelines) and liquefaction capacity (if natural gas is to be shipped and exported). Note that LNG can also be consumed by end demand sectors and be delivered by tank trucks. On the other hand, if secondary energy sectors are considered, energy-transformation agents (gas into electric power) can also be considered. This representation allows to better model the where the power sector competes with natural gas sector for final energy in the residential, commercial and industrial demand sectors (or downstream, to be discussed next). Liquefiers and regasifiers are explicitly modeled by the WGM (see the WGM reference (Egging et al. 2010) for the mathematical formulation). NANGAM-MultiMod (see reference for MultiMod (Huppmann and Egging 2014)) provides the flexibility to include these agents, although they have not been explicitly considered in Feijoo et al. (2016).

15.2.3 Downstream Sector: Use of Natural Gas by Demand Sectors

Most natural gas market models consider specific demand markets for the downstream sector. These markets are normally defined as the industrial, residential, commercial, transport, and electric power sectors. Different sectors utilize natural gas for different purposes. For instance, the industrial sector uses natural gas as a fuel for process heating (in combined heat and power systems), and as feedstocks (raw material) to produce chemical, hydrogen, or fertilizers. In the USA, natural gas in the industrial sector accounted for 31% of the total industrial energy consumption. Another example is the power sector, which directly uses natural gas to generate electricity (26% in the US power sector was generated by natural gas in 2017). The power sector is also considered to produce a secondary type of energy, since electricity is also consumed by the downstream sector, competing with other fuels, including natural gas (for instance in the residential sector, where gas and electricity can be used for cooking or heating).

We model the final demand sectors as profit maximizing agents, where their utility comes from having access to energy services, where different fuels (if natural gas is not considered to be the only energy commodity available in the market) provide different levels of a service (parameter eff^{D}_{ynde} as shown in Eq. (15.27)). When optimizing the demand problems (taking the derivative over the consumption variable of an energy commodity), Eq. (15.27) is obtained (per the NANGAM-MultiMod framework, where a similar idea and formulation is used by the WGM.), which represent the inverse demand function for a particular energy commodity. Note that this is the inverse demand function (π^{D}_{yhnde}) that appears in the supplier's problem in Sect. 15.2.1. Parameters and variable definitions are shown in Table 15.5.

$$p^{D}_{yhnde} = eff^{D}_{ynde}\left[\mathrm{int}^{D}_{yhnd} - slp^{D}_{yhnd}\left(\sum_{s\in S, f\in E} eff^{D}_{yndf}\, q^{D}_{yhsndf}\right)\right] - eucc^{D}_{yhnde}$$

$$-eucl^{D}_{yhnde}\left(\sum_{s\in S, f\in E} q^{D}_{yhsndf}\right) - \sum_{s\hat{I}S, f\hat{I}E} p^{G}_{yng}\, ems^{D}_{yedg} \tag{15.27}$$

Note that the final demand price obtained from the inverse demand function (p^{D}_{yhnde}) is composed of the sector price index in terms of energy services, weighted by the efficiency of the particular fuel in supplying the service. The price is also affected by the end-use-cost, modeled via a linear function with parameters $eucc^{D}_{yhnde}$ and $eucl^{D}_{yhnde}$ and the emissions cost (or tax imposed to consumers, if any).

We next discuss strategies to simultaneously solve the mathematical problems that have been presented for each market agent, and hence, obtain solutions (equilibria) in natural gas markets.

Table 15.5 Set of parameters and variables for the downstream (demand sector) problem in NANGAM-MultiMod

Functions and parameters for the demand sector	
Pi^{D}_{yhnde}	Inverse demand curve of sector *d* for fuel *e*
int^{D}_{yhnd}	Intercept of inverse demand curve of sector *d* at node *n*
slp^{D}_{yhnd}	Slope of inverse demand curve of sector *d* at node *n*
eff^{D}_{ynde}	Efficiency of energy service demand satisfaction of sector *d* by fuel *e* at node *n*
$eucc^{D}_{yhnde}$	Constant end use cost parameter of sector *d* regarding fuel *e*
$eucl^{D}_{yhnde}$	Linear end use cost parameter of sector *d* regarding fuel *e*
ems^{D}_{ydeg}	Emission of type *g* during consumption of fuel *e* at node *n*

15.2.4 Solution Approaches for Natural Gas Models

We formulate the problem of solving the natural gas model for Nash equilibrium as a Mixed Complementarity Problem (MCP) (Cottle and Dantzig 1970). MCP is a problem of finding $\begin{pmatrix} x \\ y \end{pmatrix} \in \mathbb{R}^g \times \mathbb{R}^f$ such that for the given $F : R^{g+f} \mapsto R^g$ and $G : R^{g+f} \mapsto R^f$, the following are satisfied.

$$x \geq 0$$

$$F(x,y) \geq 0$$

$$x^T F(x,y) = 0$$

$$G(x,y) = 0$$

$$y \text{ free}$$

We write this compactly as.

$$\begin{array}{lllllll} 0 & \leq & x & \perp & F(x,y) & \geq & 0 \\ \text{Free} & & y & , & G(x,y) & = & 0 \end{array}$$

Standard routines (e.g., the PATH algorithm (Ferris and Munson 2000) or Lemke's algorithm (Adler et al. 2016)) are available to solve large-scale MCPs, which are typically used to solve industry-relevant MCPs.

We now motivate how the problem of finding equilibrium in the natural gas market can be posed as an MCP. Convex optimization problems with linear constraints is a class of optimization problems where the first-order optimality conditions (Karush–Kuhn–Tucker conditions or KKT conditions (Karush and Albert 2008; Cottle 2012; Dreves et al. 2011)) are both necessary and sufficient for optimality. Given

$$\min_{x} f(x)$$

subject to

$$g(x) \leq 0$$

$$h(x) = 0$$

where $f : R^n \mapsto R$ is a continuously differentiable convex function and $g, h : R^n \mapsto R$ are linear functions, the necessary and sufficient first order optimality conditions (or KKT conditions) are

$$\begin{array}{rcl} \nabla_x f(x) + \lambda^T h(x) + \mu^T g(x) & = & 0 \\ g(x) & \leq & 0 \\ h(x) & = & 0 \\ \mu g(x) & = & 0 \\ \lambda & & \text{Free} \\ \mu & \geq & 0 \end{array}$$

Note that the KKT conditions can be compactly written as below, showing that it is indeed an MCP.

$$\begin{array}{lllcl} \text{Free} & x & \perp & \nabla_x f(x) + \lambda^T \nabla_x h(x) + \mu^T \nabla_X g(x) & = & 0 \\ 0 \leq & \mu & \perp & -g(x) & \geq & 0 \\ \text{Free} & \lambda & , & h(x) & = & 0 \end{array}$$

In this manner, the KKT conditions of each player can be stacked together to form an MCP which can be solved efficiently. The solution can be interpreted as a Nash equilibrium solution of the market, in which some players act as price-takers (i.e., perfectly competitive), while others may exert Cournot market power. Nevertheless, since not all players' optimization problems are strictly convex, we cannot guarantee uniqueness of the equilibrium. This is an inherent problem common to all large-scale energy models that are not entirely linear.

Another approach taken for relatively smaller models is to solve for equilibrium by posing the problem as a Mathematical programming with Equilibrium constraints (MPEC). An MPEC is a problem of the form

$$\min_{x,y} f(x,y)$$

subject to

$$g(x,y) \leq 0$$

y solves an MCP parameterized by x

While we have much less efficient algorithms to solve MPECs, they have the ability to model market power, by modeling a player as a *leader* and the rest as *followers* in the context of *Stackelberg* games (Von Stackelberg 2011). Such modeling paradigm is important to analyze policies if a player, owing to their large market share or other privileges obtained from the government, can anticipate the reaction of the market. However, the computational burden imposed in solving large, industry-relevant MPECs hurdles this modeling approach.

15.3 Data Sources for Natural Gas Markets and Case Study

In this section we present a case study based on the NANGAM model described above. NANGAM is calibrated using different data sources. Particularly, data for the USA is obtained (both historical and projections) from the Annual Energy Outlook (AEO) developed by the US Energy Information Administration (EIA) (US Energy Information Administration 2017, 2018b). US EIA provides both historical data and projections for natural gas production, consumption by sector, trade (exports and imports), among others. Projections are based on the National Energy Model System (NEMS), an integrated model of the US energy system linked to a macroeconomic model. NANGAM has also been calibrated to other projections, including natural gas consumption and production generated by the Global Change Assessment Model (GCAM) (Feijoo et al. 2018). Data for Mexico is obtained primarily from the *Prospectiva de gas natural (PGN)* developed by the Mexican Secretary of Energy, SENER (Secretaria de Energia (SENER) 2016). The PGN also provides historical data for Mexico as well as projections (regional) for expected production and consumption. PGN also provides information regarding pipelines that are under construction or have been approved or planned. Similarly, the Canadian government, via the National Energy Board (NEB), produces the *Canadian Energy Future* report, providing projections that are obtained by consults with Canadian energy experts and interested stakeholders representing industry, government, nongovernmental organizations, and academia. For global models (not focused on North America), such as the WGM described earlier, production capacity and forecast are obtained from technical literature (e.g., Oil and gas journal), European Energy models (such as the PRIMES model (Gusbin 2012)) and other global energy models for the rest of the world (e.g., POLES model (European Commission 2006)). Global Energy balances can also be obtained from the International Energy Agency (IEA) (International Energy Agency (IEA) 2015; OECD/IEA 2017; IEA 2017).

As mentioned above, NANGAM uses data from governmental entities of the USA, Canada, and Mexico, to obtain past and future data regarding production and consumption. Other important data, such as prices, investment costs, tariff for pipeline are also obtained from reports from the US government, including the Federal Energy Regulatory Commission (FERC).

Given the abovementioned data used to calibrate NANGAM, the baseline quantities for the 10 US census regions that are considered in the model are plotted in Fig. 15.2, from 2016 to 2050. Naturally, there is a mismatch between regional supply and demand, primarily because supply is driven from regional extraction costs whereas consumption is driven from different factors in the five sectors: residential, commercial, industrial, electrical, and transportation. On aggregate, supply is slightly higher than demand due to the losses of the system at different points of the supply chain (when storing or transporting natural gas).

In addition, we are focusing on the region of the USA and do not report here the trade with Canada, Mexico, and the Rest of the World, which is substantial. In our

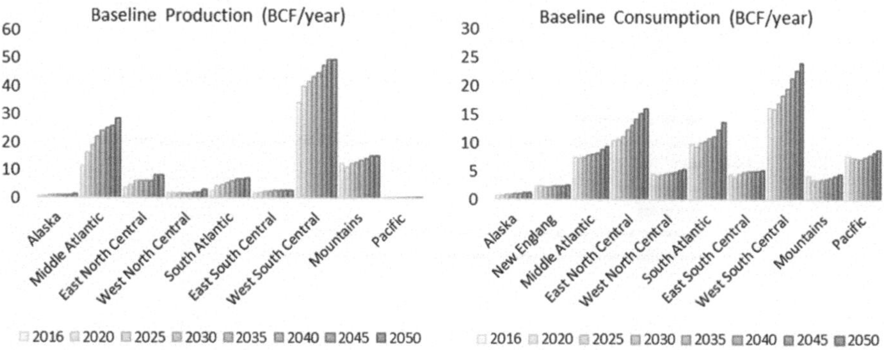

Fig. 15.2 Baseline production and consumption of US regions through 2050

calibration, following AEO2017 (US Energy Information Administration 2017b), the USA is a net exporter after 2020, which also implies that regional production will be higher than regional consumption. The mismatch between the two highlights the importance of the pipeline infrastructure for the robust operation of the system. Moreover, from a modeling perspective, it hints at the significance of using a framework that jointly accounts for both the production and the pipeline system when assessing the impact of a scenario to the natural gas sector.

To illustrate the type of analysis that can be conducted with these types of models, an increased demand scenario is implemented and studied with NANGAM. One source of increased demand for natural gas is the electricity sector. The discovery of cheap shale gas at Marcellus, coupled with the low emissions factor of natural gas, has led to the substitution of other fossil fuels, such as coal. Motivated by the already documented increasing penetration of natural gas (US Energy Information Administration 2017; Nalbandian 2015; Hayhoe et al. 2002), we will design a scenario where no coal plants are built in the future. In the design of our scenario, we will follow Feijoo et al. (2018). More specifically, we will assume that by 2050, 50% of electricity will be produced from gas-fired plants, compared to 30% in the baseline scenario. Furthermore, we assume that the increase happens gradually, that is, demand from the electricity sector increases by 0.97% every year. In what follows, we will focus our analysis to the USA. The results are shown in Fig. 15.3.

Albeit the percentage increase of demand is similar for all regions, regional production does not follow the same trend. We observe that most of the increase in consumption is covered from the production of three major regions, which, at the same time, are the three biggest producers: West South Central (Texas), Middle Atlantic, and the Mountains region. In particular, the change in cumulative consumption amounts to 20.43 BCF/year, out of which 6.19 BCF/year are covered from West South Central, 4.47 BCF/year from Middle Atlantic, and 1.36 BCF/year from Mountains. From the smaller regions, East South Central picks up 3.98 BCF/year while the increase in production of the rest of the regions is insignificant. For the rest of the regions, we observe that consumption increases more than their production. We can thus conclude that the first three regions exploit their potential to grow,

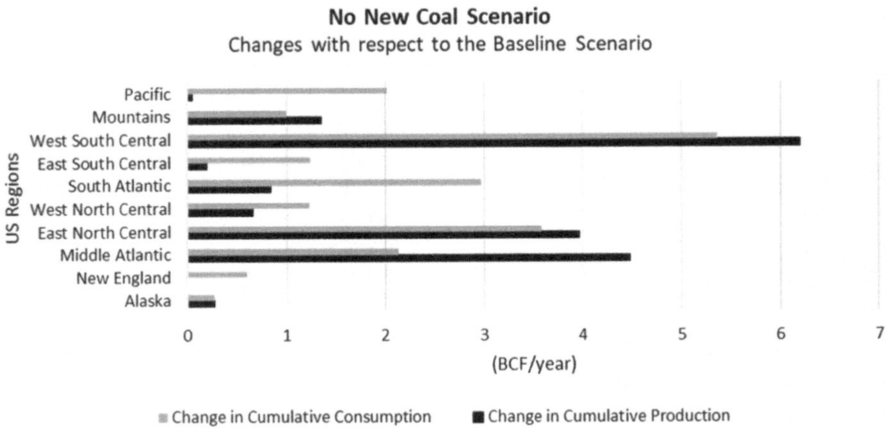

Fig. 15.3 Cumulative production & consumption changes compared to the baseline scenario Increase of cumulative regional demand leads to increase of cumulative regional production, when the existing resources allow for it. Production of certain regions grows more than others, due to their cost structure, resources available, and the dependence of their neighbors from them

since they can increase their production and at the same time remain competitive with respect to the producers of the other US regions. Regarding the three cheaper regions, their comparative growth depends on many other parameters apart from their cost, namely the demand of their neighbors. For example, Middle Atlantic also accommodates the New England and East North Central regions, whose demand grows and their production has limited potential. Hence, Middle Atlantic's production grows proportionally more than that of West South Central. For the latter region, the demand of the major exporting destination, Mexico, remains unchanged. Nevertheless, their production still grows to accommodate the increased in demand in East South Central.

In conclusion, in this stylized example, we explore the intuition behind these models. Although Game Theory provides us with a framework for understanding the interactions between market participants with conflicting objectives, it is not always clear what the driving forces are. Therefore, the usefulness of such tools is twofold. Firstly, we are able to gain a better understanding of the fundamental drivers that are at play and their significance. Secondly, they allow us to numerically assess the impact of a policy. At the hands of policy-makers, they are valuable tools with which they are able to compare different policies, for a range of scenarios, and pick the one best suited for their goal. It is also important to mention that MCPs have their challenges. The first challenge is associated to the development of solution techniques that can be scalable for large scale problems, particularly when certain market structure (e.g., Stackelberg game-MPEC or EPEC models) are considered. Authors in (Ruiz et al. 2014) also comment about the concerns, when using MCP or LCP formulations, for the representation of uncertainties, appropriate considerations of both physical and economical constraints, and the assumptions regarding rivals' behavior, or market structure (as mentioned above).

15.4 Outlook and Examples of Energy System Transformation

Burning fossil fuels such as coal, gas and oil produces more than 80% of the world's energy and more than 90% of global carbon dioxide emissions. Slowing and ultimately stopping climate change depends on decarbonization and transformation of the global energy system into one that does not dump CO2 into the atmosphere. Because gas-fired power plants emit roughly half as much CO2 per unit of energy produced as coal-fired plants, the greatly expanded gas supplies promised by new hydraulic fracturing (fracking) methods have been celebrated as a means of cutting emissions. Progressive inclusion of gas can thus decarbonize the energy sector and serve as a "bridge" to a more distant future when carbon-free, renewable-energy technologies are more affordable and reliable than they are now (Davis and Shearer 2014). However, carbon emissions (CO2) are not the only Greenhouse Gas Emissions (GHGs) contributing to climate change. Others include aerosols and methane (CH4), the main component of natural gas. Because of the high global warming potential (GWP) of methane (estimated by the EPA to have a GWP of 28–36 over 100 years (EPA 2016)), climate benefits from using natural gas as a bridge fuel will highly depend on the use system leakage rates (Brandt et al. 2014).

The Regional Greenhouse Gas Initiative (RGGI) experience supports the conventional wisdom that natural gas may act as a "bridge fuel" to move away from higher CO2 polluting coal and oil (Littell 2017). However, the bridge fuel relies in the need that there will be a follow-on transition to renewables as a primary generation resource. The Achilles' heel (as defined by Littell 2017) in the bridge-fuel hypothesis may be that overinvestment in natural gas infrastructure can undercut the transition to lower-emitting technologies (Feijoo et al. 2018; Littell 2017). In fact, authors in reference (Feijoo et al. 2016) show how the transition to low-carbon economy still relies on natural gas infrastructure investment (midstream pipelines) which is likely to be underutilized by mid-century. In the short run, there is a distinct possibility that optimistic gas-price financial models and cost allocation constructs that encourage natural gas switching result in overbuilding interstate and intrastate pipelines, local gas distribution facilities, and gas power plants. This could permanently undercut a transition to renewables, efficiency, and other advanced energy technology resources, or result in high expenditures and loses from stranded assets.

Despite these concerns, there are other international gas markets that have taken the lead in increasing natural gas consumption in order to transition to a "lower" carbon future economy. Examples include the US power sector, where natural gas surpassed coal as the main fuel used for generation. Natural gas is not only expected to play an important role in the power sector. New methanol industrial plants under construction in the USA are expected to increase natural gas consumption in the industrial sector, mainly as a feedstock for chemical production. The Energy

Information Administration (US EIA)[1] estimates that by the end of 2020 two new methanol plants on the Gulf Coast will enter service. The US EIA reports that these plants (and others under construction) will result in combined capacity of 3.3 million metric tons per year, equating to methanol production of 0.30 billion British thermal units per day, a 45% increase from the current US capacity (US Energy Information Administration 2019).

A second example is the energy reform and energy sector transformations in Mexico. Based on the 2014 Mexican Energy reform, the country opened its energy reserves to private investors, in an attempt to meet its rising demands, while also advancing an ambitious goal of carbon reduction (Veysey et al. 2014; SENER 2015). Mexico's total energy consumption in 2015 consisted mostly of petroleum (46%), followed by natural gas (40%) (US Energy Information Administration 2017a). Natural gas is increasingly replacing oil and coal in electric power generation. Projected increases in natural gas consumption are resulting in plans to many new pipelines to import natural gas from the USA (Feijoo et al. 2016, 2018), principally because the development of Mexico's shale gas resources is proceeding slowly, while consumption is projected to increase 31% from 2015 to 2029 (Secretaria de Energia (SENER) 2016; SENER 2015). Indeed, Mexico is a net importer of natural gas, mostly via pipeline from the USA. Consequently, Mexico will (be expected to) rely on increased pipeline imports of natural gas from the USA and liquefied natural gas (LNG) imports from other countries (Feijoo et al. 2016).

Lastly, another interesting example of increasing natural gas consumption is the Chinese natural gas sector. Historically, China has had a regulated and complex market, with limited pipeline access and price caps that are imposed in order to keep city-gate (downstream) prices accessible for consumers. Within this context, natural gas production in China still grew rapidly, increasing over 500% (from 27.2 bcm in 2000 to 136.9 bcm in 2016), yet being outpaced by an increasing demand of about 850% over the same period (Rioux et al. 2019). The gap between supply and demand has been projected to increase further, reaching 210 bcm by 2020 (Wang et al. 2013). The gap is in part justified by the ambitious targets set by Chinese policy makers for the natural gas industry; which consist in meeting the goal of 10% of total energy consumption by 2020, 15% by 2030, and increasing annual demand to 360 bcm by 2020 (see references in Rioux et al. 2019). This is sustained by the 13th Five-Year Plan of China for Natural Gas Development, which outlines a substantial build up in capacity, particularly in the natural gas power generation and pipeline transportation segments (Central Compilation and Translation Press 2016).

[1] https://www.eia.gov/naturalgas/weekly/. Accessed on 29/1/2019. Short-Term Energy Outlook (STEO), January 2019 (U.S. Energy Information Administration 2019).

References

Adler, I., Cottle, R. W., & Pang, J. S. (2016). Some LCPs solvable in strongly polynomial time with Lemke's algorithm. *Mathematical Programming, 160*, 477.

Aguiar, A., Narayanan, B., & McDougall, R. (2016). An overview of the GTAP 9 data base. *Journal of Global Economic Analysis, 1*, 181.

Alizadeh, F., & Goldfarb, D. (2003). Second-order cone programming. *Mathematical Programming, 95*(1), 3–51.

Andersen, E. D., Roos, C., & Terlaky, T. (2003). On implementing a primal-dual interior-point method for conic quadratic optimization. *Mathematical Programming, 95*, 249.

Beltramo, M. A., Marine, A. S., & Weyant, J. P. (2018). A North American Gas Trade Model (GTM). *The Energy Journal, 7*(3), 15–32.

Bernard, J. T., Bolduc, D., & Hardy, A. (2002). The costs of natural gas distribution pipelines: The case of SCGM, Québec. *Energy Economics, 24*(5), 425–438.

Bernstein, P., Tuladhar, S. D., & Yuan, M. (2016). Economics of U.S. natural gas exports: Should regulators limit U.S. LNG exports? *Energy Economics, 60*, 427–437.

Boots, M. G., Rijkers, F. A. M., & Hobbs, B. F. (2004). Trading in the downstream European gas market: A successive oligopoly approach. *The Energy Journal., 25*(3), 73–102.

Bosetti, V., Carraro, C., Galeotti, M., Massetti, E., & Tavoni, M. (2006). WITCH: A World Induced Technical Change Hybrid model. *The Energy Journal, 27*(SPEC. ISS. OCT), 13–37.

Brandt, A. R., Heath, G. A., Kort, E. A., O'Sullivan, F., Petron, G., Jordaan, S. M., et al. (2014). Methane leaks from North American natural gas systems. *Science, 343*, 733–735.

Brown, S. P. A., & Yücel, M. K. (2008). Deliverability and regional pricing in U.S. natural gas markets. *Energy Economics, 30*(5), 2441–2453.

Capros, P., Van Regemorter Paroussos, P. D., Karkatsoulis, P., Fragkiadakis, C., Tsani, S., Charalampidis, I., et al. (2013) *GEM-E3 model documentation. JRC technical report.* https://ec.europa.eu/jrc/en/publication/eur-scientific-and-technical-research-reports/gem-e3-model-documentation

Capros, P., Kannavou, M., Evangelopoulou, S., Petropoulos, A., Siskos, P., Tasios, N., et al. (2018). Outlook of the EU energy system up to 2050: The case of scenarios prepared for European Commission's 'clean energy for all Europeans' package using the PRIMES model. *Energy Strategy Reviews, 22*, 255–263.

Central Compilation and Translation Press. (2016). 13th five-year plan of China for natural gas development.

Cole, W. J., Medlock, K. B., & Jani, A. (2016). A view to the future of natural gas and electricity: An integrated modeling approach. *Energy Economics, 60*, 486–496.

Colombo, S., El Harrak, M., & Sartori, N. (Eds.). (2016). *The future of natural gas, markets and geopolitics*. Rome: Istituto di Affari Internazionali (IAI).

Cottle, R. W. (2012). William Karush and the KKT theorem. *Documenta Mathematica Extra Volume ISMP, 1*, 255–269.

Cottle, R. W., & Dantzig, G. B. (1970). A generalization of the linear complementarity problem. *Journal of Combinatorial Theory, 8*, 79.

Davis, S. J., & Shearer, C. (2014). Climate change: A crack in the natural-gas bridge. *Nature, 514*(7523), 436–437.

de Joode, J., & Özdemir, Ö. (2010). Demand for seasonal gas storage in Northwest Europe until 2030: Simulation results with a dynamic model. *Energy Policy, 38*(10), 5817–5829.

Dieckhöner, C., Lochner, S., & Lindenberger, D. (2013). European natural gas infrastructure: The impact of market developments on gas flows and physical market integration. *Applied Energy, 102*, 994–1003.

Dreves, A., Facchinei, F., Kanzow, C., & Sagratella, S. (2011). On the solution of the KKT conditions of generalized Nash equilibrium problems. *SIAM Journal on Optimization, 21*, 1082.

Egging, R., Gabriel, S. A., Holz, F., & Zhuang, J. (2008). A complementarity model for the European natural gas market. *Energy Policy, 36*(7), 2385–2414.

Egging, R., & Holz, F. (2016). Risks in global natural gas markets: Investment, hedging and trade. *Energy Policy, 94*, 468–479.

Egging, R., Holz, F., & Gabriel, S. A. (2010). The world gas model. A multi-period mixed complementarity model for the global natural gas market. *Energy, 35*(10), 4016–4029.

EPA. (2016). *Understanding global warming potentials*. https://www.epa.gov/ghgemissions/understanding-global-warming-potentials.

European Commission. (2006). *World energy technology outlook 2050–WETO H2*.

Feijoo, F., Huppmann, D., Sakiyama, L., & Siddiqui, S. (2016). North American natural gas model: Impact of cross-border trade with Mexico. *Energy, 112*, 1084–1095.

Feijoo, F., Iyer, G. C., Avraam, C., Siddiqui, S. A., Clarke, L. E., Sankaranarayanan, S., et al. (2018). The future of natural gas infrastructure development in the United States. *Applied Energy, 228*, 149–166.

Ferris, M. C., & Munson, T. S. (2000). Complementarity problems in GAMS and the PATH solver. *Journal of Economic Dynamics and Control, 24*, 165.

Fragkos, P., Kouvaritakis, N., & Capros, P. (2016). Incorporating uncertainty into world energy modelling: The PROMETHEUS model. *Environmental Modeling and Assessment, 20*(5), 549–569.

Gabriel, S. A., Zhuang, J., & Kiet, S. (2005). A large-scale linear complementarity model of the north American natural gas market. *Energy Economics, 27*(4), 639–665.

Golombek, R., Gjelsvik, E., & Rosendahl, K. E. (1995). Effects of liberalizing the natural gas markets in Western Europe. *The Energy Journal, 16*, 85–112.

Gusbin, D. (2012). The PRIMES energy model.

Hartley, P., & Medlock, K. B. (2006). The Baker Institute world gas trade model. In A. Jaffe, D. Victor, & M. Hayes (Eds.), *Natural gas and geopolitics: From 1970 to 2040* (pp. 357–406). Cambridge: Cambridge University Press.

Hayhoe, K., Kheshgi, H. S., Jain, A. K., & Wuebbles, D. J. (2002). Substitution of natural gas for coal: Climatic effects of utility sector emissions. *Climatic Change, 54*(1), 107–139.

Hellemo, L., Midthun, K., Tomasgard, A., & Werner, A. (2012). Natural gas infrastructure design with an operational perspective. *Energy Procedia, 26*, 67–73.

Holz, F., Richter, P. M., & Egging, R. (2016). The role of natural gas in a low-carbon Europe: Infrastructure and supply security. *The Energy Journal, 37*(SpecialIssue3), 33–59.

Holz, F., & von Hirschhausen, C. (2013). The infrastructure implications of the energy transformation in Europe until 2050—Lessons from the Emf28 modeling exercise. *Climate Change Economics, 4*(supp01), 1340006.

Holz, F., von Hirschhausen, C., & Kemfert, C. (2008). A strategic model of European gas supply (GASMOD). *Energy Economics, 30*, 766.

Huntington, H. G. (2016). Introduction: North American natural gas markets in transition. *Energy Economics, 60*, 401–404.

Huppmann, D., & Egging, R. (2014). Market power, fuel substitution and infrastructure—A large-scale equilibrium model of global energy markets. *Energy, 75*, 483–500.

International Energy Agency (IEA). (2015). World Energy Outlook 2015. *World Energy Outlook 2015 Present to Press London, 10 November 2015*.

IEA. (2017). *World energy balances: an overview of global trends*.

Jacoby, H. D., O'Sullivan, F. M., & Paltsev, S. (2012). The influence of shale gas on U.S. Energy and Environmental Policy. *Economics of Energy and Environmental Policy, 1*(1).

Karush, W., & Albert, W. (2008). Karush-Kuhn-Tucker conditions.

Kim, S. H., Edmonds, J., Lurz, J., Smith, S. J., & Wise, M. (2006). The ObjECTS framework for integrated assessment: Hybrid modeling of transportation. *The Energy Journal, 27*(SPEC. ISS. OCT.), 63–91.

Levi, M. (2013). Climate consequences of natural gas as a bridge fuel. *Climatic Change, 118*(3–4), 609–623.

Lin, B., & Wang, T. (2012). Forecasting natural gas supply in China: Production peak and import trends. *Energy Policy, 49*, 225–233.
Lise, W., Hobbs, B. F., & van Oostvoorn, F. (2008). Natural gas corridors between the EU and its main suppliers: Simulation results with the dynamic GASTALE model. *Energy Policy, 36*(6), 1890–1906.
Littell, D. (2017). Natural gas: bridge or wall in transition to low-carbon economy? *Natural Gas and Electricity, 33*(6).
McJeon, H., et al. (2014). Limited impact on decadal-scale climate change from increased use of natural gas. *Nature, 514*(7523), 482–485.
Midthun, K. T. (2007). *Optimization models for liberalized natural gas markets*. Trondheim: Norwegian University of Science and Technology.
Mignone, B. K., Showalter, S., Wood, F., McJeon, H., & Steinberg, D. (2017). Sensitivity of natural gas deployment in the US power sector to future carbon policy expectations. *Energy Policy, 110*(January), 518–524.
Nalbandian, H. (2015). Climate implications of coal-to-gas substitution in power generation. IEA Clean Coal Centre.
OECD/IEA. (2017). *World energy balances: overview 2017*. IEA.
Oliver, M. E. (2015). Economies of scale and scope in expansion of the U.S. natural gas pipeline network. *Energy Economics, 52*, 265–276.
Oliver, M. E., Mason, C. F., & Finnoff, D. (2014). Pipeline congestion and basis differentials. *Journal of Regulatory Economics, 46*(3), 261–291.
Paltsev, S., & Zhang, D. (2015). Natural gas pricing reform in China: Getting closer to a market system? *Energy Policy, 86*, 43–56.
Rioux, B., Galkin, P., Murphy, F., Feijoo, F., Pierru, A., Malov, A., Li, Y., et al. (2019). The economic impact of price controls on China's natural gas supply chain. *Energy Economis, 80(C)*, 394–410.
Ruiz, C., Conejo, A. J., Fuller, J. D., Gabriel, S. A., & Hobbs, B. F. (2014). A tutorial review of complementarity models for decision-making in energy markets. *EURO Journal on Decision Processes, 2*(1-2), 91–120.
Sankaranarayanan, S., Feijoo, F., & Siddiqui, S. (2018). Sensitivity and covariance in stochastic complementarity problems with an application to north American natural gas markets. *European Journal of Operational Research, 268*(1), 25–36.
SENER. (2015). Prospectiva de Petróleo Crudo, y Petroliferos. Prosepectivas de Gas Natural y Gas L.P 2015–2029. p. 209.
Secretaria de Energia (SENER). (2016). Prospectiva de gas natural 2016–2030.
Siddiqui, S., & Gabriel, S. A. (2016). Modeling market power in the U.S. shale gas market. *Optimization and Engineering, 18*, 203–213.
U.S. Department of Energy. (2015). *Natural gas infrastructure implications of increased demand from the electric power sector*. Southwest, Washington, DC: U.S. Department of Energy.
US Energy Information Administration. (2017a). Country analysis brief: Mexico. October, p. 32.
US Energy Information Administration. (2017b). Annual energy outlook 2017 with projections to 2050.
US Energy Information Administration. (2018a). Natural gas market module of the national energy modeling system: model documentation 2018.
US Energy Information Administration. (2018b). Annual energy outlook 2018 with projections to 2050.
US Energy Information Administration. (2019). Short-Term Energy Outlook (STEO). January.
Veysey, J., et al. (2014). Pathways to Mexico's climate change mitigation targets: A multi-model analysis. *Energy Economics, 56*, 587–599.
Von Stackelberg, H. (2011). *Market structure and equilibrium* (Vol. XII, p. 134). New York: Springer.
Wang, J., Feng, L., Zhao, L., & Snowden, S. (2013). China's natural gas: Resources, production and its impacts. *Energy Policy, 55*, 690–698.

Xu, J., Hallack, M., & Vazquez, M. (2017). Applying a third party access model for China's gas pipeline network: An independent pipeline operator and congestion rent transfer. *Journal of Regulatory Economics, 51*(1), 72–97.

Yeh, S., Cai, Y., Huppman, D., Bernstein, P., Tuladhar, S., & Huntington, H. G. (2016). North American natural gas and energy markets in transition: Insights from global models. *Energy Economics, 60*, 405–415.

Zhang, Q., Li, Z., Wang, G., & Li, H. (2016). Study on the impacts of natural gas supply cost on gas flow and infrastructure deployment in China. *Applied Energy, 162*, 1385–1398.

Chapter 16
Future Research Directions for Sustainable Natural Resource Management

Yihsu Chen, Antonio J. Conejo, and Andrew L. Liu

We discuss three directions for future research in sustainable natural resource management. These directions are drawn mainly from our research experience and expertise in the energy sector and other natural resource areas. First, we believe that future sustainability and resilience in natural resource management can be enhanced through better utilization of outcomes from process- or physical-based climate change models, such as Global Climate Models.[1] Second, directly related to the electric power sectors is a better harnessing variability and unpredictability via flexibility, transmission, and storage. Third, future natural resource management should consider interdependence of multiple systems, such as power, natural gas, and water systems, through co-optimization of these interdependent systems. We elaborate each of these three future research directions in this chapter.

[1] Much of the discussion in this section draws upon the first and third author's experience of working on a project of the California's Fourth Climate Assessment. However, the discussions here are sufficiently broad to generalize to other situations.

Y. Chen (✉)
Jack Baskin School of Engineering, University of California, Santa Cruz, Santa Cruz, CA, USA
e-mail: ychen225@ucsc.edu

A. J. Conejo
Integrated Systems Engineering Department, Electrical and Computer Engineering Department, The Ohio State University, Columbus, OH, USA

A. L. Liu
School of Industrial Engineering, Purdue University, West Lafayette, IN, USA

C. Chen et al. (eds.), *Pursuing Sustainability*, International Series in Operations Research & Management Science 301,
https://doi.org/10.1007/978-3-030-58023-0_16

16.1 Harnessing Results from Global Climate Models

A robust assessment of resource and technology management to adapt to climate change or to enhance sustainability relies on prediction of future climate conditions. One approach is to harness the results from the Global Climate Models or GCMs. GCMs are tools developed in physical science research communities to predict future climate when the earth system is subject to various climate forcing. As an example of common climate change scenarios used by the California's Fourth Climate Change Assessment, data from historical period of 1950–2005 were used for projecting 2006–2100 climate conditions under two scenarios: Representative Concentration Pathways (RCPs) 4.5 and 8.5, corresponding to medium and higher future greenhouse and aerosol loading, respectively (California Natural Resource Agency 2019).

However, outputs from GCMs are known to be with great uncertainties, mainly from three sources: models, scenarios, and randomness associated with climate variability (Hawkins and Sutton 2011). The model uncertainty stems from the fact that each GCM projects somewhat different future changes in climate even when they are in response to the same radiative forcing. The scenario uncertainty arises from unknown future changes in anthropogenic forcing. GCMs from various international modeling centers are now hosted by the World Climate Change Programme under the United Nations Intergovernmental Panel on Climate Change (IPCC). The CMIP6 (Coupled Model Intercomparison Project) archives more than 50 different GCMs with standard experimental design and organization to facilitate model comparisons (World Climate Research Programme 2019). Finally, climate itself also fluctuates, which affects the signal of anthropogenic emissions. GCMs are also subject to systematic errors and bias (Cayan et al. 2018). Examples include failure to capture the amount of precipitation and runoff given the soil moisture conditions. Therefore, output from GCMs need to be corrected before using them (Pierce et al. 2015).

Moreover, the outputs from GCMs typically have very coarse spatial resolutions with the size of grid cells in 100 km or more on each side, making them less useful for natural resource management in energy, water, and other sectors, which need much finer resolutions. To better account for the impacts of local topography on climate change, researchers typically apply different downscaled techniques in order to produce results at a finer resolution, for example, 4–6 miles or 1/16 degree. There are two types of downscaling: statistical and dynamical downscale (Pierce et al. 2014). Whereas the statistical approaches harness the historic relationships between large- and small-scale conditions, the dynamical downscaling methods rely on regional climate models at the domain boundaries of GCM's output. Finally, downscaled GCM data are typically with daily, or even hourly, temporal resolutions, which make them appropriate for coupling with various optimization-based models that we introduce in Chapters 12–15. We illustrate the usefulness of GCM data to identify infrastructure vulnerability and assist decision makers for resource allocation and investment planning. Our focus is on how to estimate the risk of sea-level-rise on the natural gas systems in Northern California.

16.1.1 Sea Level Rise Risk

According to the US National Oceanic and Atmospheric Administration (NOAA), sea level is rising and is at an increasing rate (National Oceanic and Atmospheric Administration n.d.). Causes of global sea level rise include the thermal expansion of the ocean (since water expands as it becomes warmer), melt from glaciers, extraction of surface water, and contributions from the large ice sheets on Antarctica and Greenland. In addition to global signals, regional sea level rise is affected by local or regional effects from ocean dynamics. Changes in the ocean circulation can affect local sea level, causing water to accumulate or disperse along the coast due to winds (Cayan et al. 2018).

The GCM's downscaled data from the California's Fourth Climate Change Assessment include nine coastal stations, from most south La Jolla to the Crescent City at the north near the border with Oregon. The data report information on daily secular sea level, hourly astronomical tides, and hourly ocean/atmosphere during 1950–2099, where the last component represents the short-term fluctuations caused by storm surges and the El Nino Southern Oscillation events. The total daily sea level can be calculated as the sum of secular sea level, the daily maximum tide, and the daily maximum ocean/atmosphere effect. Figure 16.1 plots the GCM's simulated total sea level in the San Francisco station during 1950–2090 of 99.9th- and 50th-percentile for RCP 4.5 and 8.5 scenarios, respectively. Since historical data during 1950–2005 are the inputs, predicted sea level prediction over this period are the same for all the scenarios. Under the RCP8.5 scenario, the sea level rises more dramatically than the RCP4.5. For each scenario, 99.9th-percentile increases at a faster rate than 50th-percentile.

Together with digital elevation model data (DEM) (U.S. Geological Survey 2019), which record the U.S. elevation data at 1-m resolution, we can identify the expected water depth under sea level rise at a given time for a natural gas asset. Since the maximum sea level under the RCP8.5 99.9th-percentile (worst case) is around 350 (centimeters) toward the end of the century, we limit our attention to those assets with an elevation below 350 cm. Figure 16.2 plots two compressor stations in the Pacific Gas & Electrics (PG&E) system that are with a greater likelihood to be submerged under the sea water. More specifically, one compressor ID1, is located near Crescent City, California; while the second one, compressor ID2, is located near the Bay Area. As the gas network is normally radial, where the pressure in the network is gradually decompressed" along the pipelines through pressure regulating stations, the impact might be limited to those consumers downstream of the damaged compressor stations. Of course, there are other types of assets (e.g., regulating stations) that might also be subject to the risk.

Tables 16.1 and 16.2 report the predicted probability that compressor stations ID1 and ID2 are subject to inundation at different water depth (cm) under RCP 8.5: 99.9th-percentile, respectively. As an illustration, the analysis is divided into four periods: 2021–2040, 2041–2060, 2061–2080, and 2080–2099. In each period, the probabilities when summing across each row is equal to 1. Overall, the risk or

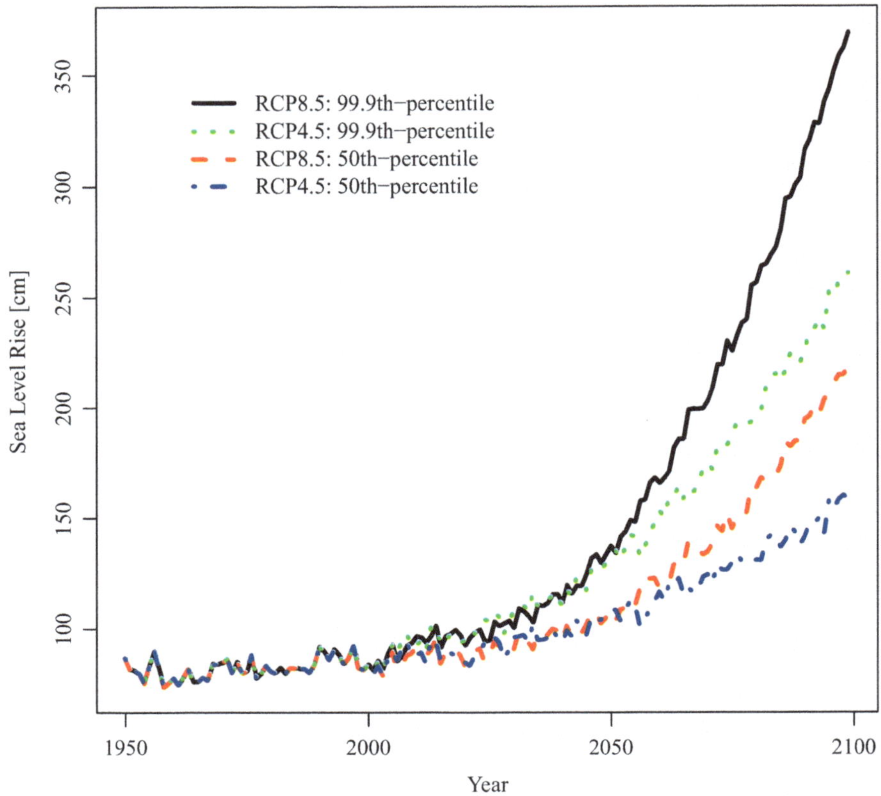

Fig. 16.1 Plot of simulated sea-level rising in [cm] during 1950–2100 for RCP 8.5 and 4.5 scenarios at the San Francisco station (Cayan et al. 2018)

probability that a compressor station is subject to inundation, as a result of sea-level rise, increases over time. For instance, station ID#1's risk associated with water depth 0–50 cm" increases from 0.01% in 2021–2040, to 0.13% in 2041–2060, to 5.39% in 2061–2080, and finally to 24.78% in last period. A similar observation is emerged in Table 16.2 for compressor station ID#2. Information in Tables 16.1 and 16.2 can then be used by stochastic operation or planning model or decision analysis tools to represent sea level rise risk and to enhance resource management.

16.2 Harnessing Variability and Unpredictability via Flexibility, Transmission and Storage

In this section, we focus on sustainable natural resource management in the electric power sector. While the electricity sector has been traditionally the biggest air pollutant emitter and is the main culprit for depleting the coal reserve on earth, the past

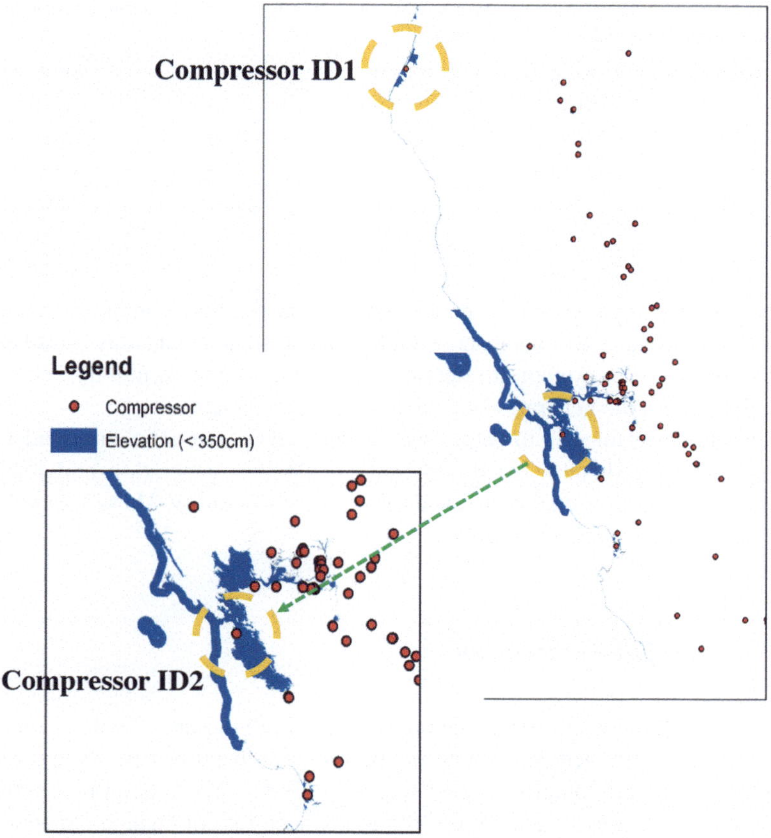

Fig. 16.2 Gas compressors near the San Francisco Bay Area that are subject to the risk of sea-level rising (Chen et al. 2020)

Table 16.1 Probability in [%] that the compressor station ID#1 is subject to inundation at different water depth (cm) under RCP 8.5: 99.9-percentile

Periods\depth	No flood	0–50	50–100	≥100
2021–2040	100.00	0.00	0.00	0.00
2041–2060	99.87	0.13	0.00	0.00
2061–2080	94.06	5.39	0.51	0.03
2081–2099	51.58	24.78	17.26	6.38

decade has witnessed tremendous growth of renewable energy resources, especially wind and solar, around the world. Weather dependent renewable sources are being increasingly integrated into power systems. At the transmission (400 kV) and sub-transmission (132 kV) levels, large utility-scale wind and solar power plants are being installed in increasing numbers. On the other hand, behind-the-meter photovoltaic installations are increasingly integrated in distribution systems making the

Table 16.2 Probability in [%] that the compressor station ID#2 is subject to inundation at different water depth (cm) under RCP 8.5: 99.9-percentile

Periods\depth	No flood	0–50	50–100	≥100
2021–2040	100.00	0.00	0.00	0.00
2041–2060	99.98	0.02	0.00	0.00
2061–2080	91.95	7.47	0.58	0.01
2081–2099	36.78	28.12	24.04	11.07

net demand of many consumers increasingly stochastic. Since the power output of both wind and solar power plants are variable and, to some extent, day-ahead unpredictable, most power system are operated under increasing variability and unpredictability at both the generation side and the demand side.

The main "instruments" to help power systems to operate efficiently under such variable and unpredictable conditions are (1) flexible generation and load, (2) expanded transmission availability, and (3) storage availability. These are explained below using three simple examples.

16.2.1 *Flexible Generation*

In this first example we illustrate the importance of relying on a flexible generation mix. Consider an isolated demand during two consecutive time periods in which its actual load is 10 and 20 units, respectively (Fig. 16.3). This demand is supplied by an inflexible unit with a capacity of 30 units, whose rate of change per time (aka ramping rate) period is just 5 units.

In such a situation, the demand is fully supplied in the first period (load is equal to 10 units), but not in the second period (load is equal to 20 units) since the inflexible unit can only increase its production by 5 units, which results in an unserved energy of 5 units.

If the inflexible unit is replaced by a flexible one, whose rate of change per time period is 10, the power supply changes as follows. The demand is fully supplied in both periods since the flexible unit has the flexibility to change its output from 10 to 20 units as the demand load increases from 10 to 20 units.

Figure 16.3 illustrates this simple example. The left-hand side corresponds to the inflexible generation case, while the right-hand side corresponds to the flexible one.

16.2.2 *Expanded Transmission Availability*

While the so-called non-wire options have been receiving the most attention to help integrate renewable resources into the grid. That is, expanding transmission availability is still very important in harnessing variability from renewable resources, which is illustrated through the following example.

Fig. 16.3 Flexibility: inflexible unit (left diagram) and flexible unit (right diagram)

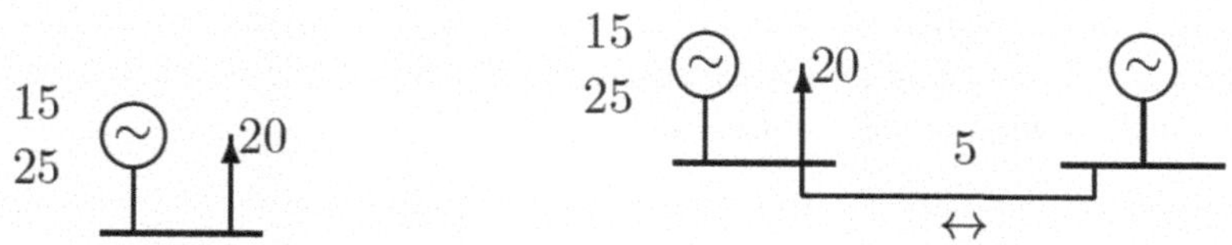

Fig. 16.4 Transmission: not available (left diagram) versus available (right diagram)

Consider an isolated demand that is supplied by a wind unit (Fig. 16.4). This demand has a constant power load of 20 units. The stochastic wind unit randomly produces between 15 and 25 units. In such a situation, if the wind unit produces 15 units, the demand is subject to an unserved energy of 5 units, which is undesirable. If, on the other hand, the wind unit produces 25 units, the demand is fully supplied, but 5 units of wind power are spilled(wasted), which is undesirable as well.

Next, we consider the same situation, but including a transmission interconnection that allows importing or exporting 5 units. In such a situation, if the wind unit produces 15 units, 5 units are imported through the transmission line and the demand is fully supplied. The (per-unit) cost of this import is generally much lower than the unserved energy cost. If, on the other hand, the wind unit produces 25, the demand is fully supplied, and 5 units are exported through the transmission line for a potential profit.

16.2.3 Storage Availability

In this third example we illustrate the relevance of storage availability.

We consider the previous example, but in this case instead of a transmission interconnection, a storage system of capacity 5 units is analyzed (see Fig. 16.5). In the case with storage, if the wind unit produces 25 units, the demand is supplied (20 units) and 5 units are stored in the storage system. If, on the other hand, the wind unit produces 15 units, the demand is fully supplied by the wind unit (15 units) and the storage system (5 units). It is important to note the storage system needs to store energy prior to using that energy. Thus, chronological effects are relevant for storage systems. Also, storage efficiency is not 1, and depends on the storage technology. However, for simplicity, we consider efficiency 1 in this example. Figure 16.5 illustrates this simple example. The left-hand side corresponds to the no storage case, while the right-hand side corresponds to the case with storage availability. We conclude this section by indicating that increasing transmission availability,

Fig. 16.5 Storage: not available (left diagram) versus available (right diagram)

increasing storage availability and an increasingly flexible thermal mix allow integrating in a power system an increasing number of renewable weather-dependent sources, such as wind or solar power units.

16.3 Interdependence of Multiple Systems

The previous section focuses on options to help integrate renewable resources into electric power systems, which account for one critical infrastructure sector in our modern society. However, based on the U.S. Department of Homeland Security, there are at least 16 such sectors that touch upon every aspect of our daily lives, and such sectors are all intrinsically linked (U.S. Department of Homeland Security 2019). The linkage of the multiple sectors, referred to as interdependence, may result in adverse effects to sustainability in one sector due to the development of the other linked sectors. For example, meeting growing electricity demand could require more fresh water for power plant cooling, causing tension on water systems, which have already witnessed historical droughts across the globe (Cook et al. 2015; Spinoni et al. 2014).

The importance to better understand and manage the interdependence of critical infrastructure systems is well recognized (Pederson et al. 2006). To realize such an aim, however, there are significant challenges. On the economic front, the more economical options in one system may strain resources in other systems. An example is the once-through cooling technology for power generation. While it is the cheapest option for cooling, it also withdraws the most water. On the policy front, policies designed to promote sustainability in one system may bring adverse environmental or reliability impacts to the other interlinked systems. For instance, policies to curb CO_2 emissions and to promote renewable energy in the power sector may bring more coal plants with carbon capture and sequestration and more concentrated solar power, both of which are among the heaviest users of water. Policies that aim to more tightly regulate power plants' water withdrawal may limit the available power generation capacity, especially during hot summer days when electricity demand is high, causing potential risks of blackouts. All these challenges arise from the reality that across different systems (and even within the same system), there are often competing objectives of lowering total system costs, improving system reliability, and bolstering sustainability, as illustrated in Fig. 16.6.

In the rest of this section, we will first present more details on the interdependence between electricity and water systems as a case study. Then we will briefly

Fig. 16.6 An illustration of competing objectives

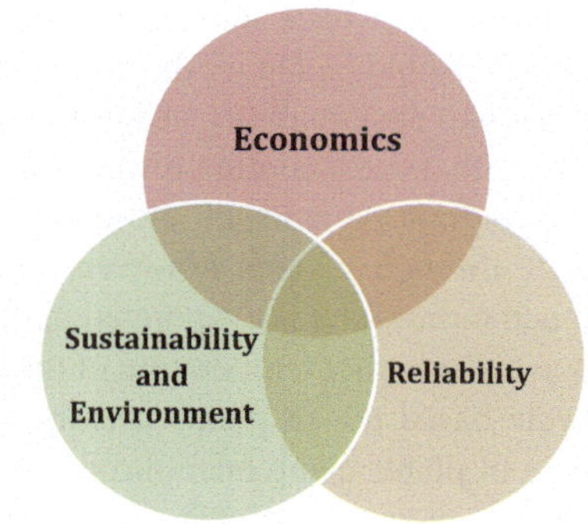

survey existing works on modeling interdependent systems and present a promising modeling and algorithm approach that may be computationally feasible to deal with the otherwise daunting task of modeling complicated systems together with high fidelity.

16.3.1 Interdependence: A Case Study of Energy and Water Systems

16.3.1.1 Water Dependency of Electric Power Systems

Water usage can generally be categorized as withdrawal and consumption. Water withdrawal refers to diverting water from its original source; while water consumption can be thought as permanent withdrawal, as the water will no longer be available due to evaporation, transpired by plants, consumed by people or livestock, etc. In the power sector, water is used in different processes during electricity generation (Kenny et al. 2009; Macknick et al. 2011), but mainly for cooling gas steam, with the liquified steam fed to a boiler to be reused to drive a steam turbine. Most of thermal-based generators, such as coal, nuclear, and natural gas plants, and certain renewable generators, such as biopower, and concentrated solar power, use water as the cooling medium to condense steam (Turchi et al. 2010). As a result, the power sector is the largest sector withdrawing fresh water in the USA, followed by the irrigation sector (Kenny et al. 2009). On the contrary, the electricity sector only accounts for a small share of water consumption, as the majority its consumptive water is through evaporation. This, however, does not mean that the power sector has little impact on the sustainability of water systems. The specific impacts highly depend on the actual cooling technologies used in the cooling process.

In the USA, commonly used cooling technologies are once-through (open cycle cooling), closed-cycle cooling (cooling towers or cooling pond), dry cooling (air-cooled condensing), and hybrid water-dry cooling. Among the least expensive and most widely used cooling options is the once-through cooling system, which withdraws water from a fresh water source (such as a river) and discharges (heated) cooling water back into the source. If the temperature of the returned water is too high, it would endanger the corresponding aquatic ecosystem, hence posing serious threats to sustainability at least in the local region. Closed-cycle cooling systems may either use a cooling tower or a (man-made) pond to recirculate the use of water for cooling. As a result, closed-cycle systems withdraw much less water than the once-through system; however, on the basis of a per unit of electricity generated, much more water is evaporated in the closed-cycle than in the once-through system, due to the higher temperature of the recirculating water. For the same reason, closed-cycle-based power plants usually have lower generation efficiency than the same type of plants using once-through cooling. While dry cooling does not use water at all, it is the most expensive cooling option. As a result, it is a complicated decision when determining which cooling options to choose for new power plants. Any sustainability-targeted policy needs to encourage or ensure the most efficient options to be adopted, when all the trade-offs of costs, reliability and environment are considered. This is precisely where the proposed modeling and computational framework can contribute significantly.

With the apparent dependency on water, electricity production can be impacted in many ways by water availability. First, prolonged drought conditions would likely limit available fresh water for cooling process. In addition, warming water with higher in-take temperature could also worsen efficiency of electricity production, especially for those who use once-through cooling technologies (Averyt et al. 2011). For instance, the Millstone nuclear plant in Connecticut had to shut down one of its two reactors in mid-July 2012, the time when electricity demand was very high, because the in-take water was too warm to cool the plant (Wald 2012). Second, limited availability of water could also impact the operations of pollution control or abatement devices (e.g., scrubbers) or other processes (e.g., flue gas desulfurization). A recent study suggests that amine-based carbon capture and storage (CCS) will increase water usage by 45–90% (Macknick et al. 2012). Third, scarce water will also constrain hydropower production (Averyt et al. 2011), which eventually will be substituted by fossil-fired fleets, leading to higher air pollution and threatening system reliability (due to the reduced available generation capacity).

16.3.1.2 Electricity Dependency of Water Systems

The opposite direction of the dependency is mainly related to the energy that is needed for abstraction, conveyance, distribution, and treatment of fresh water or wastewater (Rothausen and Declan 2011). Abstraction and conveyance include pumping groundwater and fresh water as well as transferring water from its source to treatment plants or a reservoir. Treatment and distribution refer to activities that process water into a consumptive quality, such as filtration, oxidation, and desalination. This is particularly the case for the Western United States. For instance, in California, the State Water Project (SWP), one of the major water projects, delivers water originated from the Northern California region over more than 400 miles to the demand location in the Southern California via San Joaquin Valley through a combination of natural rivers and man-made aqueducts. The delivery of water is not entirely by gravity, and a series of pumping stations were built to lift water for more than 3500 feet beyond which the gravity-ow takes place. As a result, the SWP is one of the largest electricity users in California. While the SWP consists of nine hydropower facilities to provide needed energy to deliver water, variations in weather conditions coupled with climate change shift the annual hydrographs (due to low snowpack and high rain-snow ratio) in an unexpected way that makes it difficult to manage the overall water delivery system. The shortage of surface supply, mainly due to drought, would exacerbate the conditions further as more groundwater needs to be extracted (and hence more energy needed for the extraction) to satisfy demand. Therefore, any mismanagement of the water system is likely to spill over to the energy system.

16.3.2 Interdependence Modeling and Methodologies

16.3.2.1 Existing Literature

The interdependence between power water and power systems is an instance of the interdependence of critical infrastructure systems. Methodologies to study such interdependence have been surveyed in papers (Pederson et al. 2006; Ouyang 2014). The major modeling-based approaches include agent-based simulation, economic theory–based approaches (input–output models and computable general equilibrium models), and iterative approaches. Numerous agent-based simulation (ABS) models addressing system interdependency exist in the literature, including Sandia National Lab's Aspen model, Argonne National Lab's SMART II, Idaho National Lab's CIMS (all surveyed in reference Ouyang 2014). While the ABS approach is flexible and particularly adept at capturing the interdependence of multiple systems/agents, its major weakness is its lack of solution quality guarantee. More specifically, ABS models can provide a workable solution to address the interdependency, but better solutions (in terms of lowering costs, improving reliability and promoting sustainability) may exist. Economic theory based approaches mainly consist of the input–output in operability model (IIM) (Haimes and Jiang 2001) and computable general equilibrium models (CGE) (Wing and Kolodziej 2008). The IIM is based upon the classic input-output model in economics, which is a static, linear model of all transactions between sectors of an economy based on the abstract production curves, representing technological relationships of production. It can hence model the rippling effect of one system to the others. CGE can be viewed as an extension to the input–output model, in the sense that it can incorporate nonlinear interdependencies among economic sectors, as well as consumers' and producers' behavioral responses to markets and prices subject to labor, resource, and capital constraints. The major drawback of both the IIM and CGE approach lies in their abstract representation of technology and physical systems, which does not permit high-fidelity representation of physical systems, and hence, is not suitable to study specific paths (such as which specific technologies will be built and how much will be built) that lead to future scenarios. There have been efforts in designing hybrid models to combine CGE models and a detailed sector-model. However, such hybrid models suffer computational challenges.

Iterative approach refers to the method that simply iteratively solve two or more systems in a sequential order. For example, in (Averyt et al. 2013), a power system model—Regional Energy Deployment System (ReEDS) and a water system model—Water Evaluation and Planning (WEAP) model are employed to study the potential impact of the power sector under various climate scenarios. In their study, one model is run first, say, ReEDS, then computing the water usage offline and passing that to WEAP. WEAP then solves for water system's resource allocation, while keeping the power sector's water usage fixed. The iteration may continue till a set of consistent solutions is achieved. However, such an approach is only heuristic as convergence is not guaranteed. Even if it converges, this simple iterative approach does not lay down a specific path or mechanism to guide the multiple systems to converge to the jointly optimal solution.

16.3.2.2 Distributed Framework for Integrated System Modeling

With the shortcomings of the abovementioned approaches, it is desirable to have a single model that can encompass all the interconnected infrastructure systems with detailed modeling of each system's components. Just using water and energy systems as an example, from a very high-level perspective, it is ideal to have an integrated model in the following form:

$$\begin{aligned} &\underset{(x\in X, y\in Y)}{\textit{minimize}} \quad f(x)+g(y) \\ &\text{subject to} \quad Ax+By=c. \end{aligned} \tag{16.1}$$

Where x and y, both vectors, represent the activities in a water and power system, respectively. The function $\boldsymbol{f}$ represents some generic objective function in a water system model (such as total operation cost), with the abstract set $\boldsymbol{X}$ denoting all the constraints on the water side. Similarly, $\boldsymbol{g}$ and $\mathbf{Y}$ respectively represent the objective function and constraint set in a power system model.

The constraint $\boldsymbol{A}x + \boldsymbol{B}y = \boldsymbol{c}$ specifies the linkage between the two systems; more specifically, it describes the supply and demand balancing requirement in both the water system and the power system. In the water system, $\boldsymbol{B}y$ can be interpreted as water usage from the power sector, and $\boldsymbol{A}x$ then represents all other water usage. Together they must equal available water resources. In the power system, the joint constraint means that electricity demand from both the water and non-water sectors need to be equal to total power generated. A recent study by NREL has provided the water withdrawal and consumption rates (gal/MWh) for all types of powerplants and cooling technologies, which will facilitate the modeling of water usage from the power sector (Macknick et al. 2011).

Though structurally simple, the joint model (16.1) can be of extreme scales in terms of the problem size, which cannot be solved as a whole even by the most advanced algorithms or computers. A natural idea in this situation is to decompose the problem and to solve it iteratively; that is, we can fix y first and solve for x; then fix x at that solution level and solve for y. Such an approach belongs to the more general class of Gauss–Seidel algorithms, and is well-known to be not provably convergent (Powell 1973). More specifically, it means that with an arbitrary starting point, the iterations may not lead to an optimal solution to the original problem.

To overcome the abovementioned technical difficulties, one particular algorithmic approach with provable convergence to an optimal solution (under convexity assumptions) is the alternating direction method of moment (ADMM) (Boyd et al. 2011). In such an approach, the decomposition is done on the augmented Lagrangian function of (16.1), defined as follows.

$$L_\rho(x,y,\lambda) := f(x)+g(y)+\lambda^T(Ax+by-c)+\frac{\rho}{2}\|Ax+by-c\|_2^2, \quad (16.2)$$

where $\boldsymbol{\lambda}$ is the Lagrangian multiplier (aka shadow price) associated with the linkage equation, and the l_2 norm is a penalty term that penalizes any deviation if $\boldsymbol{A}x + \boldsymbol{B}y\text{-}c$ is not zero, with a chosen penalty parameter. The ADMM algorithm then sequentially solves for x, y, and $\boldsymbol{\lambda}$ in an iterative fashion as in Fig. 16.7.

The ADMM algorithm as so presented is not new and can be dated back to 1980s (Eckstein and Yao 2012). Its convergence (with respect an arbitrary starting point) when all the functions involved are convex is well established (Boyd et al. 2011). The advantage of such an algorithm to model interdependent systems, in addition to its distributed fashion and convergence property, is that the step for updating the multiplier $\boldsymbol{\lambda}$ has a natural economic meaning, with $\boldsymbol{\lambda}$ providing the pricing information as how much a particular item or service is valued at the current stage. Based on the algorithm description in Fig. 16.7, two systems can achieve the joint optimality through sharing the current system solution (x^k and y^k) and the pricing information $\boldsymbol{\lambda}$.

The iterative procedure in Fig. 16.7, however, is still in a sequential fashion and is not amenable to parallel computing. Recent research has proposed several ways

Fig. 16.7 ADMM iteration

Algorithm 1 Generic ADMM algorithm

Initialize x^0, y^0, λ^0, $\rho > 0$

for $k = 0, 1, \ldots$ **do**

$$x^{k+1} \in \operatorname*{argmin}_{x \in X} L_\rho(x, y^k, \lambda^k)$$

$$y^{k+1} \in \operatorname*{argmin}_{y \in Y} L_\rho(x^{k+1}, y, \lambda^k)$$

$$\lambda^{k+1} = \lambda^k + (Ax^{k+1} + By^{k+1} - c).$$

to make the algorithm parallelable (Deng et al. 2017). One easy approach is to introduce auxiliary variables (z_1; z_2) and an indicator function $I_Z(z_1; z_2)$, which equals 1, if (z_1; z_2) belongs to Z, which is defined as $\{f(z_1; z_2): z_1 + z_2 = 0\}$, and equals infinity otherwise. The original problem (16.1) can then be equivalently rewritten as follows.

$$\begin{aligned} &\underset{(x \in X, y \in Y, z)}{minimize} \quad f(x) + g(y) + \mathcal{I}_Z(z_1, z_2) \\ &\text{subject to} \quad Ax - z_1 = c/2, B_y - z_2 = c/2. \end{aligned} \tag{16.3}$$

Then applying the ADMM algorithm to (16.3), with (z_1; z_2) fixed, the remaining problem can be decomposed into two completely separable problems in x and y, respectively.

The interdependent electric power and natural gas systems have seen the most applications of distributed optimization in general, and the ADMM-type algorithms in particular. For short-term operations, the reference (He et al. 2018a) provided a formulation and the corresponding ADMM algorithm for co-optimizing power and natural gas flow in a distributed fashion. There, simulation results show that considering the natural gas systems' operations can indeed change energy system's unit commitment/dispatch schedules (when compared to without), and correspondingly, the power system's operational costs. While the power systems' operation costs could increase when considering the uncertainties of natural gas availability, such as under extreme cold weather conditions, according to their results, the overall costs of the two systems combined can indeed be reduced, when compared to optimizing each system separately, and both systems reliability can be improved. Another paper (He et al. 2018b) extend such a co-optimization framework into long-term planning for both systems, especially considering power-to-gas (PtG), which is a technology to use excessive electricity from wind power plants to generate natural gas. Their results show that when PtG is considered in the long-term planning process, it can facilitate a deeper penetration of wind energy and hence, postpone the construction of transmission lines. While there have been few applications of ADMM-type algorithms outside of the power-natural gas co-optimization, the nexus between power and water systems has received increasing attention recently, and a co-optimization model on the short-term operations of power and water distribution systems is proposed in (Zuloaga et al. 2019). It is envisioned that distributed optimization algorithms in general will experience increased applications in interdependent systems to help improve operation and planning efficiencies, enhance system reliabilities, and help promote sustainable development among all the inter linked systems.

16.4 Conclusions

This chapter concludes the applications of operations research in sustainable management of natural resources with three future research directions. These include better utilization of the outcomes from physical- or process-based climate models to represent uncertainties, research related to harnessing variability and unpredictability via flexibility, transmission, and storage in the energy sectors, and finally recognizing the interdependence among multiple natural resource systems. These three future research directions improve the representation of natural systems in the models and account for interdependence among systems, and the resulting research outcomes are expected to advance our understanding of sustainable natural resource management.

References

Averyt, K., Fisher, J., Huber-Lee, A., Lewis, A., Macknick, J., Madden, N., Rogers, J., & Tellinghuisen, S. (2011). *Freshwater use by U.S. power plants: electricity's thirst for a precious resource: A report of the energy and water in a warming world initiative*. Cambridge, MA: Union of Concerned Scientists.

Averyt, K., Fisher, J., Huber-Lee, A., Lewis, A., Macknick, J., Madden, N., Rogers, J., & Tellinghuisen, S. (2013). *Water-smart power: Strengthening the U.S. electricity system in a warming world*. Cambridge, MA: Union of Concerned Scientists.

Boyd, S., Parikh, N., Chu, E., Peleato, B., Eckstein, J. F. K., Barber, N. L., Hutson, S. S., Linsey, K. S., Lovelace, J. K., & Maupin, M. A. (2011). Distributed optimization and statistical learning via the alternating direction method of multipliers. *Foundations and Trends in Machine Learning, 3*(1), 1–122.

California Natural Resource Agency. (2019). *Introduction to California's fourth climate change assessment*. Retrieved from http://resources.ca.gov/climate/safeguarding/research/.

Cayan, D. R., Pierce, D., & Kalansky, J. (2018). *Climate scenarios for the California 4th climate change assessment model selection, downscaling, drought scenario and sea level rise*. Retrieved from http://www.climateassessment.ca.gov/techreports/docs/20180827-Projections_CCCA4-CEC-2018-006.pdf.

Chen, Y., Chen, N., Guldmann, J.-M., Liu, A. L., & Wang, C. H. (2020). *Investigating climate-change-induced vulnerability of the northern California natural gas energy system and identifying resilience options*. Unpublished report, University of California Santa Cruz.

Cook, B. I., Ault, T. R., & Smerdon, J. E. (2015). Unprecedented 21st century drought risk in the American Southwest and Central Plains. *Science Advances, 1*, 1–7.

Deng, W., Lai, M. J., Peng, Z., & Yin, W. (2017). Parallel multi-block ADMM with o(1=k) convergence. *Journal of Scientic Computing, 71*(2), 712–736.

Eckstein, J., & Yao, W. (2012). *Augmented Lagrangian and alternating direction methods for convex optimization: A tutorial and some illustrative computational results*. RUTCOR research reports 32-3.

Haimes, Y. Y., & Jiang, P. (2001). Leontief-based model of risk in complex interconnected infrastructures. *Journal of Infrastructure Systems, 7*(1), 1–12.

Hawkins, E., & Sutton, R. T. (2011). The potential to narrow uncertainty in projections of regional precipitation change. *Climate Dynamics, 37*, 407–418.

He, C., Wu, L., Liu, T., & Shahidehpour, M. (2018a). Robust co-optimization scheduling of electricity and natural gas systems via ADMM. *IEEE Transactions on Sustainable Energy, 8*(2), 658–670.

He, C., Wu, L., Liu, T., & Bie, Z. (2018b). Robust co-optimization planning of interdependent electricity and natural gas systems with a joint N-1 and probabilistic reliability criterion. *IEEE Transactions on Power Systems, 33*(2), 2140–2154.

Kenny, J. F., Barber, N. L., Hutson, S. S., Linsey, K. S., Lovelace, J. K., & Maupin, M. A. (2009). *A 2009 estimated use of water in the United States in 2005*. Reston, VA: The U.S. Geological Survey.

Macknick, J., Newmark, R., Heath, G., & Hallett, K. C. (2011). *A review of operational water consumption and withdrawal factors for electricity generating technologies*. National Renewable Energy Laboratory, technical report: NREL/TP-6A20–50900.

Macknick, J., Newmark, R., Heath, G., & Hallett, K. C. (2012). Operational water consumption and withdrawal factors for electricity generating technologies: A review of existing literature. *Environmental Research Letters, 7*(4), 1–109.

National Oceanic and Atmospheric Administration. (n.d.) *Is sea level rising? Yes, sea level is rising at an increasing rate.* Retrieved from https://oceanservice.noaa.gov/facts/sealevel.html.

Ouyang, M. (2014). Review on modeling and simulation of interdependent critical infrastructure systems. *Reliability Engineering and System Safety, 121*, 43–60.

Pederson, P., Dudenhoeer, D., Hartley, S., & Permann, M. (2006). *Critical infrastructure interdependency modeling: A survey of the U.S. and international research.* Idaho Falls, ID: Idaho National Laboratory.

Pierce, D. W., Cayan, D. R., Maurer, E. P., Abatzoglou, J. T., & Hegewisch, K. C. (2015). Improved bias correction techniques for hydrological simulations of climate change. *Journal of Hydrometerology, 16*, 2421–2422.

Pierce, D. W., Cayan, D. R., & Thrasher, B. L. (2014). Statistical downscaling using localized constructed analogs (LOCA). *Journal of Hydrometerology, 15*, 2558–2585.

Powell, M. J. (1973). On search directions for minimization algorithms. *Mathematical Programming, 4*(1), 193–201.

Rothausen, S. G. S. A., & Declan, C. (2011). Greenhouse-gas emissions from energy use in the water sector. *Nature: Climate Change, 7*(4), 210–219.

Spinoni, J., Naumann, G., Carrao, H., Barbosa, P., & Vogt, J. (2014). World drought frequency, duration, and severity for 1951-2010. *International Journal of Climatology, 34*(8), 2792–2804.

Turchi, C. S., Wagner, M. J., & Kutscher, C. F. (2010). *Water use in parabolic trough power plants: Summary results from Worley Parsons' analyses.* National Renewable Energy Laboratory technical report NREL/TP-5500-49468.

U.S. Department of Homeland Security. (2019). *Critical infrastructure sectors.* Retrieved from https://www.cisa.gov/critical-infrastructure-sectors

U.S. Geological Survey. (2019). *1 meter digital elevation models (DEMs)—USGS National map 3DEP downloadable data collection.* Retrieved from https://catalog.data.gov/dataset/usgs-national-elevation-dataset-ned-1-meter-downloadable-data-collection-from-the-national-map.

Wald, M. L. (2012). *So, how hot was it? The New York Times.* Retrieved from https://green.blogs.nytimes.com/2012/07/17/so-how-hot-was-it/?_r=0.

Wing, I. S., & Kolodziej, M. (2008). *The regional greenhouse gas initiative: emissions leakage and the effectiveness of interstate border adjustments.* John F. Kennedy School of Government, Harvard University, policy program working paper RPP-2008-03.

World Climate Research Programme. (2019). *Overview CMIP6 experimental design and organization.* Retrieved from https://www.wcrp-climate.org/wgcm-cmip/wgcm-cmip6.

Zuloaga, S., Khatavkar, P., Vittal, V., Vittal, V., & Mays, L. W. (2019). Interdependent electric and water infrastructure modelling, optimisation and control. *IET Energy Systems Integration, 2*(1), 9–21.

Index

A
ABS, *see* Agent-based simulation (ABS)
ADMM, *see* Alternating direction method of moment (ADMM)
AEO, *see* Annual Energy Outlook (AEO)
After-sales service networks
  analytical modeling
    approach, 210–213
    future research directions, 213–214
    multi-echelon repaired component, 208, 209
  item inventory situation, 5
  product-centric view, 199
  RIIS, 201–204
  TAT, 200
Agent-based simulation (ABS), 397
Alignment, 47, 61, 233–234
Alternating direction method of moment (ADMM), 399–400
Although game theory, 380
Annual Energy Outlook (AEO), 378
Anthropogenic emissions, 388
Augmented ε-constraint method, 350, 352

B
Bellman equations, 286–287, 291, 296
Biofuels
  demand and mass-balance, 351–352
  switchgrass, 342
Biorefineries, 344, 354–356
Brownian motion, 293–294

C
California's Fourth Climate Change Assessment, 388, 389
Capabilities
  DCV, 172, 175
  green supply chain, 247
  Hou-Lung River Basin in Taiwan, 8
  innovation, 187
  location, 269
  loop reinforcing, 251
  PLE mode, 270–271
  shared value, 240
  supply chain, 265
  and support mechanisms, 55
  sustainability practices, 181–182
  technological, 32
Carbon capture and storage (CCS), 396
Case study
  industrial, 130–135
  NANGAM, 9
  natural gas markets, 378–380
  polypropylene, 4
  sugar manufacturing, 3
  and water systems, 395–397
CCS, *see* Carbon capture and storage (CCS)
CGE, *see* Computable general equilibrium (CGE)
China
  agricultural energy efficiency, 98
  GG, 246
  manufacturing industry, 93
  natural gas production, 360
  slacks-based efficiency, 83
  vehicle seating systems, 179
  water efficiency, 96

C. Chen et al. (eds.), *Pursuing Sustainability*, International Series in Operations Research & Management Science 301,
https://doi.org/10.1007/978-3-030-58023-0

CI, *see* Composite indicator (CI)
Circular economy
applications
business case development, 162–163
feasibility assessment, 160–162
framing, 157–158
idea generation, 160
scoping, 158–159
business models, 149–151
case company profiles, 177–180
challenges, 6
green human capital, 250
green supply chain capabilities, 247
integration across network partners, 247, 249–250
networked architecture, 250
organizational culture, 250
product and recovery design, 248–249
product "return" model, 250–251
value chain considerations, 248
concepts, 142–145
data analysis, 180–182
decision-making processes, 154–156
evaluation, 154–156
goals and strategies
governmental strategy, 146–147
operation mechanism, 147–149
target setting, 145–146
MFA, 151–154
principles, 142–145
product life cycle, 240, 244–246
researchers and practitioners, 4
and sustainability, 141–142
Citation analysis, 73, 97, 100
Citation chronological graph
findings, 80–81
literature development, 3
local main path analysis, 82–86
Climate change
business leaders, 273
natural environment, 6
natural gas, 9
political level, 76
resource and energy depletion, 162
shift, 397
Closed-cycle cooling systems, 396
Closed-loop supply chains
cost-effective remanufacturing, 227–233
elements, 221–222
management, 4–6
remanufacturing, 222–227
strategic and organizational alignment, 233–234
sustainability, 220–221
waste reduction, 222–227
Communication, 64, 91, 113, 118, 121, 133, 158, 181, 184–186, 234, 235, 258
Composite indicator (CI)
construction, 3, 86, 87
current status of the literature, 91–92
future directions, 92–93
Computable general equilibrium (CGE) models, 363, 397
Corporate social responsibility (CSR), 116–119
decision-making process, 87
implications, 3
pressures of, 44
triple bottom line concepts, 240
Corporate sustainability (CS)
assessment, 3, 87–90
current status, 87–90
DEA methods, 84
empirical studies, 176
evaluation, 113
external drivers
institutional theory, 173
stakeholder theory, 172
future directions, 90
hierarchical framework, 131
internal motivations
DCV, 175
NRBV, 174
PBV, 174–175
RBV, 173
performance analysis techniques, 3
procedure, 124
propositions, 193
research clusters, 98
topic clusters, 87
Coupled Model Intercomparison Project (CMIP6), 388
Cradle-to-cradle operations, 5, 143, 220–222, 227, 232, 234, 236
Critical period technique, 316
Cropping system, 353
Cross-case comparison analysis
auto end assemblers *vs.* parts suppliers, 188–189, 194
large *vs.* small companies, 189, 195
older *vs.* younger companies, 192–193
CS, *see* Corporate sustainability (CS)
CSR, *see* Corporate social responsibility (CSR)
Cumulative regional demand, 379, 380
Current-value Hamiltonian, 285

D

Da Lu Keng hydrologic monitoring station, 321, 322
Data envelopment analysis (DEA)
  advantages and limitations, 126
  application area, 77–78
  citation-based review methods, 74–75
  collection, 74
  data source, 74
  DMUs, 73
  methodologies employed, 78–80
  publications
    outlets, 76–77
    over time, 75–76
  research analysis unit, 77–78
  scholarly community, 76–77
  sustainability, 3, 73
DCV, *see* Dynamic capabilities view (DCV)
DEA, *see* Data envelopment analysis (DEA)
DEM, *see* Digital elevation model (DEM)
Decarbonization scenarios, 361
Decision-making units (DMUs), 73
Decision-making processes, 7, 17, 24, 47, 53, 54, 62, 73, 87, 112, 120, 127, 154–156, 301–303
Decision support
  article structure, 45
  motivation, 44–45
  multiobjective, 356
  objectives, 44–45
  research requirements, 51–53
  in resource management, 151
  SC, 43
  SCM, 44
  SSCM, 43
  tools, 2
  transformation, SSCM, 54–57
Deterministic equivalent formulation, 302
Digital elevation model (DEM), 389
Discounted value Hamiltonian, 308–310
Discrete problem, 304
Distributed optimization algorithms, 400
DMUs, *see* Decision-making units (DMUs)
DOE, *see* US Department of Energy (DOE)
Downscaled techniques, 388
Dynamic capabilities view (DCV), 172, 175
Dynamic optimization
  deterministic continuous time, 288–290
  deterministic discrete time
    Bellman equations, 286–287
    Hamiltonian, 286–287
    Lagrangean method, 279–285
    nomenclature, 281
    Pontryagin's maximum principle, 288
  electric power sector, 279
  optimal replacement
    Bellman equation, 291
    chemical batteries, 291
    optimality conditions, 291–292
    results, 292
    sensitivity, 292

E

Econometric models, 362
Economic theory, 397
Eco-risk mitigation strategies, 273
Electricity dependency, 396–397
Emissions prediction and policy analysis (EPPA), 363
Empirical evidence, 4, 174, 175, 194, 222–227
Energy modeling forum (EMF) 31, 361
Energy storage system (ESS)
  analytical solution, 298–300
  continuous-time process, 296
  energy capacity, 297
  HJB equation, 297
  optimal feedback policy, 298
  optimality conditions, 298
  time trajectory, 300
Energy supply chain, 86, 100, 360
Energy system transformation, 381–382
Environmental management programs, 317
EPPA, *see* Emissions prediction and policy analysis (EPPA)
EPR, *see* Extended producer responsibility (EPR)
European Electronic Take Back Legislation, 272
Extended producer responsibility (EPR), 259

F

Federal Energy Regulatory Commission (FERC), 378
Firm activity, 265
First order conditions (FOCs), 281–282, 284, 286, 290, 298
FOCs, *see* First order conditions (FOCs)

G

Game-theoretic models, 362
GaMMES, *see* Gas market modeling with energy substitution (GaMMES)
Gas market modeling with energy substitution (GaMMES), 362–363
Gas trade model (GTM), 362

Gauss–Seidel algorithms, 399
GCAM, *see* Global change assessment model (GCAM)
GCMs, *see* Global climate models (GCMs)
Geographic information system (GIS), 343
Geometric Brownian Motion with drift, 294
GIS, *see* Geographic information system (GIS)
Global change assessment model (GCAM), 363, 378
Global climate models (GCMs)
  harnessing results, 388–390
  sea level rise risk, 389–390
Global Trade Analysis Project (GTAP), 363
GM, *see* Green manufacturing (GM)
Golombek-type cost function, 368
Green design, 2, 18, 119, 120
Green manufacturing (GM), 112, 114–116, 119, 121, 135, 177, 247
Green product development
  competitive advantages, 15
  consumers, 16
  demand, 273
  downstream processes, 119
  environmental benefits, 15
  literature review, 16–18
  model structure, 18–22
  sequential games, 26–34
  simultaneous games, 22–26
Green supply chain (GSC), 2, 112, 114, 119–122, 135, 176, 247
GSC, *see* Green supply chain (GSC)
GTAP, *see* Global Trade Analysis Project (GTAP)
GTM, *see* Gas trade model (GTM)

H
Hamilton-Jacobi-Bellman (HJB) equation, 297
Here-and-now decisions, 302
Hotelling's rule, 310–311
Hou-Lung River system, 319–323
Hydraulic fracturing, 361
Hydro-meteorological forecasting technique, 317

I
IHS Markit macroeconomic model, 362
IIM, *see* Input–output in operability model (IIM)
Infinite horizon problem, 284
Input–output in operability model (IIM), 397
Integrated system modeling, 398–400
Integrative modeling framework, 345–346
Interdependence of multiple systems
  competing objectives, 394, 395
  distributed framework, integrated system modeling, 398–400
  electricity dependency, water systems, 396–397
  electric power systems, 394
  existing literature, 397–398
  once-through cooling technology, 394, 396
  water dependency, electric power systems, 395–396
Inventory systems
  multi-echelon, 209
  product returns, 204–206
  remanufacturing, 206–208
IPCC, *see* United Nations Intergovernmental Panel on Climate Change (IPCC)
Itô process, 294–295

K
Karush–Kuhn–Tucker conditions (KKT conditions), 376–377

L
Lagrangean method, 279–285
LCS, *see* Local citation scores (LCS)
Lemke's algorithm, 376
Lignocellulosic biomass (LCB)
  assumptions, 344–345
  biofuel supply chain configurations, 344–345
  computational power and capacity, 356
  data sources, 351
  economic and environmental objectives, 342
  empirical results, 352–355
  feedstock, 342, 356
  geospatial technologies, 343
  indices, 347–348
  integrative modeling framework, 345–346
  model structure, 346–352
  multiobjective optimization model, 343
  parameters, 347–348
  variables, 347–348
Linear complementarity framework, 362
Liquefied natural gas (LNG), 360, 374, 382
Local citation scores (LCS), 3, 76, 77, 80, 81
Local main path analysis, 73, 80–86, 100

M

Markov Chain Monte Carlo (MCMC), 318, 330
Markov decision processes, 306–307
Material flow analysis (MFA), 4, 151–156
Materials requirements planning (MRP), 263
Mathematical programming with equilibrium constraints (MPEC), 377
MCMC, *see* Markov Chain Monte Carlo (MCMC)
MCP, *see* Mixed complementarity problem (MCP)
MFA, *see* Material flow analysis (MFA)
Mixed complementarity problem (MCP), 376–377
Monte Carlo techniques, 305
MPEC, *see* Mathematical programming with equilibrium constraints (MPEC)
MRP, *see* Materials requirements planning (MRP)

N

NANGAM-MultiMod framework
- downstream (demand sector) problem, 375
- sets and parameters, 365
- storage operators' problem, 371–372
- suppliers' problem, 366

National energy modeling system (NEMS), 363, 378
National Oceanic and Atmospheric Administration (NOAA), 389
Natural gas market module (NGMM), 363
Natural gas markets
- bridge fuel, 360
- data sources, 378–380
- energy system transformation, 381–382
- evolution, 361–362
- global and regional studies, 362–363
- global unconventional resources, 360
- infrastructure development and interregional flows, 361
- network congestion, 361–362
- supply chain (*see* Natural gas supply chain)
- technological development, 360

Natural gas supply chain
- downstream sector, 374–375
- midstream sector, 369–374
- representation, 364
- solution approaches, 376–377
- upstream sector, 364–368

Natural resource-based view (NRBV), 172, 174
Natural resource management, 388
NEMS, *see* National energy modeling system (NEMS)
Network congestion, 361–362
NGMM, *see* Natural gas market module (NGMM)
NOAA, *see* National Oceanic and Atmospheric Administration (NOAA)
NRBV, *see* Natural resource-based view (NRBV)

O

Once-through cooling system, 395–396
Operational challenges, 261–263

P

Pacific gas & electrics (PG&E) system, 389
PBV, *see* Practice-based view (PBV)
PGN, *see Prospectiva de gas natural* (PGN)
Poisson distribution, 297
Pontryagin's maximum principle, 288
Practice-based view (PBV), 172, 174–175
Predictive maintenance, 258
Preventative maintenance (PM), 257
Price cut immune, 20, 23, 34, 36
Price-induced market equilibrium system (PRIMES), 363
PRIMES, *see* Price-induced market equilibrium system (PRIMES)
Probability matrix method, 316
Product life cycle (PLC) extensions
- circular economy, 240
- and circular economy, 244–246
- CSR framework, 5
- environmental innovations, 48
- extension/reuse, 151
- extension strategies, 242
- internal processes, 114
- remanufactured, 231
- sustainability performance, 120

Product life extension (PLE)
- challenges, 256
- functional areas, 263
- managerial implications
  - firm activity, 265
  - firms outside your supply chain, 267–268
  - periodic reassessment, 272
  - six-step process, 264
  - strategic redirection, 269–271

Product life extension (PLE) (*cont.*)
supply chain partner activity, 265–267
supply side, 264
SWOT, 268–269
policy implications, 272
recall, 256–257
recover, 259
recycle, 259
redirect, 260–263
reduce, 260
remanufacturing, 259
repair
improve existing product, 258
maintain post-failure, 257
maintain pre-failure predictive maintenance, 258
maintain pre-failure preventative maintenance, 257
restore (refurbish), 258
reuse, 258
12 R framework, 256
Prospectiva de gas natural (PGN), 378

R
RBV, *see* Resource-based view (RBV)
RCPs, *see* Representative concentration pathways (RCPs)
Recondition, 245
Recovery, 48, 144, 149–151, 158, 160, 205, 220, 223, 241, 245, 248–251
Recursive algorithm, 282
Recyclable packaging, 266–267
Recycling
closed-loop supply chains, 6
policy makers, 272
PP-based car bumpers, 159
remanufacturing, 120
targets, 146
wastewater, 99
Redirect, 256, 260–263, 267, 269–272
Refurbishing, 223, 248, 258
Regional energy deployment system (ReEDS), 398
Regional greenhouse gas initiative (RGGI), 381
Remanufacturing, 2, 4–6, 17, 99, 143, 144, 163, 200, 204–208, 222–229, 231–235, 248, 259
Remanufacturing consumer valuation
consumer leasing and renting scenario, 232–233
consumer owned products scenario
buyback acquisition, 229–230
consumer valuation, 228–229
new product sales, 228–229
profitability, 229–231
remanufactured product demand, 228–229
trade-in acquisition, 230–231
trade-in and buyback alternatives, 231–232
volume, 229–231
strategic and organizational alignment, 233–234
Remote-sensing techniques, 343
Renewable energy sources (RES), 307
Renewable Fuel Standard 2 (RFS 2), 342
Repair
analytical modeling approach, 210–213
improve existing product, 258
maintain post-failure, 257
maintain pre-failure predictive maintenance, 258
maintain pre-failure preventative maintenance, 257
retreatment of products and materials, 146
RIIS, 201–204
Repairable item inventory systems (RIIS), 200–204
Representative concentration pathways (RCPs), 388
Research topic clusters
corporate sustainability assessment, 87–90
regional sustainable development assessment, 95–97
sustainability
CI, 91–93
performance analysis, 93–95
Reservoir capacity planning
background, 318–320
critical period technique, 316
data analysis
cost and benefit databases, 325–327
water quality monitoring and modeling, 324–325
water supply and demand, 321–323
economy development, 316
framework, 320–321
human and natural system, 318
optimal pollution removal strategy, 334, 336
optimization analysis
chance-constrained multiobjective optimization model, 327
sensitivity analysis, 330–331
sizing and reservoir operation, 328
water quality, 329, 330
water resources balance, 327–328

organic pollutant, 332
pollution control cost, 332
probability matrix method, 316
sizing, 316
systems analysis, 316
water quality standard, 319
RES, *see* Renewable energy sources (RES)
Restore, 256, 258, 262, 269–271
Resource-based view (RBV), 172–175
Returns acquisition, 228
Reuse, 258
cost-effective remanufacturing, 227–233
PP, 157
and recycle, 115
virgin resource usage, 223
wastewater applications, 129
RFS 2, *see* Renewable Fuel Standard 2 (RFS 2)
RGGI, *see* Regional greenhouse gas initiative (RGGI)
Rice World Gas Trade Model (RWGTM), 362
RIIS, *see* Repairable item inventory systems (RIIS)
RWGTM, *see* Rice World Gas Trade Model (RWGTM)

S
SC, *see* Supply chains (SC)
SCM, *see* Supply chain management (SCM)
Sea level rise risk, 389–390
Second order cone programming (SOCP), 371
Sequential games
first mover advantage, 31–34
green product development, 16, 31–34
price competition, 27–31
role-choosing games, 27–31
Shared value, 244
capabilities loop reinforcing, 251
challenges to circular economy, 246–251
circular economy, 244–246
conceptual model, 241
financial profits, 131
firm level value creation, 242–243
future research agenda, 246–251
product life cycle extensions, 240, 242, 244–246
stakeholder value creation, 240, 243–244
strategies and actions, 239
Short-term meteorological processes, 317
Simultaneous games
entry decisions, 23–25
environmental quality, 25–26
Nash–Bertrand price equilibrium, 20
pricing, 23–25
quality, 23–25
subgame perfect equilibrium, 2
Simulation/optimization approaches, 316–317
Small and medium-sized enterprises (SMEs), 4, 53
SOCP, *see* Second order cone programming (SOCP)
SSCM, *see* Sustainable supply chain management (SSCM)
State Water Project (SWP), 396
Stochastic dynamic programming, 295–296
Stochastic optimization
dynamic programming, 295–296
Itô process, 294–295
optimal ESS management, 296–300
representative agent
analytical solution, 299–300
continuous-time process, 296
energy capacity, 297
HJB equation, 297
optimal feedback policy, 298
optimality conditions, 298
time trajectory, 300
Weiner process, 292
Stochastic programming model, 317, 318
decision-making process, 301–303
inter-temporal constraints, 308
Markov decision processes, 306–307
nonlinear constraints, 307
parameter realizations, 301
RES, 307
storage/battery, 300
transition probability matrix, 308
unit commitment with economic dispatch, 303–306
Strengths, weaknesses, opportunities and threats (SWOT), 268–269
Streeter–Phelps model, 325, 329
Supply chain management (SCM), 44, 45, 47, 48, 50, 59
Supply chain partner activity, 265–267
Supply chains (SC)
activities, 51
orientation, 49
practitioners, 44
profitability, 47
stakeholders, 56, 63
strategic and operational perspectives, 54
sustainable development, 43
transformation, 3

Sustainability
closed loop, 220–221
considerations
economic, 47–48
environmental, 48
social, 49
defined, 1
design, 2–4
ecology, 72
innovation, 2–4
logistics, 4–6
manufacturing, 4–6
natural resource management, 6–10
system analysis, 72
technology, 2–4
value creation, 4–6
Sustainability performance analysis
current status, 93–94
future directions, 94–95
Sustainable manufacturing
components, 3, 112–114
CSR, 116–119
definitions, 3, 112–114
evaluation
criteria, 122–124
indicators, 122–124
performance analysis techniques, 126–129
steps, 124–126
GM, 114–116
GSC, 119–122
industrial case study, 130–135
performance evaluation, 112
and technology management, 2
wastewater recycling, 99
Sustainable natural resource management
GCMs, 387–390
harnessing variability and unpredictability
electric power sector, 390
expanded transmission availability, 392–393
flexible generation, 392
gas compressors, 391
instruments, 392
renewable energy resources, 391
storage availability, 393–394
interdependence of multiple systems, 394–400
physical-/process-based climate models, 400–401
Sustainable practices, 4, 53, 64, 175, 182, 248
Sustainable reservoir sizing, 317
Sustainable supply chain management (SSCM)
building blocks, 50–51
case illustration, 63–65
characteristics, 3
CS, Corporate sustainability (CS)
definitions, 45–47
focus, 49–50
maturity models, 57–61
motivation, 44–45
objectives, 44–45
progression, 61–63
success of, 250
sustainability considerations, 47–49
transformation models, 54–57
Switchgrass biofuel supply chains, 344–346, 353
SWOT, *see* Strengths, weaknesses, opportunities and threats (SWOT)
Systems analysis, 316–317

T
Taiwan's Environmental Protection Administration, 319
TAT, *see* Turnaround time (TAT)
Taylor series, 295
Thermal-based generators, 395
TMDL, *see* Total maximum daily loading (TMDL)
Total maximum daily loading (TMDL), 329
Transversality condition, 284, 290
Turnaround time (TAT), 200

U
Uncertainty
models, 388
randomness, 388
scenarios, 388
Unit commitment with economic dispatch, 303–306
United Nations Intergovernmental Panel on Climate Change (IPCC), 388
US Department of Energy (DOE), 361
U.S. Department of Homeland Security, 394
US Energy Information Administration (US EIA), 381–382

V
Value creation
firm level, 242–243
logistics, 4–6
manufacturing, 4–6
optimal cost, 240
principles, 144
stakeholder, 243–244

W

Wait-and-see decisions, 302
Waste load reduction rate, 335
Wastewater treatment cost, 327
Water dependency, 395–396
Water evaluation and planning (WEAP), 398
Water quality
    modeling, 324–325
    monitoring, 324–325
    parameter, 324
    river reaches, 329–330
    simulation model, 332
    standard, 319
Water resources
    balance, 327–328
    cumulated possibility distribution, 322
    management, 316–318, 335
    planning, 317–318
Watershed management, 318
Weather dependent renewable sources, 391
Weiner process, 292
Weymouth equation, 370, 371
WGM, *see* World Gas Model (WGM)
WITCH, *see* World Induced Technical Change Hybrid (WITCH)
Within-case analysis, 182–188, 190–192
World Gas Model (WGM), 362
World Induced Technical Change Hybrid (WITCH), 363

The manufacturer's authorised representative in the EU is Springer Nature Customer Service Centre GmbH, Europaplatz 3, 69115 Heidelberg, Germany. If you have any concerns regarding our products, please contact ProductSafety@springernature.com

Printed and bound by CPI Group (UK) Ltd, Croydon, CR0 4YY
30/09/2026
02225508-0017